Hydroclimate Changes: Implications of the Middle East Region

Mashael M. Al Saud
Editor

Hydroclimate Changes: Implications of the Middle East Region

Springer

Editor
Mashael M. Al Saud
King Saud University
Riyadh, Saudi Arabia

ISBN 978-3-032-20052-5 ISBN 978-3-032-20053-2 (eBook)
https://doi.org/10.1007/978-3-032-20053-2

This Springer imprint is published by the registered company Springer Nature Switzerland AG
The registered company address is: Gewerbestrasse 11, 6330 Cham, Switzerland

Preface

Water scarcity and drought become an issue for many regions which witness a deficit in water availability and then supply for various communities. These is obvious concern by regional and international organizations to enhance water availability and to promote transnational cooperation between different countries for upscaling the management of water resources. Thus, conferences, assemblies, symposia, and workshops are continuously performed under the patronage of global agencies such as World Bank, FAO, UNEP, WHO, UNFCCD, and many other concerned entities who made many initiatives and programmes to promote sufficient and pure water supply. This has been also put as a global target in the Sustainable Development Goals, and more specifically SDG6.

Water scarcity is well pronounced in arid and semiarid regions, notably those with average precipitation rate less than 100 mm/year and scarce water resources. This is almost coinciding with the Middle East Region where drylands and deserts comprise about 1/3 of its total area, while it occupies countries with unprecedented water deficit where water demand exceeding 40 times water availability; and thus, complain for water supply becomes a daily issue that has been reflected on other vital sectors including mainly agriculture and food security, plus energy, and socioeconomic.

In this respect, the problems and challenges on water resources in the Middle East Region are many and diverse by regions. They can be on small and large scale starting from point source to large water bodies or vast dry lands, but these problems are not well addressed in the largest part of the region. The management of water resources is still a crucial issue, and it was resulted that the economy plays a role in managing these resources. Hence, there are Middle East countries with available water resources but poor supply; besides other arid countries with acceptable supply due to the economic-induced management. Another point of concern implies the geographic distribution of water sources and more certainly the large rivers (e.g., Nile, Tigris, etc.); especially in vast regions where these sources are remote enough to feed water to urban and agricultural clusters. The issue of transboundary water resources is also unmanaged in the entire Middle East Region and geopolitical conflicts existed behind this challenge. Moreover, pollution of water sources became common, notably in the low-income countries with the absence of environmental controls. While, water-related risks are also taking place, notably floods which have been lately pronounced even in dry regions. The list of problems and challenges is long enough to be illustrated in this preface.

For the mentioned problems and challenges, climate exists as the main generator and control for water and its journey from the atmosphere—Earth's body—atmosphere, and it is unreachable to study water resources and analyze their regime (i.e., flow/storage) unless climate is in-depth involved. The largest number of recent studies distinctly integrate water cycle (i.e., hydrology) and climatic variables together. However, the problem of data is still a constraint confronting these studies, notably where historical datasets are absent. This imposed researcher to seek for innovative tools where they can roughly retrieve these datasets. The development of remote sensing and satellite/climate-based data sources became more common source to fill gaps in ground-based data, and this can be the sole solution to establish hydrological and climatic models and scenarios. Yet, there must be a concern about the implementation of these models notably with data uncertainty. For example, when the IPCC scenarios are declaring that by 2100 the temperature will be increased by about 5 °C; this mean that by 2200 the temperature will increase by about 12 °C, and this is merely unrealistic concept.

In the view of the current situation, there must be more attention and realism, and this requires collaborative effort and cooperation between various geographic region and experts; especially if cross-border monitoring networks are established, the climate-related geographic regions are classified and ground-based measuring stations are installed, and the interrelationship between climate and hydrology should be well understood considering the acting variables and well defining the implications. Thus, the existing global initiatives (e.g., SDG Agenda, Conference of Parties-COP, etc.).

Tackling water resources and scarcity is undoubtedly linked with the existing climatic conditions, and they can never be separately taken. This is obviously found in many terminologies, such as drought, aridity, precipitation, etc. Therefore, water and climate are a dual term for a meteorology and water resources. From this point of view, "Hydroclimate" is a term joining the relationship between both where it also involves the sense of study and research. It becomes a necessity to emphasis on hydroclimate issues, notably when it is tackled in a regional level, and this require cooperative research work to achieve the success in identifying the temporal and geographic dimension of climate-induced water problems in a region like the Middle East, one of the driest region worldwide.

The author initiated the idea of this book in order to response for many questions and to highlight a number of concepts and understandings. Therefore, "Hydroclimate Changes: Implications of the Middle East Region" was the most indicative and subjective title for this book which constitutes 14 chapters by a number of distinguished authors from various geographic regions who put their knowledge and cases studies as indicators for diagnosing the current hydroclimate conditions in the Middle East. It is a collective document for various audience including students, researchers, academics, stakeholders and decision makers.

Riyadh, Saudi Arabia Mashael M. Al Saud

Contents

The Middle East Region: Facts on Climate and Water Resources

Mashael M. Al Saud

1 Introduction

The world's population is projected to continue growing for the next fifty to sixty years, climaxing at about 10.3 billion by the mid-2080. After reaching this climax, it is anticipated to slightly decline to around 10.2 billion by the end of the century. Presently, one in four people lives in a country where the population has already peaked (UN 2024). The global population growth has been increasingly over time, and this is well concentrated among the world's poorest countries. This incases in population rate is associated with higher demand on resources; especially the natural ones which are almost under stress and have therefore become depleted in many countries. The level of impact from the natural resources depletion is well pronounced in low-income regions, even though these countries may include civilized and urbanized clusters associated with industrial development. In addition, the increased population rate exacerbated the stress on fossil resources which are exploited without an account in order to upscaling the development projects, resulting in several negative impacts on human and the environment.

Newly global problem has been lately existed and affects all regions of the Globe. It is the climate change, the physical process which is acting with unpreceded level on the human being and life style as well as its impact is extending to the environment and many other vital sectors (e.g., agriculture, energy, etc.). Thus, facts on climate change are frightful. For example, it has been stated by NOAA (2024) that the rate of warming since 1982 is more than is approximately three times as fast: 0.20 °C) per decade; while, 2024 was the warmest year since global records began in 1850, and the ten warmest years in the historical record have been occurred in the decade of 2015–2024. Besides, the global precipitation patterns have shifted significantly over the past century, and the precipitation rate has been increased in some regions and

M. M. Al Saud (✉)
King Saud University, Riyadh, Kingdom of Saudi Arabia
e-mail: geomm5.mashael@gmail.com

M. M. Al Saud (ed.), *Hydroclimate Changes: Implications of the Middle East Region*,
https://doi.org/10.1007/978-3-032-20053-2_1

decreased in others. It has been estimated that the global precipitation has increased by about 0.76 mm per decade since 1900.

People are experiencing and treating climate change in different ways. It impacts human health, agricultural productivity, housing, safety and work. Some communities are more vulnerable to climate change impacts. Besides, some other communities have had to relocate because of sea-level rise and saltwater intrusion. The problem has many arms at diverse level of impact and uneven geographic and temporal dimensions. In the future, the number of people displaced due to climate change is expected to rise. The story of climate change is still in its beginning, especially the increasing greenhouse gas (GHGs) emissions from industrial and other human activities act like a blanket wrapped around the earth, trapping the sun's heat and raising temperatures.

GHGs emissions, causing changes in our climate, come from various regions of the World, but they affect all people; however, some countries produce GHGs much more than others. The G20, the group of the World's largest 20 economies, is responsible for about 77% of global GHGs emissions (UN, 25). This motivates considering "Climate Action", as it was illustrated by sustainable Development goal (SDG14), while countries and large-scale industrial facilities who are creating more of the problem have a greater responsibility to act first and take the responsibility of adopting protection measures before time become late to address this problem.

Perhaps, water are the most affected resources by the changing climate. It is not a limited impact to these resources, but it is a chain of influences starts by the decrease water availability and then developed and acting the other vital sectors, human life, the environment and the circular economy of the entire globe. In many regions worldwide, climate change is likely to increase people's demand for water while also shrinking water supplies. While, in humid and mountainous, people depend on snowpack for drinking water, agriculture, and other aspects of water consumption.

The changing climatic conditions influence water quality by various processes (e.g., the increased rainfall rate may lead to more runoff of sediments, nutrients, pathogens, and other substances into water bodies. Increases in nutrient runoff, along with warming water temperatures, can also lead to harmful algal blooms). Moreover, climate change enhances the processes of salinity of water bodies and more specifically in coastal groundwater aquifer through saltwater intrusions. While, rising sea level and increased drought can enable saline water to advance farther upstream and inland in estuaries, wetlands, and aquifers. Increased salinity can contaminate freshwater supplies and harm aquatic plants and animals.

Recently, water deficit has been raised in more than 60% of the World's countries, with a special emphasis in the arid and semiarid regions, like the Middle East. The paradox is that, water problems are extending in some instances to other regions with slight humid climate, such as in some northern parts of the Mediterranean Basin evidencing that the least precipitation is not the only reason behind water deficit and the related problems. In this respect, there is no clear discrimination to determine the degree of impact on water resources between climate causes and human interference, and yet water deficit and quality deterioration is always attributed to the changes in the climatic conditions.

2 Water Problem

The Middle East Region and even North Africa (MENA) Region are the World's most vulnerable to the impacts of climate change, witnessing most higher temperatures, pollution, drought, floods and sea level rising. The need to address the climate challenges in vulnerable communities became a must and precautionary and rapid implementations should be taken. In this regards, regional and international organizations, plus national governance bodies, give concern to the issue of climate change and more specifically its impact on water and agriculture. for example, World Bank has integrated climate into its tasks for the MENA Region and is on track to align 100% of new operations with the goals of the Paris Agreement, such as the World Bank's MENA Climate Change Roadmap runs between 2021 and 2025; emphasizing on four key transformation areas: food systems, water security and resilient natural capital; energy transition and low-carbon mobility; carbon-smart cities and resilient coastal economies; and sustainable finance. Despite this big support of the World Bank, The Middle East Region, as a part of the MENA Region is still the smallest recipient of climate change financial support worldwide; while the existing vulnerabilities and continuous conflicts have put additional stress on the resources of many countries. WBG (2025) stated that these regions will need to build on climate action progress of recent years and focus on long-term reforms and prioritize resilient and low carbon investments to protect against climate impacts and reduce poverty on a livable planet. Of course, this is not only the responsibility of these countries located in the MENA Region as a whole, but industrial countries must also take the action and be responsible on reducing carbon footprint.

With respect to the increased global temperature increase of 1.1 °C since pre-industrial times (IPCC 2021), droughts and heatwaves have been intensified, and these changes disrupt crop yields and increase the vulnerability of agricultural systems (Lobell et al. 2011); resulting in threatening global food systems. The global population is expected to rise to nearly 9.7 billion by 2050 (UN 2019), which means that food systems are under increasing pressure; particularly in regions already facing structural weaknesses (Elkhalfi et al. 2025). Recently, around 820 million people remain affected by hunger, and this figure is likely to rise due to the accelerating impacts of climate change and its impact on water resources (FAO 2020).

More than 4 billion people, equal to ½ of the World population, are witnessing water deficit over at least one month of the year, and this number will be increased between 4.8 and 5.7 billion in 2050 (Mekonnen and Hoekstra 2016). In this regard, countries consuming more than 25% of its renewable freshwater resources is considered as "water-stressed" country. According to the UN Water (2021), there is approximately 2.3 billion people live in water-stressed countries, among which 733 million live in high and critically water-stressed regions. In this respect, agriculture consumes more than 60% and it is expected to increase by 20,250 as per the developed number of urban clusters and the need to feed more than 9 billion people. This will be associated with a remarkable increase in water abstraction estimated at 15% increase from different surface water and groundwater).

Located in the middle part of the Globe, the Middle East Region is known term for the geographic area including southwest Asia and, usually, at least part of North Africa. It is the home to many countries with cultural, historical and religious significance. The region is characterized by its diverse landscapes, including deserts, mountains, and coasts. The Middle East Region represents the crossroads of the world, because it connects the three continents of Asia, Africa, and Europe. This makes the Middle East with World's oldest civilizations; especially it is the junction of trade routes connecting Europe and China, India and Africa, and all the cultures of the Mediterranean basin. Years has had a profound effect on the region's culture, politics, and economy.

The landscape of the Middle East Region is more diverse and it encompasses all natural landscapes including deserts, rivers, plains, long coasts, mountains with snow, cultivated and dry lands. Population of the Middle East Region is mainly situated along the coastal zone (e.g., Dubai, Beirut); and there are many dense urban clusters (e.g., Cairo, Riyadh), but at the same time empty lands and deserts are dominant (e.g., Western desert, Empty Quarter desert). The central region of the globe is being rapidly developed and many giant and global projects are implemented along with the increased water demand. For this reason, water became a scarce resource and competitive commodity notably precipitation is uneven distributed between various geographic regions, resulting in water stress that has been exacerbated lately on the national level, and even geo-political conflicts on shared water resources between riparian countries occurred, and lately, the sixteen countries of the Middle East Region are witnessing unprecedented water deficit.

According to Shaban (2022), the largest part of the Middle East Region is located within arid and semiarid regions, with few parts situated in sub-humid climate zones, and the average precipitation is often less than 200 mm while the evapotranspiration exceeds 55%. Among the Middle East Region, there are 13 Arab countries within the (Fig. 1), occupying more than 62% of the total population and 66.5% in surface area. Generally, the Arab Region is mainly an arid or semi-arid; and thus, water availability is lower than the threshold of the World Health Organization (1000 m^3/capita/year) for water poverty.

3 The Middle East Region

The Middle East Region is a transcontinental region including parts of Eurasia and Africa, and it includes countries that almost share common factors like ethnic groups, geographic features, political history, etc. The Middle East Region includes 16 countries. These are Bahrain, Cyprus, Egypt, Iran, Iraq, Jordan, Kuwait, Lebanon, Oman, Palestine, Qatar, Saudi Arabia, the Syrian, Turkey, the United Arab Emirates and Yemen (Fig. 1). In many instances, the name "Greater Middle East" is raised to include, in addition to the 16 identified countries, Afghanistan, the Comoros, Djibouti, Maghreb, Pakistan, Sudan, and Somalia. Even though, countries in Central Asia and Transcaucasia is sometime linked into the Middle East countries.

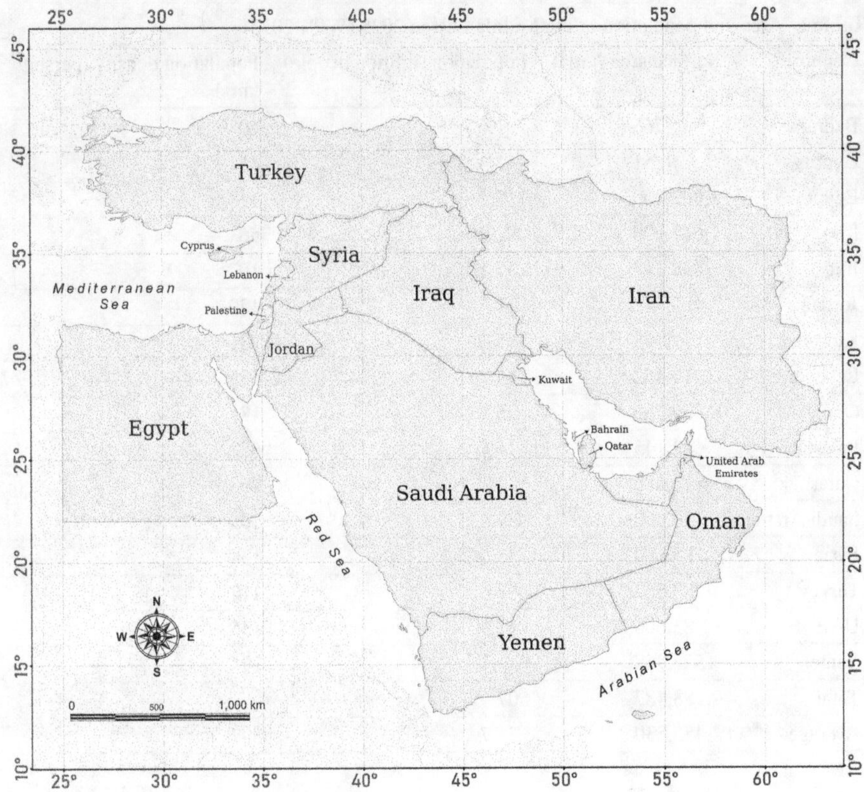

Fig. 1 Map of the Middle East countries

The middle East Region is located between the following geographic coordinates: 42° 00′ 01″ N & 12° 35′ 50″ N and 24° 59′ 58″ E and 63° 11′ 04″ E. It has a total surface area of the Middle East Region is 7,288,487 km^2 (2,814,100 miles^2), with total population of about 507 million people (Table 1) which is equivalent to about 6.2% of the total global population. This also implies an average population density of 307 person/km^2 which is five time larger than the average global population density (62 person/km^2). For the surface area, it is obvious that the Kingdom of Saudi Arabia (2.15 million km^2) occupies the largest one; while the biggest population is in Egypt (118.4 million people). However, there is always contradictory in mentioning the surface area of many Middle Easter countries, and this can be attributed to the existing geopolitical conflicts.

The Middle East Region has a varied physical geography in that it includes the lowest altitude on Earth's surface which is in the Dead Sea exceeding 440 m below sea level; whilst there are many crests with remarkable altitudes such as the crests of Kuh-e Damavand in Iran with 5671 m, and mountain of Ararat in Turkey with 5137 m, in addition to other mountain peaks such as Kornet Esawda in Lebanon with 3088 m. The mountains regions are characterized by humid climate with dominant

Table 1 Area and population of the Middle East in 2025 (WPR 2025)

Country	Surface area (km^2)	Population (million people)	Population density (person/ km^2)
Bahrain	766	1.6	2089
Cyprus	9251	1.4	151
Egypt	1,002,000	118.4	118
Iran	1,647,000	92.4	56
Iraq	438,317	47.1	107
Jordan	89,342	11.5	129
Kuwait	17,818	5.0	281
Lebanon	10,452	5.8	558
Oman	309,500	5.5	18
Palestine	22,145	14.4	652
Qatar	11,586	3.1	268
Saudi Arabia	2,150,000	34.6	16
Syria	185,180	25.6	138
Turkey	783,562	87.6	112
UAE	83,600	11.3	135
Yemen	527,968	41.7	79
Total	7,288,487	507	–
Average	455,530	31.7	307

snowpack remains for few months over the year. Besides, desert lands are spread in the Middle East Region, and they comprise about 2.39 million km^2, equivalent to 1/3 of the total area, and the majority of the desert land is in the Arabian Peninsula. Moreover, the Middle East Region has a long coastline that exceed 11,000 km where more than 60% of urban settlements are situated.

4 Climate of the Middle East Region

To characterize the climate of the Middle East Region, it is sometimes difficult, because of it has large surface area, with diverse topographic features which in turn control the climatic parameters. However, the general location of the Middle East Region makes it characterized by diverse meteorological description including sub-humid, semiarid or arid climate zone. Typically, the Middle East Region has a warm and dry climate, with an oscillating arid and semi-arid environment characterising most territories. These climatic conditions result distinct seasons and seasonal variability, with winds blowing into the interior in summer and blowing towards the coasts in winter. However, low temperatures and snow are common in the mountainous regions which are dominant with vegetation cover such as Turkey, northern Iran and Lebanon where winter is cold, especially at elevated regions. Moreover, much of the Middle East is desert (about 33%), and the daytime temperatures can reach over 52 °C, but it cools quickly and sometimes it reaches below 0 °C. The deserts of the Middle East are very dry with less than 50 mm/year and sometimes less. Besides, some regions of the Middle East have a cooler and wetter climatic conditions where precipitation exceeds 150 mm/year.

All Middle East countries lack to complete and long-term climate data that enable establishing creditable climatic trends. Even though, the reason is still undefined, but concerns have been lately raised and many of these countries installed new meteorological stations and reactivate the existing and old ones. The obvious diversity in climate even within small areas makes it necessary to increase the number of meteorological stations with uneven geographic distribution. The existed climate and remote sensing-based platforms played a role in filling climatic gaps. While, for climate change, most recent studies are tackling future scenarios and models (e.g., IPCC, NGFS, CMCC-CM, etc.) rather than using historical data for performing trend analysis for assessing climate change, while, historical trend analysis is most credible since it is based on measured datasets; even though, this data might be incomplete or does not cover uneven geographic extent.

4.1 Precipitation

Perhaps precipitation is one of the most significant climatic variables; especially it is used to evaluate climate change and the input of water volume into the hydrological system; and therefore, it is always a numeric values used while making strategic plans for water and agriculture. Globally, the volume of precipitated water is always considered in decrease, but this is not quite accurate and the many studies and interpretations have been lately raised. The largest number of these studies declare precipitation rate decrease (e.g., Karami 2019), other studies indicate increase in precipitation rate (e.g., Hasanean and Almazroui 2015), and few assure existence of changing climatic patterns rather than increase/decrease (e.g., Shaban 2020). This contradictory might be attributed to the diverse regions where various studies were applied, data inaccuracy, nature of data source, misleading of data analysis.

Rainfall in the Middle East Region often falls in boreal winter due to the eastward pathway of Mediterranean cyclones over the region. While, the summer is completely dry and humid in the coastal zones, with sporadic rainy spells triggered by anomalous synoptic conditions. Although Mediterranean cyclones are the main factor behind rainfall, yet other meteorological factors may play significant role. In addition, the Middle East experiences remarkable inter-annual variability in annual rainfall rate. In addition, the geographic proximity of the Middle East to Europe, and the similarity latitudes, as well as the open seas (e.g., Arab Sea-Indian Ocean) raises the possibility that the modes of inter-annual variability that affect rainfall in these regions also affect the Middle East; especially that rarely a year pass by without wet storm(s) in the Middle East Region with the increased frequency of both snow storms and heavy rains.

Good examples are the snow storm affected Europe and the Middle East in December 2010, as well as the winter storm existed in 2023, referred to locally as Storm Joyce, affected the eastern Mediterranean. In addition, the cyclone Gonu, which hit Oman and the east coast of the UAE in 2007, and the cyclone Asna in August 2024 over Oman, as well as the frequent torrential rain over the Arabian Peninsula that resulted in flash floods hitting mainly Saudi Arabia on several occasions in the 1970s, 1990s and more lately in 2009 and 2011.

The average rainfall rate in the Middle East Region is approximately 295 mm/year and standard deviation (σ) of 220.5; where it ranges between 34 mm/year as the minimum and 1760 mm/year as the maximum recorded values. The rainfall gradient is abruptly changed between different topographies, but it is normally increase with the mountainous areas. For example, in Lebanon the precipitation gradient reaches 48 mm/km; and its calculated at 34 mm/km in Turkey.

However, the geographic distribution of average annual rainfall rate is located in dry and desert lands. Thus, about 82% (5,976,560 km^2) of the Middle East Region is characterized by average annual rainfall rate of less than 205 mm/year. The geographic distribution of the average annual rainfall rate between different Middle East countries for the last few decades is shown in Fig. 2. It shows that the least average is 62 mm/year in Kuwait besides 834 mm/year in Lebanon.

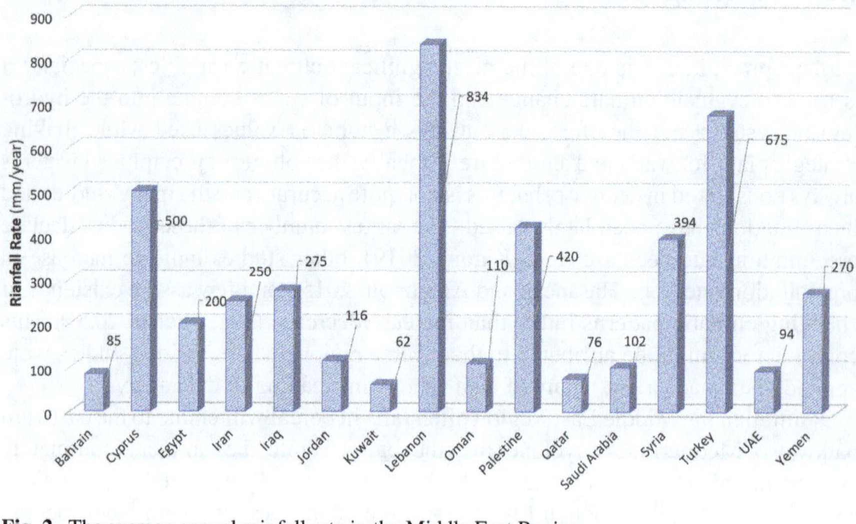

Fig. 2 The average annual rainfall rate in the Middle East Region

4.2 Temperature

Global warming is a temperature-induced process that represented by long-term heating of Earth's surface which has been noticed since the pre-industrial period due to exacerbated human activities with a special emphasis to the use of fossil fuel burning, which increases heat-trapping greenhouse gas levels in Earth's atmosphere. The temperature has been estimated to have increased Earth's global average temperature by about 1 °C (1.8 °F), a number that is currently increasing by more than 0.2 °C (0.36 °F) per decade (NASA 2025).

Similar to rainfall, air temperature is also the another major meteorological variable, and it may have much consideration in the Middle East Region because it is almost a dryland and characterized by warm weather. This is normal notably that the Middle East Region has the highest temperature rate worldwide. Probably, the increased air temperature rates in the Middle East Region is a result of its global location on the upper portion of the equator forming titled surface facing the sunlight and makes it with effective aspect to capture the heat from the sunlight radiation, and this is enhanced by the presence of dry and desert lands with low-lands and blowing wind.

In opposite to rainfall, which is almost fluctuating between decrease and increase, temperature is in ascending trend; and the spatially-averaged annual air temperature for the Middle East Region increased at a rate of 0.36 °C/decade during 1981–2020 with an even more marked increase of 0.45 °C/decade in the summer (Hadjinicolaou et al. 2023), and air temperature extremes have also become more frequent. The increase in temperature in the Middle East Region is uneven notably between different latitudes, altitude and morphological characteristics (e.g., low-land, closed land, coastal area, etc.). Therefore, trend analysis for temperature of the Middle East Region has been established for the last few decades (Fig. 3).

The average temperature rate in the Middle East Region is approximately 23 °C, which is above the global average air temperature by about 8 °C for the mid-latitudes regions. The calculated standard deviation (σ) is 3.2, and the difference between biggest and smallest average is 10 °C; while it ranges between 17.5 °C and 27.5 °C for the minimum and maximum rates as recorded for Cyprus and Kuwait; respectively.

It is obvious that the latitude it ranges between 17.5 and 27.5 °C for the minimum and maximum and topography play a significant role in controlling the temperature rate which averages it almost below 20 °C; and therefore, this exists in the western part of the Middle East Region including Cyprus, Lebanon and Turkey, while Iran is also averaging below 20 °C, but this extends even to latitude of about 25°, but the existence of mountain chains acts in lowering its temperature rates well.

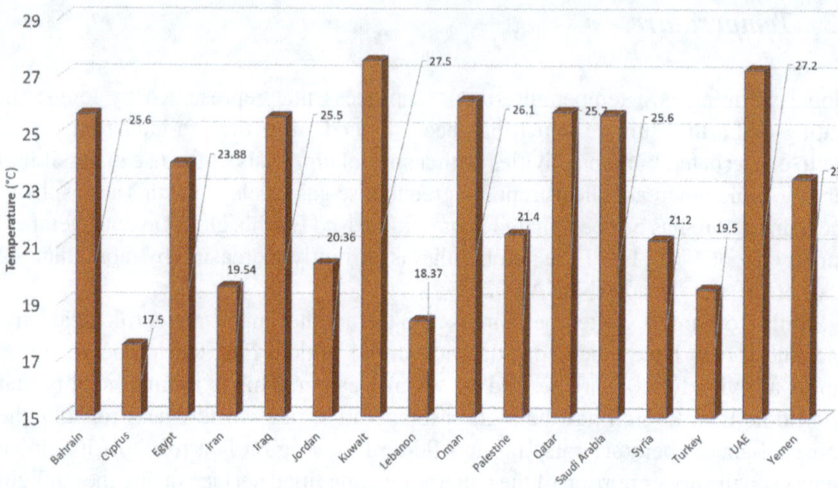

Fig. 3 The average annual air temperature rate in the Middle East Region

5 Water Resources

The Middle East Region is naturally dry zone with dominant aridity climate at different levels and this is the primary reason that characterizes it with water scarcity. If it is coupled with limited freshwater supplies resources and the growing water demand, virtually the Middle East Region is facing exacerbated water stress. The increased impact of climate variability and oscillation widens the imbalance in water supply/demand and this is exacerbating by amplified drought spells.

The Middle East, and even North Africa Region, have been widely acknowledged as the most water-stressed regions in the world. In fact, according to data from 2019, sixteen of twenty-five most water-stressed countries in the world are located in the Middle East Region, with Bahrain ranked as the world's most water-stressed country (Mahmoud 2024). Nevertheless, this region occupies large-scale water bodies including river and reservoirs. The main ones are: Nile River (about 1500 km in Egypt), Euphrates River (2800 km), Tigris River (1850 km), Kızılırmak River (1355 km), Karun River (950 km), Sakarya River (824 km), Büyük Menderes River (584 km), Orontes River (571 km), Seyhan River (560 km), Diyala River (445 km) and Jordan River (251 km). In addition, there are large-scale open water bodies (reservoirs) including: Lake Nasser (132 billion m^3), Atatürk Reservoir (48.7 billion m^3), Keban Reservoir (31 billion m^3), Mosul Reservoir (11 billion m^3), Ilısu Reservoir (10.4 billion m^3), Karakaya Reservoir (9.5 billion m^3), King Fahd Reservoir (325 million m^3), Qaraaoun Reservoir (220 million m^3) and Wadi Dayqah Reservoir (100 million m^3). Some of these large-scale water bodies are located in drylands such as Nile River, Lake Nasser and King Fahd Reservoir.

Even though, water availability in the Middle East Region is still inaccurately defined notably with the rapid changing in population and their needs, climatic conditions and water storage. The total average annual volume of water from precipitation is about 1,760,879 billion m^3, where more than 55% of the volume is returned back to the atmosphere by evaporation and evapotranspiration. The average water availability in the Middle East Region is estimated as 485 m^3/capita/year, which is almost of the limit of the threshold of water-scarce countries. While, water demand is estimated as 527 m^3/capita/year; and this normal that water demand exceeds the available water (Table 2). However, this ratio is viewed from the mathematical perspective, and there must be made clear that some countries occupy large areal extend, such as Saudi Arabia where the precipitation rate is only 96 mm/year but it has a surface area of 2,150,000 km^2; and therefore, this increases water availability or it may create an erroneous result. In addition, the total available water might be high if the population is low. In addition, rainfall distribution is uneven in most countries, and rainfall water may be flow or store in remote areas where human consumption is negligible.

There is extreme difference in the ratio between water demand and the available water resources; and thus, some countries have a huge deficit such as in Bahrain where the deficit between the available and demand water exceeding 41 times; while other has a deficit representing several times such as Kuwait, Qatar, Saudi Arabia and UAE.

Table 2 Water availability and demand in the Middle East Region

Country	Precipitation (mm)	Total water volume (million m^3)	Water availability (m^3/capita/year)	Water demand (m^3/capita/year)	Water demand/availability ratio
Bahrain	85	65	18	746	41.5
Cyprus	544	5032	656	228	0.35
Egypt	200	200,400	465	757	1.63
Iran	250	411,750	1675	1110	0.66
Iraq	275	120,544	915	1140	1.25
Jordan	116	10,363	182	108	0.59
Kuwait	62	1104	51	293	5.75
Lebanon	834	8720	1027	220	0.21
Oman	110	34,045	300	375	1.25
Palestine	420	9301	307	88	0.28
Qatar	76	881	27	307	11.37
Saudi Arabia	96	206,400	73	747	10.23
Syria	394	72,961	417	958	2.29
Turkey	675	528,904	1500	738	0.49
UAE	94	7858	26	504	19.38
Yemen	270	142,551	125	120	0.97
Average	281	–	485	527	6.14

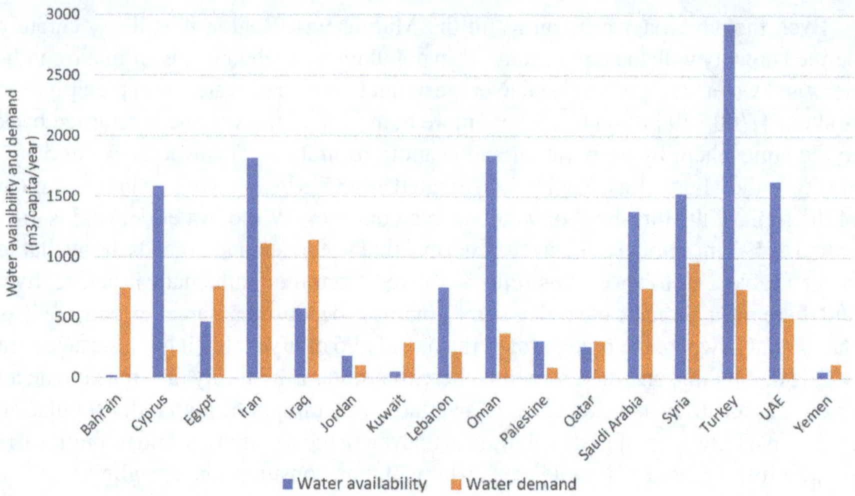

Fig. 4 Water availability and demand in the Middle East Region

Besides, there is surplus in water availability such as in Cyprus, Lebanon, Palestine and Turkey where the available water is about 3, 4.5, 3.5, 4 and time; respectively (Fig. 4). In this regard, it must be made clear that water availability/demand is not directly controlled by the precipitation rate but by other factors including mainly, surface area, demand rate and population size; and perhaps the Kingdom of Saudi Arabia and United Arab Emirates are good examples.

For water supply, it is almost different in terms of calculations in that water supply is often considered as the volume of water reached to consumers from the public sector, and it is also calculated as m^3/capita/year. In other words, in many countries maybe water availability exceeds water demand, but the supply can be poor and this supply is controlled mainly by the management approaches such as adoption of integrated water resources management (IWRM), or the use of new technologies as well as the implementation of the adaptive measures required. Unfortunately, estimations on water supply in the Middle East Region are few and uncertain and this might be attributed to many reasons, such as water partioning, and supply intermittency, water leakage, lack of water-meters for controlling the consumption as well as the uncalculated water that managed by individuals (e.g., construction of ponds and boreholes). However, some examples can be mentioned such as for Syria where water supply from the public sector does not exceed 130 m^3/capita/year, and this value represents only about 14% of water demand; while from Lebanon water supply is only 80 m^3/capita/year i.e., about 36% of water demand.

6 Precipitation/Availability Index

Precipitation is often considered as the main indicator for water availability for any region whether there is a surplus or deficit. However, in many instances, this conflicts with the available water in that region. In other words, some regions receive high amount of precipitated of water, but the available water volume remains low. In this regard, water availability has many interpretations, for example Meran et al. (2021) considered it as the total stored water in various water sources. While, Welsh and Pinder-Darling (2022) de-scribed it as the quantity or amount of water available for human use. By taking into consideration both definitions and the overall understanding about available water, it can be described as the amount of stored water in various forms and resources, with possibility to utilize this water per each person for different uses; and thus, it will be the feasible-used volume of water per capita per unit time, and that is usually coded as m^3/capita/year.

Therefore, if the volume of the available water equals to about ½ of the precipitated water that mean the management of precipitated water is good enough. This can be justified according to the fact that about ½ of precipitated water is returned back to the atmosphere and the rest ½ is stored in various water sources. In this case, the Precipitation/Availability Index (PAI) will be about 2, and it is simply calculated as precipitated water volume divided by available water volume. Therefore, Table 3 characterize the determined PAI (Shaban 2025).

For the Middle East Region, the PAI has been calculated and it shows significant findings (Fig. 5); especially that the studied Middle East countries revel diverse results even countries with similar hydroclimatic characteristics. In this regard, 8 (equivalent to 50%) of the Middle East countries reveal PAI below 1 which is extremely optimized index, 4 countries with PAI between 1 and 1.5, two countries with PAI between 2

Table 3 Precipitation/availability Index (PAI) classes and characteristics (Shaban 2025)

PAI	Description	Interpretation
1.0	Extremely optimized	Rarely reached index, and it means that almost all precipitated water is stored. It evidenced either a successful water management (e.g., large-scale water harvesting projects), hydrogeological characteristics enabling good water retention, least climatic variability and renewable water sources derived from the neighbouring regions
1.0–1.5	Optimized	A substantial portion of precipitated water (> 65%) is stored, almost due to the above mentioned evidences in the above PAI
1.5–2.0	Promising	This is the most sought index in many countries that suffering water challenges; especially climate change
2.0–2.5	Good to normal	It is almost the normal case in regular water storage in different water resources and mainly as groundwater
2.5–3.0	Normal to bad	It represents a percentage between 33 and 40% of precipitated to available water
> 3.0	Bad	When less than 33% of precipitated water is stored, and it is normally found in arid regions with no management plans

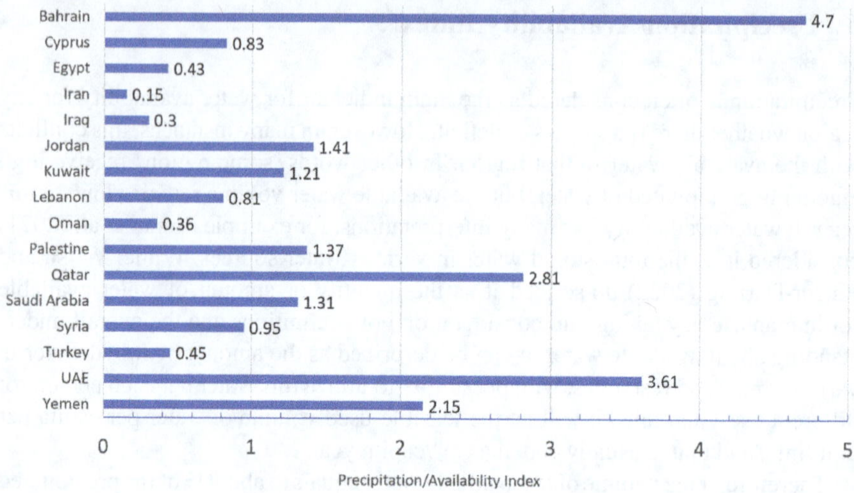

Fig. 5 Precipitation/availability Index in the Arab Region (Shaban 2025)

and 3, and the rest two countries are above 3. According to Shaban (2025), the PAIs of these countries have many interpretations (Table 4).

There are potential factors increasing the availability of water with respect to the precipitation volume. This largely includes the presence of snow cover rivers and dams as water feeding sources, plus the existence of karstic rock with high infiltration rage and high permeability representing storage characteristics. From the other side, the relatively low or moderate water demand is also a factor for more water availability. These factors are found in countries with PAI less than 1 (Table 4). Besides, there are also potential factors affecting the least availability and this is obvious in countries with low precipitation, open coastal aquifers seeping water to the sea, plus the relatively high water demand as in countries with PAI above 3 (Table 4). However, the combination of these factors, in many instances, play an opposite role, such as the low precipitation in Saudi Arabia bedsides more than 950 dams that harvest considerable volume of water, or the presence of large rivers such as in Egypt and Iraq. In this respect, water availability is changeable between years and the management of water resources remains the main factor.

These are: Cyprus, Egypt, Iran, Iraq, Lebanon, Oman, Syria and Turkey. Some of these countries already have considerable precipitation rate exceeding 500 mm/year (i.e., Iran, Lebanon, Turkey); and perhaps these three countries store more water due to the presence of snowpack as well as the presence of karstic conduits and aquiferous rock formations that can groundwater, but the question is still for the rest counties with PAI below 1 (Egypt, Iraq, Oman and Syria).

Table 4 Precipitation/availability Index (PAI) for the Middle East Region (Shaban 2025)

PAI class	Country	PAI	Potential factors
< 1 (Extremely optimized)	Iran	0.15	Snow cover, five major rivers, more than 500 dams
	Iraq	0.30	Euphrates and Tigris Rivers
	Oman	0.36	Karstification, more than 200 dams
	Egypt	0.43	Nile River, Aswan dam, Delta and Asuit Barrages
	Turkey	0.45	Karstification, more than 900 dams
	Lebanon	0.81	Karstification, snow cover, moderate demand
	Cyprus	0.83	Karstification, moderate demand and low-dense population
	Syria	0.95	Euphrates and Tigris Rivers, more than 140 dams
1-1.5 (Optimized)	Kuwait	1.21	Relatively low demand
	KSA	1.31	Huge sand aquifers, more than 950 dams
	Palestine	1.37	Least water demand and karstification
	Jordan	1.41	Least water demand, Jordan River
2.0–3.0 (Almost normal)	Yemen	2.15	More than 300 dams, least water demand, but relatively low precipitation
	Qatar	2.81	Very low precipitation, open coastal aquifers
> 3.0 (Bad)	UAE	3.61	Low precipitation, high demand
	Bahrain	4.72	Low precipitation, open coastal aquifers

7 Climate Scenarios

Global warming as a result of the increased activities of carbon emission activities, (i.e., specifically the energy supply (34%), industries (24%), agriculture (22%) energy supply, transportation (15%), and building structures (6%)) in the most recent decade constrains mitigation efforts to limit global warming. Estimates indicate that the sum of carbon emissions between 2010 and 2019 is about 4/5 of the remaining carbon budget post-2020 needed to keep global warming below 1.5 °C. Moreover, carbon emissions during this period amount to 1/3 of the remaining carbon budget for 2020 onward required to keep global warming below 2 °C. While, in order to curb future global warming for the period between 2020 and 2050, carbon intensity would need to decrease globally by 3.5% per year to limit warming to 2 °C and by 7.7% per year to limit warming to 1.5 °C (IPCC 2022). The low-carbon-emissions is likely correspond with the least developed countries, which contributed with fossil fuel and industry carbon emissions from 1850 to 2019 (Fig. 6), with north America as

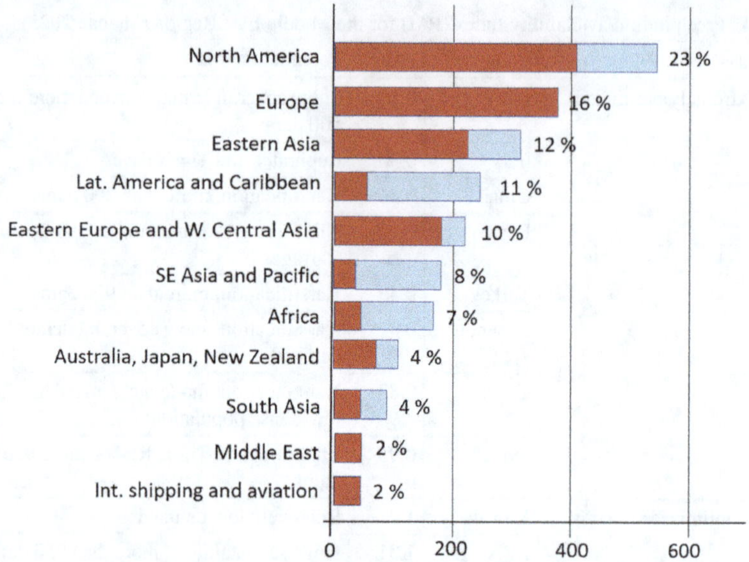

Fig. 6 Historical cumulative net anthropogenic CO_2 emissions per region between 1850 and 2019 (IPCC 2022)

the most carbon emission producer (23%) and Europe 16%, while the Middle East Region is still with least carbon amount of about 2% (IPCC 2022).

Climate scenarios have been lately raised as numerical projections of potential change in future climate, and they are generation is based on various assumptions about greenhouse gas (GHG) emissions integrated with land use and socio-economic development. These scenarios support scientists, stakeholders, policymakers, and planners for predicating the incoming risks and plan for adaptation or mitigation plans and measures. In this respect, there is a numbers of climate scenarios, the Intergovernmental Panel on Climate Change (IPCC 2014) scenarios as the most used and reliable one. IPCC scenarios is implied by the Representative Concentration Pathways including RCP2.6 (strong mitigation, warming about 1.5–2.0 °C); RCP4.5 (stabilization, warming about 2.5–3.0 °C), RCP6.0 (moderate emission, warming about 3.0 °C) and RCP8.5 (high emission, business as usual, waring > 4 °C). However, RCP 4.5 and RCP 8.5 are the most used ones. These scenarios, with different rates and magnitudes of climate change, provide a basis for assessing the risk of crossing identifiable thresholds in both physical change and impacts on biological and human systems; and therefore, they can predict changes in temperature, precipitation and sea level. From these variables, RCPs can introduce data on relevant components (e.g., number of warm days, number of snowing days, etc.).

The United Nations Framework Convention on Climate Change (UNFCCC) estimates that current commitments will increase emissions by 10.6% by 2030 compared to 2010, and that current nationally determined contributions (NDCs) may only limit global temperature rise to around 2.7–3.2 °C by the end of the century (UNFCCC

2022). Moreover, there will be differences between what is experienced on average at the global scale, compared to what might be realized at the local and regional scale. That is, even under a 2 °C warmer climate, there will be variations around the average response, with the Middle East Region expecting to see amplified warming relative to other regions (McCabe et al. 2023). Generally, the climate change in the Middle East Region is anticipated to significantly exceed global mean values. Given current projections, this will equally apply to most of the other Middle East and North Africa Region and countries of southwest Asia, and it will be accompanied with increases in the frequency and intensity of drought spells and warm days (Stocker 2013; Pal and Eltahir 2015). For this purpose, the results from the global climate models are utilized as boundary conditions for regional climate models.

By considering results of the regional climate modeling, an increase in temperature is anticipated in the Middle East Region. Results with wider scale can be concluded as follows (Bucchignani et al. 2018).

- Changes in summer temperatures will be uniformly above those observed for the winter months, which are more spatially uniform.
- The increases in temperature for the period between 2070 and 2099 (3.5–7 °C) time slots are significantly higher than for the mid-century period between 2040 and 2069 with 3–5 °C)
- The highest temperature rises are expected for region in the northeastern Mediterranean at latitudes north of 36°–38° N across the Balkan Peninsula and Turkey.

Ground-based measurements including the near past indicate consistently an increase in heat extremes in the Middle East Region and even North Africa, and the climate model results project a continuation of this trend and further rising mean temperatures until the end of the twenty-first century (Lelieveld et al. 2014, 2016). This implies that the number of heat waves (i.e., a prolonged period of more than five hot days and warm nights), may increase remarkably over the next decades. However, the average maximum temperature during the hottest days in the near past was about 43 °C, and it may increase to about 46 °C by the middle of the century and also may reach almost 50 °C by the end of the century (Lelieveld et al. 2016).

For the anticipated change in precipitation, many projections have made from regional climate models, but they are less robust (Lelieveld et al. 2012; Kostopoulou et al. 2014). However, there is a general decreasing precipitation trend and this is consistent with global climate models. According to Lange (2019), the period between 2040 and 2069, precipitation will be decreased by about 10–20% relative to the period 1961–1990 period. Besides, there will be increase in precipitation projected for Bahrain, Kuwait, and Qatar exceeds 30%, but since absolute amounts are extremely small, these changes are considered insignificant. While, for the Arabian Peninsula, extreme precipitation events are expected to occur up to three or four times per year by mid-term (ESCWA. 2023). While most Middle East countries are expected to witness reductions in annual mean precipitation by about 10%, while, for both Cyprus and Lebanon the decrease in precipitation will be in excess of 20% (Lange 2019). Even though precipitation will continue to decline until the end of the century, the overall decrease will be only moderately larger relative to mid-century for most of the Middle East Region (Lelieveld et al. 2012).

The changing climate will result in several implications on water availability and security in the Middle East Region, and they can be concluded as follows:

- Dry seasons with be widened for most Middle East countries (Lange 2019).
- An anticipated reduction in recharge of aquifers in the Mediterranean Coast between 30–70%, impacting the quantity and quality of ground water (Döll and Flörke 2005).
- There is expected reduction in surface and groundwater availability which will affect rivers, springs and the soil water reservoirs and resulting in adverse consequences on food production and security (Mohtar 2017).
- Runoff into the Middle East region will decrease by between 20 and 30% by the end of the century (Karami 2019).
- By 2030, the annual evaporation rate will be 40% cent higher than the precipitation rate, resulting in increased groundwater losses and increased salinity (ESCWA 2023).

8 Conclusion and Discussion

Talking about the climate and water of the Middle East Region always extends an expression of warm climate, dry land and deserts as well as scarce water resources. This is well pronounced in most of this region except those at the most upper latitude and notably those located at elevated areas. However, the real status is almost different; especially that the countries with driest lands in of the Middle East Region are known with sufficient water supply. In fact, the reason behind this is that these countries are either occupying large-scale water bodies, such as Egypt and Iraq (i.e., Nile River and Euphrates-Tigris River); or they made giant adaptive measures to secure water supply (e.g., desalination plants) and this is the case for most Golf Cooperation Council (GCC) countries. Therefore, mild and humid climate is not always reflected in water supply such as for Lebanon and Syria.

All performed studies and projections on hydroclimate of the Middle East Region confirmed that the region is characterized by a changing climate between arid, semi-arid and sub-humid and this has been characterizing the region since a long time. Nevertheless, the trending towards more aridity is obvious and the reduction in water availability is also touched, but also the current case is alarming and the climatic conditions are being rapidly changing and this is accompanied with an accelerating in population increase linked with new human aspects of water consumption. This will overload the stress on water availability, reduces water supply, water insecurity and make a confusion in socioeconomic status as a whole.

Based on trend analysis, field observations and tangible cases from the Middle East Region, the future is not imaginable for the region and the degree of changing towards water scarcity is cumulatively accelerated with time. In this respect, the issue is well pronounced and rapidly exacerbated in the low-income countries, while countries with better income are executing projects on the national level and it can be said that they are still working in the right way. While, the international and regional organizations are supporting different-scale water projects almost in the low-income countries, but the status quo is still undesirable and getting worse.

By the next few decades, the Middle East Region will witness unpreceded warm climate, least precipitation and the volume of water resources will be under their long-term average. Thus, water availability in many countries of the Middle East will be decreased to about 50% by 2040, making harmful consequence on agriculture, food security and energy and this will be associated with demographic changes and social and economic impact. Since the Middle East Region does not produce influential percentage of carbon emissions, maybe solutions are not tackled through the clime itself rather than the adaptive measures for water scarcity. This requires serious management, polices, strategies and awareness approaches; especially that the trend water consumption increase is moving in the opposite direction with renewable water decrease, and perhaps they now exactly on the intersection point between both.

The adaptive measures must be distinctly identified and the should be based on scientific approaches and adopting new and smart technologies. This is very crucial notably that many projects illustrated a number of measures in a generic sense regardless for the region characteristics, and this might be a reason for the failure of these measures in many cases. In addition, adaptive measures on the individual basis should be upscaled and improved notably in low-income countries, as well as the nature-based and ecosystem solutions must be adopted as a supportive tool to improve the status of water availability and supply in the Middle East Region.

References

Bucchignani E, Mercogliano P, Panitz J, Montesarchio M (2018) Climate change projections for the Middle East-North Africa domain with COSMO-CLM at different spatial resolutions. Adv Clim Change Res 9:66–80

Döll P, Flörke M (2005) Global-scale estimation of Di_use groundwater recharge; Institute of Physical Geography, Frankfurt University: Frankfurt am Main, Germany. Available online: https://www.uni-frankfurt.de/

Elkhalfi O, Chaabita R, Ghoujdam M, Zehraoui K, EL Alaoui H, Belhaj I (2025) The impact of climate change on food security in the Middle East and North Africa: challenges and adaptation strategies. J Agric Food Res 21:101963. https://doi.org/10.1016/j.jafr.2025.101963

ESCWA (2023) Regional climate modelling outputs for Saudi Arabia: key findings. Technical Report. Regional Initiative for the Assessment of Climate Change Impacts on Water Resources and Socio-Economic Vulnerability in the Arab Region (RICCAR). 25 pp

FAO (2020) The state of food security and nutrition in the World 2020. Available at: https://openknowledge.fao.org/items/08c592f2-1962-4e1a-a541-695f9404b26d

Hadjinicolaou P, Tzyrkalli A, Zittis G, Lelieveld J (2023) Urbanization and geographical signatures in observed air temperature station trends over the Mediterranean and the Middle East-North Africa. Earth Syst Environ 7:649–659. https://doi.org/10.1007/s41748-023-00348-y

Hasanean H, Almazroui M (2015) Rainfall: features and variations over Saudi Arabia, a review. Climate 3(3):578-626. https://doi.org/10.3390/cli3030578. https://www.un.org/en/global-issues/population

IPCC (2014) Climate change 2014: impacts, adaptation and vulnerability. In: Working group II contribution to the fifth assessment report of the intergovernmental panel on climate change, 151 pp

IPCC (Intergovernmental Panel on Climate Change0 (2021) Climate change 21: the physical science basis. IPCC 6th assessment report. Available at: https://www.ipcc.ch/report/ar6/wg1/

IPCC (2022) Climate change 2022: impacts, adaptation and vulnerability. In: Working group II contribution to the sixth assessment report of the intergovernmental panel on climate change, 3056 pp

Karami N (2019) The modality of climate change in the middle east: drought or drying up? J Interrup Stud 2(1):118–140. https://doi.org/10.1163/25430149-00201003

Kostopoulou E, Giannakopoulos C, Hatzaki M, Karali A, Hadjinicolaou P, Lelieveld J, Lange M (2014) Spatio-temporal patterns of recent and future climate extremes in the eastern Mediterranean and MiddleEast region. Nat Hazards Earth Syst Sci 14:1565–1577

Lange M (2019) Mediterranean and the Middle East and North Africa region and the water–energy nexus. Atmosphere 10:455. https://doi.org/10.3390/atmos10080455

Lelieveld J, Hadjinicolaou P, Kostopoulou E, Chenoweth J, El Maayar M, Giannakopoulos C, Hannides C, Lange M, Tanarhte M, Tyrlis E (2012) Climate change and impacts in the Eastern Mediterranean and the Middle East. Clim Chang 114:667–687

Lelieveld J, Hadjinicolaou P, Kostopoulou E, Giannakopoulos C, Pozzer A, Tanarhte M, Tyrlis E (2014) Model projected heat extremes and air pollution in the eastern Mediterranean and Middle East in the twenty-first century. Reg Environ Chang 14:1937–1949

Lelieveld J, Proestos Y, Hadjinicolaou P, Tanarhte M, Tyrlis E, Zittis G (2016) Strongly increasing heat extremes in the Middle East and North Africa (MENA) in the 21st century. Clim Change 137:245–260

Lobell D, Schlenker W, Costa-Roberts J (2011) Climate trends and global crop production since 1980. JSTOR 333(6042):616–620

Mahmoud M (2024) The looming climate and water crisis in the Middle East and North Africa. Carnegie Endowment for International Peace. Available at: https://carnegieendowment.org/research/2024/04/

McCabe M, Al Shalan M, Al Saud N, Hejazi M, Wada W, Peixoto R, Duarte C, Beck H, Maestre F, Guirado E (2023) Climate futures report: Saudi Arabia in a 3-degrees warmer world. KAUST King Abdullah University of Science and Technology, AEON Collective, p 133

Mekonnen M, Hoekstra A (2016) Four billion people facing severe water scarcity. Sci Adv 2(2):e1500323. https://doi.org/10.1126/sciadv.1500323

Meran G, Siehlow M, von Hirschhausen C (2021) Water availability: a hydrological view. In: The economics of water. Springer Water. Springer, Cham. https://doi.org/10.1007/978-3-030-48485-9_2

Mohtar R (2017) Climate change and the water-energy-food nexus in the MENA region. Available online: https://econpapers.repec.org/paper/ocpppaper/pb-1739.htm

NASA (2025) What is a climate change? Available at: https://science.nasa.gov/climate-change/what-is-climate-change/

NOAA (2024) Climate change: global temperature. Available at: https://www.climate.gov/news-features/understanding-climate/climate-change-global-temperature

Pal S, Eltahir E (2015) Future temperature in southwest Asia projected to exceed a threshold for human adaptability. National Climate Change 6:197

Shaban A (2020) Water resources of Lebanon. Springer Science Publisher, p 229. https://doi.org/10.1007/978-3-030-48717-1

Shaban A (2022) Satellite monitoring of water resources in the Middle East. Springer Science Publisher, 413 pp. https://doi.org/10.1007/978-3-031-15549-9

Shaban A (2025) Challenges on water resources of Lebanon. Presentation on simian: water resources in the Middle East: Challenges and evolutions. Agence Universitaire de la Francophonie, AUF

Stocker T (2013) Climate change: the physical science basis. In: Working Group I contribution to the fifth assessment report of the intergovernmental panel on climate change. Cambridge University Press, Cambridge (2014)
UN (2019) World population prospects 2019. Available at: https://www.un.org/development/desa/pd/news/world-population-prospects-2019-0
UN (2024) Peace, dignity and equality. Population. Global Issue. Available at:
UNFCCC (2022) Nationally determined contributions under the Paris Agreement. Synthesis report by the secretariat. https://unfccc.int/documents/619180
WBG (World Bank Group) (2025) Climate and development in the Middle East and North Africa. Available at: https://www.worldbank.org/en/region/mena/brief/climate-and-development-in-the-middle-east-and-north-africa
Welsh K, Pinder-Darling (2022) Water Availability and Access. In: Leal Filho W, Azul AM, Brandli L, Lange Salvia A, Wall T (eds) Clean water and sanitation. Encyclopedia of the UN Sustainable Development Goals. Springer, Cham. https://doi.org/10.1007/978-3-319-95846-0_165
WPR (World Population Review) (2025) Middle East countries 2025. Available at: https://worldpopulationreview.com/country-rankings/middle-east-countries

Hydroclimate Risks in the Middle East Region

Mashael M. Al Saud

1 Introduction

Climate change is influencing millions of people everywhere, and it does not affect only on water resources and other vital sector (e.g., agriculture, energy, food security), but also it affects human security, infrastructures and the environment by a number of recurrent hazards resulted from the extreme weather events, such as floods, mass movements and erosion, sea level rise, saltwater intrusion and a loss of biodiversity which led to more volatile, increasing competition for dwindling natural resources. The implications of climate change on human security as a whole have been tackled by the UN General Assembly and the Human Rights Council; and then, they were first recognized by the Security Council (Resolution 2349/2017) on the cri-sis in the Lake Chad Basin (Schaar 2023). Therefore, the Council has recognized the opposite effects of climate change on human life in the context of seven other regions. While, in 2018, UNDPPA, UNDP and UNEP jointly established the Climate Security Mechanism to strengthen the UN's ability to address climate-induced security risks.

The year 2022 revealed the 5th warmest year on record, with the average land surface temperature 1.11 °C warmer than pre-industrial period. According to the World Meteorological Organization, the increase in global air temperature is likely to reach 1.5 °C between 2023 and 2027. This slight increases in temperature can result a large non-linear impact on a wide range of catastrophic events, including drought, flood and tropical cyclone and many other climate risks. Without a radical reduction in Greenhouse gasses emissions, 3.5 to 7 °C of warming is likely before the end of the century, which would bring more severe consequences (Lange 2019).

An outstanding initiative has been carried out by the Regional Initiative for the Assessment of Climate Change Impacts in the Arab Region (RICCAR 2017), prepare by UNESCWA with a number of regional and international partners. The middle

M. M. Al Saud (✉)
King Saud University, Riyadh, Kingdom of Saudi Arabia
e-mail: geomm5.mashael@gmail.com

M. M. Al Saud (ed.), *Hydroclimate Changes: Implications of the Middle East Region*,
https://doi.org/10.1007/978-3-032-20053-2_2

east Region has been given a geographic domain providing more granular detail than the Intergovernmental Panel on Climate Change's (IPCC) regional domain, focusing on the assessment of the impact of climate change on the region's water resources. RICCAR, based on n the climate and hydrological models has carried out an integrated assessment for identifying vulnerability of various sectors including water, agriculture, biodiversity and ecosystems, infrastructure and urbanization, plus the availability of potable water, health, agricultural employment and the overall vulnerability.

The hydrologic cycle is affected by climate change and increasing the frequency and intensity of instable and upnormal climatic conditions. UNEP (2024) attributed over 90% of natural" disasters to weather-water-related ones, including drought and aridity, wildfires, pollution and floods. These disasters lead to death, injury, loss of livelihoods and displacement and place a huge burden on societies, economies and the environment. Climate risks often refer to an extreme weather or high-impact climate event whose characteristics, such as recurrence, intensity and duration, are strongly influenced or worsened by climate change induced human interference. Perhaps, climate risks are distinguished from other natural risks, which are naturally occurring hazards that cause harm. Even though, climatic events such as droughts, floods, hurricanes, etc. occur naturally, they become climate disasters when their severity is amplified by long-term shifts in global climate patterns. For instant, a typical drought might be part of a natural cycle, but when it is prolonged and intensified by deficit precipitation and warming temperatures; it is therefore, represents a climate risk.

The warmest year in the 175-year observational record was 2024 where alarming escalation in climate-related disasters existed with more than 150 significant events recorded worldwide (WMO 2024). This evidences that the world is on a dangerous path, driven by greenhouse gas emissions and on-going dependence on fossil fuels, and it is on a perilous path from record-breaking heat to destructive floods and storms. This ascending trend in temperature is accompanied with instability is precipitation and mainly the torrential rainfall and increase in cyclonic events along with prolonged drought spells. Therefore, implications have been existed on water resources and agriculture, plus other relevant sectors.

Climate risks have been lately existed on the top of the list for all natural hazards, and the meteorological anomalies have been exacerbated indicating a clear difference between the long-term average of precipitation, temperature, wind, etc. and the fluctuated average whether in ascending or descending trends. However, in some instance, these anomalies are slight enough with respect to the average, but the exposure (e.g., urban areas, water courses, excavated areas, etc.) has been changed and became more fragile and sensitive to these relatively small anomalies. For example, a watercourse is normally runoff surface water along channels, but the distortion in channel embankments or urban encroachments on these channels may hinder water flow and cause floods, which means that the existed flood is a result of human interference on land exposure. Therefore, it worth considering the changes made mainly by human of these exposure before making the judge of climate regime, and this can be a first-hand information for implementing adaptation and mitigation measures to what is attributed as climate risks.

In the Middle East Region, rapid urbanization along with new land planning projects as well as the development of industrial facilities are well pronounced and they increase the exposure of people and economic assets to disaster events; while risk reduction and controls are not significantly accounted and this increases the impact level of natural risk notably the climatic ones. The impact of urbanization development is mainly influencing in the coastal areas where the largest cities and economic hubs are located and representing zones for outlets of rivers and large streams resulting in collusion between the runoff water downstream with coastal urban settlement, or confronting coastlines for marine risks (e.g., coastal flooding, sea level rise, etc.). There are also many reasons acting on the impact of the existed disasters including insufficient resources in the low-income countries, lack or poor infrastructure, chaotic urban sprawl, lack of monitoring tools, inadequate risk management and lack to Early Warning Systems (EWSs) which became a successful tool for risk alter and it contributed in the reduction of risk impact. In the Middle East Region, EWSs have been established with scales and by adopting a number of Anticipatory Climate Action Models (MEACAMs) where satellite images are the main source of geospatial data (Fig. 1). This will help in making pprojections where economies and livelihoods of the Middle East Region is anticipated to be the second most affected by climate change (WB 2018).

Natural hazards resulted from climate several aspects of physical systems including the atmospheric, hydrologic and marine systems where the impact is often catastrophic. According to Al Saud (2022), these hazards can make implications on:

- People and human activities (decease, injury, disease, economic, life style, etc.),
- Property and environment (property damage, ecosystem, economic loss).

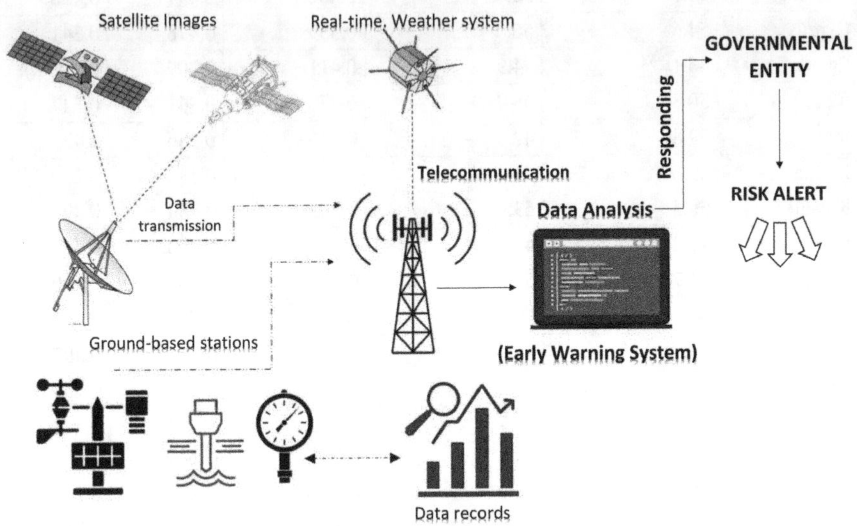

Fig. 1 Schematic figure for proposed Early Warning System

2 Hydroclimate Risks

According to EM-DAT (2023), natural disasters were classified as: 4%, 5.5%, 39%, 5%, 31%, 3%, 2% and 10.5% for extreme temperature, drought, floods, landslides, storms, wildfire, volcanoes and earthquakes; respectively. This indicates that about 48% of natural hazards are directly connected with climatic conditions, and 39% can be connected to climate, while the rest 12% are mainly due to geological hazards. While current estimates suggest the global economic impact of natural hazards (e.g., floods, earthquakes, etc.) costs to $2.3 trillion annually (UN-Geneva 2024), yet this was about $125 billion in 2013 (MRC 2013) which show an in-creased cost of about 18 time over about one decade.

Many countries in the Middle East Region are currently in a state of fragility or in conflicts making them the most severely exposed to climate change impacts and its hazardous risks (Peter et al. 2019). According to the Notre Dame Global Adaptation Initiative index (ND-GAIN 2023), which combines measures of readiness and vulnerability to climate change, there are Syria, and Yemen are at the bottom of the ranking, besides good ND-GAIN index in GCC countries including mainly UAE, Qatar and Saudi Arabia (Table 1).

This consistence between upper-rank income countries and ND-GAIN index reflects the role of financial resources in developing adaptation measures needed.

Table 1 Notre Dame Global Adaptation Initiative index for the Middle East (ND-GAIN 2023)

Country	ND-GAIN Index (2020)	ND-GAIN Index (2023)	Vulnerability	Readiness	Exposure
Bahrain	48.7	52.8	0.436	0.491	0.418
Cyprus	54.6	56.0	0.359	0.478	0.341
Egypt	46.1	44.9	0.411	0.309	0.360
Iran	50.8	48.2	0.376	0.341	0.347
Iraq	39.8	42.5	0.436	0.286	0.437
Jordan	49.6	51.0	0.366	0.387	0.289
Kuwait	50.3	53.9	0.362	0.439	0.397
Lebanon	45.2	42.5	0.399	0.250	0.332
Oman	54.7	52.3	0.420	0.465	0.410
Palestine	ND[a]	ND	ND	ND	ND
Qatar	53.0	58.0	0.351	0.512	0.397
Saudi Arabia	54.3	57.9	0.408	0.566	0.358
Syria	39.0	35.7	0.486	0.200	0.372
Turkey	51.4	53.9	0.375	0.453	0.415
UAE	58.9	59.9	0.383	0.582	0.367
Yemen	33.5	34.2	0.534	0.219	0.452

[a]No data available

It is also obvious from Table 1 that the ND-GAIN index has been abruptly changes between 2020 and 2023 where some Middle Eastern countries increased its index such as in Qatar, Saudi Arabia, Bahrain; besides a decrease in other countries such as Lebanon, Iran, and Oman. In addition, the Notre Dame Global Adaptation Initiative has illustrated the vulnerability, readiness and exposure for the climate impact worldwide.

Table 1 shows the results for the Middle East countries. Perhaps the vulnerability and exposure, which are almost naturally existed are not controlled by human and development of adaptive measures, but readiness must be. For example, this interprets the scoring as very low in vulnerability and exposure, but very high in readiness for the UAE (0.383, 0.367 for vulnerability and exposure) and 0.582 for readiness, and this is the case for other countries, such as Saudi Arabia, Qatar, Cyprus, etc.; while other many other Middle Eastern countries show very low readiness, such as Syria, Lebanon, Iraq and Yemen, and this might be linked with the geo-political situation in these countries.

Risks and hazards are always two terms used to explain dangerous impacts on human life, infrastructures and the environment (terrestrial and marine), while the risk is the chance of that harm actually happening. CBSE (2006) defined natural hazard as a dangerous condition or event, that threat or have the potential for causing injury to life or damage to property or the environment. Therefore, the term “risk” is more alarming to the potential natural hazards to be occurred. Risks have various origin(s) and mechanism of occurrence and they have diverse impact depending on the exposure and land they occur in. Risks can be natural/ or man-made and in many instances both are combined and then result different level of damages. Thus, human might directly or indirectly trigging the disastrous processes (Al Saud 2020).

Both variability in physical processes and human-induced climate change and extremes have driven considerable shifts in global weather patterns, profoundly impacting water resources, agriculture, and ecosystems (Fawaz et al. 2020). However, growing evidence highlights that anthropogenic factors including mainly greenhouse gas (GHG) emissions and the changed land cover/use are key contributors to the increasing frequency and intensity of meteorological disasters and climatic extremes (Raymond et al. 2020). This is well pronounced in arid and semi-arid regions, including the Middle East Region, where drought and water scarcity pose serious threats; especially food security, natural resources and even the geopolitical situation. The increased temperature rate does not only impact hydrological processes, but also drives profound changes in regional climates, leading to more frequent and intense extreme events such as droughts and heat stress (Ouarda et al. 2014). According to the Fifth Assessment Report (AR5) of the Intergovernmental Panel on Climate Change, a global mean temperature increase of 0.85 °C between 1880 and 2012 has already heightened the frequency of meteorological disasters including droughts, floods, cyclone, etc. (Stocker 2013).

3 Hydroclimate Risks in the Middle East

Hydroclimate risks is a terminology used to integrate the combined hydrological and meteorological risk; especially they both have an integrated mechanism of occurrence and act in the same process with similar impact. Hydroclimate risks are physical conditions that they are slightly upnormal or even they may occur under normal meteorological/hydrological conditions if human interference play a significant role in the hydroclimatic cycle, such as expanding human settlements in the pathways of water flow. Perhaps, hydroclimatic risks differ from other natural risk in that they can be predicated even within a period of few days and sometimes hours; especially that satellite-based images and products can significantly help in this concern. While, monitoring the risk occurrence can be on the basis of near-real time or even at the real-time. Lately, the impact of hydroclimate risks has been preceded the impact of other risks notably the recurrence and number of climate risks are numerous. Moreover, natural risks, other than those of hydroclimate ones, are still occurring almost constantly in terms of number and geographic zone, such as earthquakes are usually in tectonic-active zones, volcanos in zone with volcanic craters and eruption. Moreover, the hydroclimate risks can be creeping events that may extend to decades such as desertification and drought. All these behaviors and mechanisms of risks characterize hydroclimate risks.

It must be made clear that there are many aspects of risks with different mechanism and level of impact and they are harmful on diverse exposures and biological elements while they are even caused by different factors and numerous sources including human impact, notably the recently occurred human activities with a special emphasis to industries, use of energy sources and the development project. Among these risks, hydroclimate may be intruded with other risks even at a slight level of impact; and thus, forming a combined aspect of risks. Table 2 shows the natural and man-made risks listed by CBSE (2006), and the relevance to the hydroclimate impact has been added. risks. Some of the above listed risks can be controlled or at least reduced or can be confront when they occur, and these can be from various rick types (e.g., mass movement, sea erosion, epidemics, etc.); while, some other are not notably those located under the geological and hydrological hazards, such as volcanoes, earthquakes, tsunamis, cyclones, Cloudburst and other climate extremes.

The physical and biological impacts of climate change will be extended to other socioeconomic pressures and the lack of resilience, posing a threat to economic development in the region (Ezzeldin et al. 2023). The climate change impact will be associated with financial loss, which is estimated between 1.2 and 1.5% of GDP for the MENA Region, as well as this is anticipated to be increased several times in the absence of appropriate mitigation and adaptation measures and controls. The economic impacts for the projected climate change are highly interlinked with drought and water security, reduced agricultural productivity, increased migration, and demographic and land cover changes. Thus, it is normal that the impact of climate change in low-income countries is more affecting and this is attributed to

Table 2 Natural hazards and their risks and their relevance to climate (adapted after CBSE 2006)

Natural hazard	Type of risk	Hydroclimate relevance
Geological hazards	Earthquakes	Nil
	Tsunami	Very weak
	Volcanics	Nil
	Mass movement	Moderate to strong
	Dam burst	
	Mine collapse	Very weak
Hydrologic hazards	Tropical cyclone	Very Strongly
	Cloudburst	
	Tornado and hurricane	
	Floods	
	Heat and cold waves	
	Drought	
	Snow avalanche	
	Hailstorm	
	Sea flood	Strong
Environmental hazards	Environmental pollutions	Moderate
	Desertification	Strong
	Deforestation	Nil
	Pest infection	Weak
	Dust storms	Strong
Biological hazards	Human/animal epidemics	Weak to moderate
	Food poisoning	Very weak
	Pest attacks	Weak to moderate
	Weapons of mass destruction	Nil
Chemical, industrial and nuclear accidents	Chemical disasters	Very weak
	Oil spills	Weak
	Industrial disasters	Nil
	Nuclear disasters	Weak
Accident related hazards	Boat/road/train/air crash/rural and urban fires	Moderate
	Building collapse	
	Mine flooding	Strong
	Forest fires	Moderate to strong

[a]Maximum impact

several factors including the lack of mitigation and protection controls, poor education and awareness, economic failure, as well as the geopolitical conflicts. All these factors weakened the resilience of low-income and remote communities even in the same country; especially that urbanized areas are always given the priority. There are many countries vulnerable to climate-induced risks, but they follow a number of management approaches and implement risk controls; and there, risk impact will be negligible and innocuous.

3.1 Drought and Desertification

Drought and desertification are two complementary terms used to describe aridity and the loss of natural resources; and both are highly affected by climate extremes and climate change. Drought is simply the regression in precipitation rate for significant time period with respect to the long-term average (LTA). Nevertheless, even if this rate declines below the LTA, yet some countries remain its water supply steady; and thus, in a volumetric sense there is no drought. this is why drought is also interlined with water supply. While desertification has a human contribution and not only climatic impact and it represents the loss of natural resources; almost focuses on the green cover, which might be attributed to climate change (and extremes) or human interference (e.g., urban expansion, wood logging, etc.), or combination of both climate and human impacts. Dry lands occupy more than 2.5 billion people, cover more than 40% of the world's surface land and they retain less than 8% of the world's renewable water resources (EcoMENA 2025).

In the Middle East Region, drought is among the most devastating consequences of climate extremes, and ranking second only to floods in its global impact and affecting about 7.5% of the population. Its repercussions include shrinking and loss of natural resources, reduced agricultural productivity, constrained water availability, disrupted ecological functions, and compromised economic resilience in local communities (Jiang et al. 2022). The growing complexity and uncertainty of droughts under climate change challenge traditional monitoring and prediction methods (Khosravi and Ouarda 2025). There are lately many traditional drought indices, such as the Palmer Drought Severity Index (PDSI), the Standardized Precipitation Index (SPI), the Standardized Precipitation Evapotranspiration Index (SPEI), etc. provide global-scale drought assessments. There are many studies done in the Middle East Region using these indices (e.g., USAID 2021; Elsayed et al. 2024).

Results, from various studies, indicate a significant shift from wet climate before 2000 to more frequent drought occurrences in the Middle East Region; especially in the Levant and western parts of the Arabian Peninsula, northern Iraq, and western Iran. Both the SPI and SPEI indicate significant positive changes in drought frequency across 19.4% of the MENA Region during the period 1998–2021, compared to 12.9% in the earlier period of 1981–1997 (Kenawy et al. 2025). In addition, the analyzed SPEI, which accounts the evapotranspiration, showed a recently increase in drought frequency and severity than the SPI, which reflects a growing impact of temperatures increase. There are also discrepancies between SPI and SPEI in hyper-arid regions like the Sahara and Arabian Peninsula. The region experienced more frequent and prolonged, but less intense, droughts in the past two decades. Spatially, the SPEI highlighted significant drought intensification in temperate and cold desert.

The temporal changes of drought risk under the Shared Socioeconomic Pathways (SSP2.6, SSP4.5, and SSP5.8) shows the extent of drought intensification for the projected years 2040 and 2070. According to Khosravi and Ouarda (2025), in order to the link between the spatial expansion of drought and the discrete high-risk drought zones (Fig. 2), a threshold-based categorization method was applied, where the Geographically Weighted Temperature Vegetation Dryness Index (GWTVDI) trends representing cumulative changes in dryness, were refined through specific percentile-based criteria to extract areas where climate stressors have converged to critical levels. Figure 2 shows that locations with recurrent extremes in temperature sensitivity, precipitation decline, and trend significance were systematically separate as drought hotspots. This systematic linkage means that high-risk zones are spatial expressions of persistent and intensifying climate-driven degradation captured through GWTVDI evolution and not as isolated anomalies.

Desertification, interlinked with land degradation, is widespread in the Middle East Region and falls mainly within the boundaries of dry lands. Problems linked to desertification are found as: soil erosion, urban encroachment, forest fire, wood logging, overgrazing, shrub clearing and salinization. These have been affecting Food and water security, Economic Implications, Loss of Biodiversity and human displacement. For the Arab region alone, 70% of the drylands are desertified (ACSAD 2008). Around 48.6% of the land area in the Mashreq, 12.4% in the Nile Basin and 9% in the Arabian Peninsula are desertified. Among the Middle East countries, significant desertification occurs in Egypt and Jordan, Bahrain, Kuwait, Qatar, and the UAE (EcoMENA 2025).

The Mashreq Region, dependent on winter rainfall which is almost less than 350 mm/year and short rainy periods, cultivate low-productivity cereal, legume and tree crops, with sheep and goat herding on fragile, low-yielding rangelands which have been degraded from overgrazing, increasing demand for meat and milk by growing populations have compelled herders to buy more and more of the feed and fodder that their animals need. Moreover, in the Mashreq occupies forests on high-land such as in Turkey, Syria and Lebanon and other elevated areas where forest fire and wood logging is pronounced and increased the rate of land degradation. While, the unplanned urban encroachment is also common and resulting in the loss of large parts of natural resources. While, the agriculture in the Nile River Basin, is based on intensively-managed irrigated systems. The exceeded population pressure and least rainfall makes it significantly dependent on maximizing food production per unit of land, which increases salinity risks, water contamination, water shortages and ecological damage (Winslow and Thomas 2016).

The Arabian Peninsula, one of most arid regions, depends significantly on irrigation, but it lacks river systems and perennial watercourses. The Arabian Peninsula depends on groundwater abstraction; which groundwater is being depleted faster than the rainfall can replenish it. The growing urban areas are taking priority for

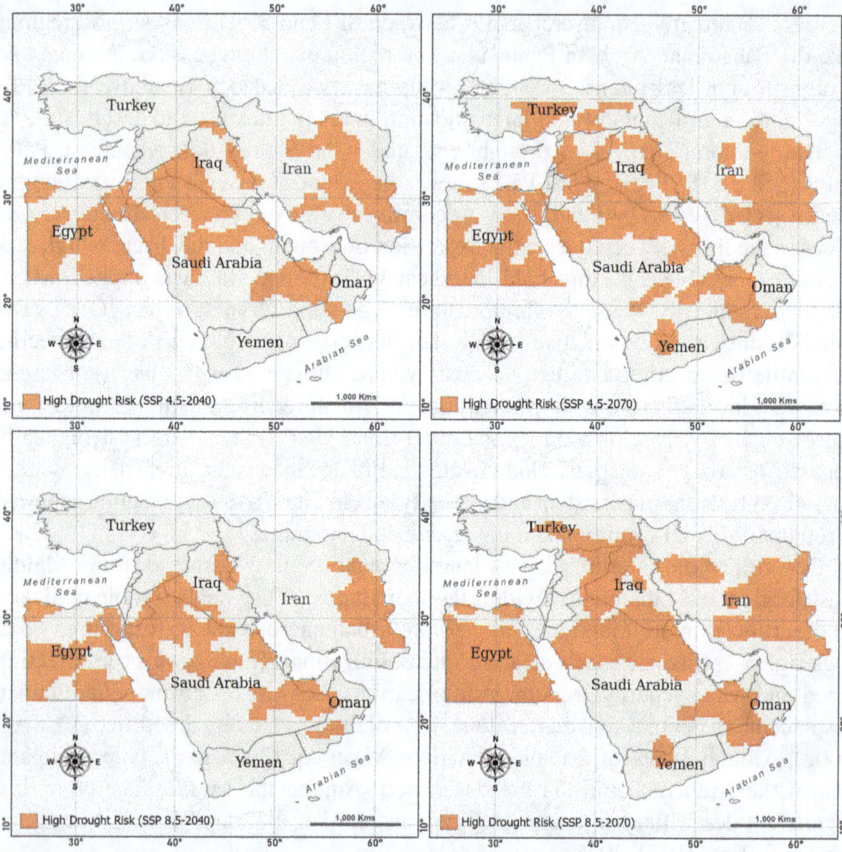

Fig. 2 Projected drought hotspots for the Middle East Region under SSP scenarios for the projected years 2040 and 2070 (Modified after Khosravi and Ouarda 2025)

the scarce freshwater, leaving agriculture to use low-value brackish or salty water that can increase the risk of salinizing the soil (Winslow and Thomas 2016). Besides, some regions in the Peninsula adopt desalinization of seawater and agriculture cannot afford this costly source. While, to the east, aridity in Iran is interrupted by rugged topography and extreme temperature variability; notably the colder winter. Therefore, irrigation in inter-mountain basins usually makes soil salinity; while irrigation along sloping terrain often scars the land with erosion gullies.

3.2 Floods

Even though floods are more damaging and they exist in short time and without any alert, yet it was ranked as the second after desertification and land degradation due to its areal coverage which is often for specific regions and not like the case of desertification covering almost the entire Middle East Region. Over the last four decades, floods have become the most recurrent natural disasters as it was reported by EM-DAT who recorded that at least 300 flood events constituting about 53% of the total number of natural disasters (Al Saud 2022).

Several factors causing floods, and these factors differ between regions. However, the general understanding is that floods are caused by heavy rainfall (i.e., torrential), but what about if torrential rainfall occurs in an empty region with no urban areas? No risk will be reported and the flood will be totally ignored. Therefore, the rick is linked with its impact, and this is exactly what is happening in many regions of the Middle East where surface water along large-scale wadis was uniformly runs over the past time, but the existence of human settlement along these wadis hinders the flow and resulted in damaging floods. From this point of view, flood occurs and makes impact from combined factors which can be classified into three main pillars: the meteorological, terrain components and human interference. For the meteorological pillar, extreme weather as recent aspect of the changing climate is the main reason, and this may be linked to other meteorological processes, such as the atmospheric rivers which have significant responsibility for some of the hydroclimatic extremes around the world (Dezfuli 2020). While, for the terrain components, Al Saud (2015), integrated ten factor causing floods and these are: drainage density, stream intersection, meandering ratio, stream order, narrowness on valleys, slopes, depressions, lithology, rock fractures and sand accumulation. These factor, other terrain components, represent the vulnerability of a region to flooding. However, for the human interference, it is rarely accounted and it can be attributed to negative human activities (e.g., excavation of valley embankment, dumping wastes and debris along valley, etc.), or the construction in the Wadis' courses. The Middle East Region has been recently witnessing frequent damaging floods and it can be said that they have uneven geographic distribution and impact.

Floods in the Middle East Region has a long history and it is almost changing between regions and dates during the years. A survey has been performed by the author for the recently occurred floods between 2020 and 2024 (Fig. 3). It is obvious that 45 damaging floods have been occurred over the projected five years in 10 Middle East countries. The number of floods per each country is a result of either meteorological, terrain components or human interference, and sometime the combination of two all of them. However, there are defined characterization for the number and distribution of these flood events. For example, the increased number of floods in Turkey (11 floods) and Iran (8 floods) can be attributed to the upper latitude plus the mountainous topography in both countries, and this concept perhaps slightly affecting Syria. The presence of floods in Yemen (6 floods) and Oman (6 floods) can be attributed to the cyclones created from the Arabian Sea. While, from the management point of view, notably the execution of flood controls, it can say that Saudi Arabia and UAE with relatively low number of floods, 4 and 3 floods; respectively is attributed to the flood reduction controls, and this is clear because these two countries were witnessing damaging floods between 2009 and 2015. From the other hand, Middle East countries who did not report floods during the projected period

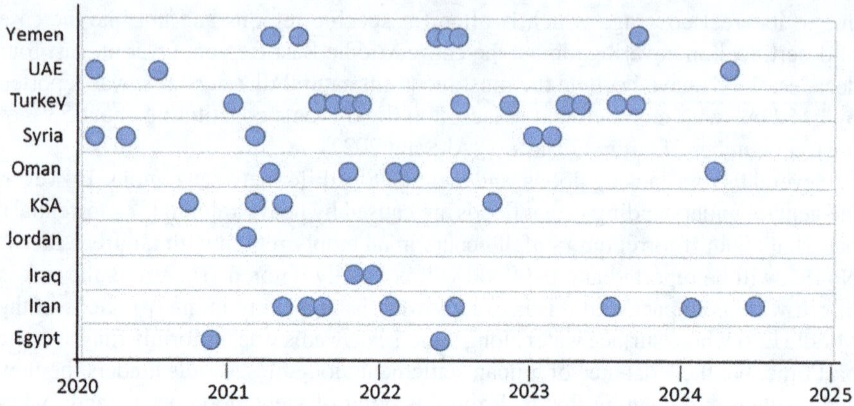

Fig. 3 Flood events occurred between 2020 and 2024 in Middle East countries

can be also attributed to various reasons. For example, Lebanon and Cyprus have narrow valleys where water directly diverted along these valley and outlet into the sea. While, Kuwait, Bahrain and Qatar have few mountainous topographies.

3.3 Cyclones

Cyclones, as atmospheric disturbances swirling winds around a low-pressure center, always one of the most dangers hydroclimatic risks. They can be of: tropical cyclones (hurricanes and typhoons), polar cyclone, mesocyclone and extratropical cycles. The Middle East region is subjected to a number of cyclones which are even dry cyclones (usually called dust storms) with no precipitation, and these are common in Iraq, Syria, Egypt and Arabian Peninsula with local names as Khamsin in Egypt and Shamal winds in the Arabian Peninsula. There are also wet cyclones and typhoons where they can be as snow storms affecting parts of Eastern Mediterranean such as Cyprus and Lebanon; or heavy rain storms frequently occur in the southern part of the Arabian Peninsula and the Arab Sea. Recently, these cyclones and their movement/speed can be predicated using advanced models such as Hurricane Weather Research and Forecasting (HWRF); example is shown in Fig. 4.

3.4 Wildfire

Climate warming and heat waves have caused a widespread increase in fire weather, making forest fires more frequent and larger. From the other side, forest fires have return impact to climate by modulating land–atmospheric car-bon, nitrogen, aerosol, energy and water fluxes. At this point, combined satellite-based datasets,

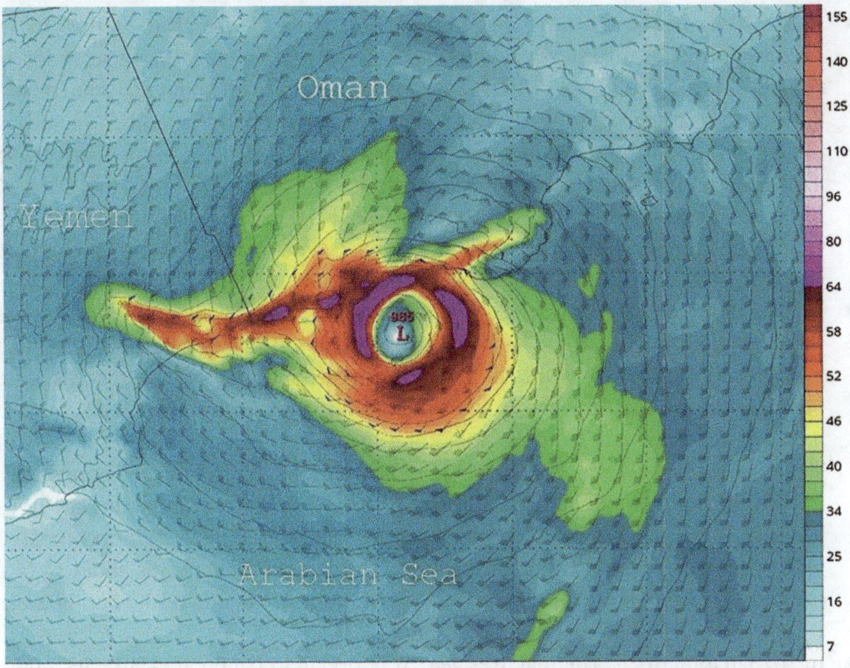

Fig. 4 Wind speed (knots) for Mekunu on May 25, 2018 as predicted by HWRF model near Salalah, Oman. It was categorized at a Category 1 hurricane with 135 km/h winds (Levi Cowan: https://www.tropicaltidbits.com/)

extended by NASA and many other space agencies, are helpful for global wildfire events (example: https://www.firemap.live/), surface radiometric temperature, surface albedo, ecosystem evapotranspiration and incoming shortwave radiation, and many other meteorological parameters. Therefore, large-scale wildfires have been associated with higher burn severity, as measured by greater post-fire decrease in vegetation greenness (Cansler and McKenzie 2014) or larger combustion of surface organic carbon resulting in higher carbon dioxide (CO_2) emissions per unit area; and therefore, have a greater biogeochemical climate warming impact.

Possibly, wildfire does not have much concern in the Middle East Region due to the limited geographic extent of forests and dense vegetation cover. However, its harmful extends primarily to the environment and ecosystem and then to human life. In addition, wildfire is due either to human inference or meteorological factors represented mainly by exceeded the heat. Therefore, the vulnerability to wildfire is firstly controlled by dense vegetation cover, and in the Middle East Region, this almost exists only in Turkey Cyprus, Lebanon, Syria and Iran which are almost, expect a part of Iran, above latitude of 33°.

Examples from the Middle East Region can be illustrated as follows:

- In 2023, Cyprus witnessed "mega-fires", one of them was close to the city of Alexandroupolis of East Macedonia, was the largest single wildfire recorded in the EU since 2000. Thus, 19 fires were mapped, burning a about 2010 ha. About 82% of the damage occurred in the months of August and September, including the largest fire of the season which covered over 650 ha in Limassol district. 545 ha burned on Natura 2000 sites, accounting for a quarter of the total and 0.229% of the total protected land of the country. The 2010 ha was distributed to 572 ha Coniferous forest, 691 ha Sclerophyllous vegetation, 614 ha Transitional and the rest were miscellaneous of agricultural lands (Joint Research Center 2024).
- In 2023, the average temperature in Turkey was 1.2 °C above the long-term average (between 1991 and 2020). Thus, in August 2023, a heat wave was occurred and it exceeded the average by about 5 °C and above; and temperature records were broken at many stations. This heatwave continued between 5 and 10 days and it has been recorded in 22 stations including the central and inner-western parts of Turkey. Due to the diverse meteorological conditions, there has been an increase in both the number and land area of forest fires in 2023 compared to the previous years. In 2023, the amount of lost area per fire increased by 0.09 ha/fire compared to the previous year, reaching 6.02 ha/fire. The total burnt area was about 15 520 hectares. The number of fires was 2579 in the same year, one of the burnet areas was about 7039 ha (Joint Research Center 2024).
- In 2023, a total of 23 principal fires were reported in Lebanon, they affected about 132.4 ha. The fire season started in 3rd March, and ended in 23rd September. The peak month for the number of fires was in August (totaling 5 large fires) and the peak month of burned areas was in June (totaling about 40.5 ha). Lebanon experienced a notable decrease in the number and extent of wildfires, which is attributed largely attributed to the successful implementation of a nationwide emergency coordination plan for fire preparedness, risk reduction (Joint Research Center 2024).

3.5 Other Hydroclimate Risks

There are many other hydroclimate risks frequently reported in the Middle East Region. These risks are directly or indirectly linked to meteorological and hydrological extremes, but the terrain characteristics (e.g., geology, soil, slope, etc.) and human interference (e.g., excavation, water abstraction, etc.) play significant roles in this respect. Among these risks are:

1. Mass movement (e.g., rock fall, landslides, mud flow, etc.) and erosion of surficial materials. This type of risk integrates intensive rainfall with other components likes clayey soil, excavated slopes, deforestation, etc. this is well pronounced in many Middle Easter regions, but they often occur locally with small-scale area. Al Saud (2022) reported several aspects of mass movement and land instability in Saudi Arabia representing many mass movement risks in the entire Arabian Peninsula.

2. Pollution of natural resources (e.g., soil, water, and air) is another type of hydroclimate risk is the and the ecosystem which is mainly created by human by it is sometimes enhanced by weather and water impact. For example, heavy rain results turbidity of water along river courses outlet into the sea and dump marine ecosystem. Another aspect of hydroclimate pollution risk is indirectly connected with exceeded evaporation and evapotranspiration which results a stress on groundwater and then leads to the exhausting of coastal aquifers and then the saltwater intrusions. This is well pronounced in many coastal regions of the Middle East such as Kuwait, Bahrain and Qatar (Shomar and Sankaran 2024).
3. Aeolian erosion in the Middle East is a major contributor to global dust emissions, accounting for 20–25% annually. This is equivalent to a staggering 400–500 million tons of dust (CMES 2025). This dust act in polluting air and then human health.
4. Sea level rise is another alarming hydroclimate risk in the Middle East Region; however, studies are still few in this concern, but the change detection of coastal morphology using satellite images evidenced remarkable invasion seawater on-land, and this is often associated with coastal erosion. The extreme temperatures could make parts of the Gulf Region unlivable by the end of the century. For example, sea levels rise in Bahraini ranges between 1.6 to 3.4 mm per year. An extreme sea level rise between 0.5 and 2 m could submerge 5–18% of the total area of Bahrain Aljenaid (2024). The extreme temperatures could make parts of the Gulf Region unlivable by the end of the century. For example, sea levels rise in Bahraini ranges between 1.6 to 3.4 mm per year. An extreme sea level rise between 0.5 and 2 m could submerge 5–18% of the total area.

4 Hydroclimate Insights in the Middle East Region

Usually the hydroclimate risks occur at different phases of impact; and thus, they may have direct impact at the time of their occurrence, and also an impact after sequential phases. For example, extreme temperature can directly hurt human forest fire or health, but it also works in evaporation processes over long time periods and leading to drought in another sequential phases and so on. This is also the case for torrential rainfall which may destruct the infrastructures at its beginning but it also reduces the infiltration rate and result in floods days later on after the initial event.

In a report produced by the Met Office-Hadley Centers climate science (Hadley Center 2021) and combined expertise from the MENA Region provided by Overseas Development Institute (ODI), a number country reports were produced, and they can be summarized/ adapted as follows:

- Bahrain: [*Average precipitation = 85 mm/year, average temperature = 26.5 °C, water availability = 18 m*^{3}*/capita/year, water demand = 746 m*^{3}*/capita/year*].

It is an island located in the Arabian Gulf where most people live in the northern part and in main urbanized cities such as Manama and Riffa, but it has very little agricultural production. The country experiences a warm desert climate with rain tends to fall in heavy spells of downpours. It is already experiencing a significant warming even before pre-industrial times. There is high confidence that Bahrain in the 2050s, will witness a temperature increase between 2 and 4 °C, with widened increase in summer and autumn time. Projections did not point out to any increase in precipitation, but downpours (i.e., torrential rains) are projected to become more intense. In addition, sea levels will rise and this will be associated with rising sea surface temperatures and increased number of intense storms. Regional level risks affecting Bahrain include heat stress and coastal impacts.

- Cyprus: [*Average precipitation = 500 mm/year, average temperature = 17.5 °C, water availability = 656 m*^{3}*/capita/year, water demand = 228 m*^{3}*/capita/year*].

Cyprus, represents an island along the Eastern Mediterranean, is known by its mild Mediterranean climate, which is warm in summer and wet and rainy in winter. Due to its limited areal extent (9251 km^2), the agricultural lands are almost limited. The country, like other Mediterranean counties, is witnessing a warm weather with an increased trend, and temperature is anticipated to increase up to about 4 °C by 2050. The country is witnessing the changing climatic conditions and risks, including heat waves, and declined precipitation rate which that affect its water availability and increasing saltwater intrusion. For the regional level risks, Cyprus will be affected, in addition to the reduced water availability and by the increased evaporation that reduces groundwater productivity.

- Egypt: [*Average precipitation = 200 mm/year, average temperature = 23.8 °C, water availability = 465 m*^{3}*/capita/year, water demand = 757 m*^{3}*/capita/year*].

It is the MENA country that borders the Mediterranean and Red Sea coasts. Most urbanism is located along the Nile River (transboundary water resource) and Nile Delta and in main cities such as Cairo and Alexandria. The majority of agricultural productivity exists along the Nile floodplain and in the Delta. Egypt has already experienced a significant warming since the pre-industrial period. In addition, projections estimated an increase in temperature between 2 and 4 °C in 2050, with larger increases during the hottest months. While, rainfall will not/or will be slightly changed. The sea levels will rise with increased sea surface temperatures and more intense Mediterranean storms. Regional level risks affecting Egypt include heat stress; mainly in cities like Cairo where humidity is high; while water availability will be affected by the increase in evaporation; in addition, and coastal inundation and erosion in the Nile Delta are anticipated.

- Iran: [*Average precipitation = 250 mm/year, average temperature = 19.5 °C, water availability = 1675 m*^{3}*/capita/year, water demand = 1110 m*^{3}*/capita/year*].

This is a Middle Eastern country which is bordered by the Caspian Sea from the north and the Arabian Gulf and Gulf of Oman from the south. It includes a number of highland and populated cities such as Tehran, Mashhad, Isfahan. Mixed farming systems are well known in the highlands where snowpack is almost found at altitudes above 3200 m, and giving a chance for snowmelt to feed river and support irrigation systems. However, some low-land areas to the south-east have been experiencing warm weather since pre-industrial period. Temperature is anticipated to increase by 2025 and it is estimated to rise 2–3 °C, and this will be accompanied with remarkable increase during the hottest seasons. While projections evidenced an increase between 2 and 4 °C in the low-land areas by 2025. However, precipitation change is still uncertain whether to increase or decrease. The Caspian Sea level is projected to drop, and increasingly intense storms will impact coastal areas. For the coastal zones along the Arabian Gulf and Gulf of Oman, they will be influenced by sea levels rise, plus frequent intense storms. Regional level risks impacting Iran include a reduced agricultural productivity and deficit in water availability as a result of exceeded evaporation.

- Iraq: [*Average precipitation = 275 mm/year, average temperature = 25.5 °C, water availability = 915 m*^{3}*/capita/year, water demand = 1140 m*^{3}*/capita/year*].

It is located across two geographic zones, the highland region of Iraq-Iran and the south-eastern Turkey zone. In the highland area, includes a number of major cities including Erbil, Kirkuk, and Mosul. Both geographic zones are rainfed and drylands with dissipated farming systems; while irrigated areas are dominant along the flood-plain of Tigris River. The highland areas are already a warm region, and it is expected to witness temperature increase between 2 and 3 C by 2050. While, precipitation has not anticipated yet and it is still uncertain if it will be changed. However, the southern part of Iraq, which includes Baghdad and Basrah, is more warm and rain is limited to torrential rainfall spells. This geographic zone is also warm region since pre-industrial times. By 2050, it is anticipated that this region will have an increase in temperature between 2 and 4 °C with extended increases in summer and autumn. While, precipitation is uncertain, but torrents are anticipated. In addition, a regression in the discharge of the Tigris and Euphrates rivers will be existed, and the sea level will rise as well as sea surface temperatures and this will impact the small coastline on the Arabian Gulf. Regional level risks that will impact Iraq are mainly reduced water availability and increased demand.

- Jordan: [*Average precipitation = 116 mm/year, average temperature = 20.4 °C, water availability = 182 m*^{3}*/capita/year, water demand = 108 m*^{3}*/capita/year*].

It is situated in the Levant region with small coastline on the Red Sea, and dense population is living in Amman. Jordan experiences a warm climate derived from the deserts to the east, and a temperate Mediterranean climate from the west. The country has already known by significant warming. Projections indicated that by 2050, there is high confidence that the temperature will increase between 2 and 4 °C. precipitation trend is not well determined but an overall decrease is most plausible. In addition, sea levels will rise along with increased sea surface temperatures plus frequent intense storms which will impact coastal regions. The regional level risks influencing Jordan include decreased water supply and decrease in the Jordan River discharge, as well as and groundwater will be depleted and this will result in food security.

- Kuwait: [*Average precipitation = 62 mm/year, average temperature = 27.5 °C, water availability = 51 m*^{3}*/capita/year, water demand = 293 m*^{3}*/capita/year*]. Kuwait is situated in the north-eastern side of the Arabian Peninsula with a coastline along the Arabian Gulf to its east. Kuwait is characterized by warm weather, which is a hot desert climate where rain is rare and tends to be as intermittent torrential downpours, and recently heat waves have been increased. The population of Kuwait is mainly along the coast, such as the Kuwait City. The agricultural activities are few due to water deficit and dominant desert lands. It is well known by extreme warming since pre-industrial period. by 2050, Kuwait is anticipated to witness temperature increase between 2 and 4 °C and this will be mainly in summer and autumn. Projected rainfall is not well determined and it is expected that annual rainfall may not change, besides more intensive torrential rain. While, sea levels will rise and sea surface temperatures will increase; in addition, the frequency of storms will increase and affects the coastal regions. The regional level of risks influencing Kuwait include heat stress and coastal impacts (e.g., erosion, saltwater intrusion, etc.).
- Lebanon: [*Average precipitation = 834 mm/year, average temperature = 18.4 °C, water availability = 1027 m*^{3}*/capita/year, water demand = 220 m*^{3}*/capita/year*]. Lebanon is in the Levant country along the Eastern Mediterranean coast. It is one of the humid regions in the Middle East with its a Mediterranean mild climate where snow covers more than ¼ of its territory (Shaban 2020), but it is characterised by dry summer and humid winters. This has been reflected on water availability which has been lately affected. Lebanon has been lately experiencing warm weather and more aridity. By 2050, the temperature is anticipated to increase between 2 and 4 °C. meanwhile, rainfall is anticipated to decrease by about 10–15% in the next few decades which will significantly affect the agricultural sector. Sea levels is anticipated to rise and associated with increasing sea surface temperatures, plus more frequent storms. For the regional level risks, Lebanon will be affected, in addition to the reduced water availability, by increased evaporation and changes in snowmelt rate which will reduce river recharge and groundwater productivity.
- Palestine (Occupied Palestinian Territories): [*Average precipitation = 420 mm/year, average temperature = 21.4 °C, water availability = 307 m*^{3}*/capita/year, water demand = 88 m*^{3}*/capita/year*].

It is located along the Levant region, at the Eastern Mediterranean coast and it is almost characterized by Mediterranean climate. The temperature is mainly warm with dry summer and sometimes mild and humid winters. It has already experienced a significant amount of warming since pre-industrial period. It is anticipated that by 2050, a temperature increase will be witnessed between 2 and 4 °C. There is less confidence around rainfall amounts, but a decrease is most probable. While, the sea levels will rise as well as the sea surface temperatures with and increased number of storms that acting mainly the coastal zone. The main regional level risk influencing

Palestine includes reduction in water availability. This is a result of less stream flow in the River Jordan, plus the increased temperature and evaporation.

- Oman: [*Average precipitation = 110 mm/year, average temperature = 26.1 °C, water availability = 300 m*^{3}*/capita/year, water demand = 375 m*^{3}*/capita/year*].

It is located at the south-eastern side of the Arabian Peninsula with long coastlines along the Gulf of Oman to the north-east and the Arabian Sea to the south-east. Oman is known by the hot desert climate where rain often falls heavy downpours almost resulted from Pacific Ocean storms. Recent events in Oman is represented by tropical cyclonic such as Gonu in 2007.Population density id mainly in the north-east and in coastal zone including the Capital Muscat. There is dissipated agricultural land even in mountainous regions. Similar to many Middle East countries, Oman has a significant warming. By 2050, the country temperature be increased between 2 and 4 °C. while, precipitation does not show any future trend and it may not change, but intensive downpours is projected. There is local climate zones for Oman's coast along the Arabian Sea, with small geographic variations in temperature and rainfall. Sea levels will rise continue to rise, as well as the sea surface temperatures, and more intense cyclones are expected. For the regional level risks, Oman is affected b heat stress, acts in increasing water temperatures in the Gulf of Oman and the existing ecosystems and fisheries.

- Qatar: [*Average precipitation = 76 mm/year, average temperature = 25.7 °C, water availability = 27 m*^{3}*/capita/year, water demand = 307 m*^{3}*/capita/year*].

Qatar is located along the eastern coast of the Arabian Peninsula and it is surrounded by the Arabian Gulf. It has a hot desert climate with little precipitation rate, but torrential rain often exists. Sine Qatar is a small-scale peninsula, thus it is almost a coastal zone where major cities, such as Doha, are located. Agricultural lands are few in the country due to water deficit as well as the arid climate with dominant desert lands. It is witnessing a significant amount of warming since pre-industrial period. In 2050, Qatar is anticipated to raise between 2 and 4 °C; while, no remarkable change in rainfall is expected, expect the increased torrential train patterns. Sea levels will continue to rise with rising sea surface temperatures, plus more intense cyclones that affecting the coastal zones. The regional level risks affecting Qatar is the heat stress along with impact on the coastal ecosystem.

- Turkey: [*Average precipitation = 675 mm/year, average temperature = 19.5 °C, water availability = 1500 m*^{3}*/capita/year, water demand = 738 m*^{3}*/capita/year*].

Turkey is situated at the northern part of the Middle East Region with coastlines on the Black Sea to the north and Mediterranean Sea to the south. It is one of the humid regions in the middle East, and it is characterized by Mediterranean climate with warm weather in summer and cool, milde and wet in winter. There are dominant farming systems include a mix of rainfed and dryland as well as highland farming in the mountains to the east. The country is experienced a remarkable warming since pre-industrial period. By 2050, temperature of Turkey will be at least 2–4 °C higher than the current climate, with larger increases in the number of warm months. Turkey and the surrounding regions will be more arid. Sea levels will continue to rise; plus, an increase in sea surface temperatures and increased number of storms including dry and wet ones, and this will impact both the Black Sea and Mediterranean coasts. Thus, discharge from the Tigris and Euphrates rivers, which are fed by spring-time from snowmelt on the Turkish mountains, is anticipated decline by about 12%. Regional level risks threatening Turkey are the reductions in river flow causing reduced water availability and impacting agriculture and food security.

- Saudi Arabia: [*Average precipitation* = *96 mm/year, average temperature* = *25.6 °C, water availability* = *73 m*^{3}*/capita/year, water demand* = *747 m*^{3}*/capita/year*].

Saudi Arabia constitutes the main territory of the Arabian Peninsula, with coastlines along the Arabian Gulf to the east and the Red Sea to the west. It is an arid region with hot desert climate and little rainfall (< 100 mm/year) recently associated with torrential rain causing floods and torrents. There are many major cities including Riyadh and Jeddah, while agricultural lands are being increased. Saudi Arabia has already experienced a significant amount of warming even before the industrial period. by 2050, temperature is anticipated to rise between 2 and 4 °C than the current climate. For rainfall, there are diverse projections which point out for probable increase and even decrease between various geographic region. This also will be accompanied with heavy rainfall spells; and therefore, groundwater will be largely affected. Sea levels will be higher and even the sea surface temperatures will increase with more intense storms. Regional level risks affecting Saudi Arabia include heat stress and coastal impacts.

- Syria: [*Average precipitation* = *394 mm/year, average temperature* = *21.2 °C, water availability* = *417 m*^{3}*/capita/year, water demand* = *958 m*^{3}*/capita/year*].

Syria is a Levantine country along the Eastern Mediterranean is characterized by a mix of hot desert climate to the east and a temperate Mediterranean climate to the west. There are many dense populated cities in various regions of the country such as Damascus, Aleppo and Homs. There is a diverse of rainfed and dryland farming systems, as well as irrigated agriculture along the floodplain of Euphrates River. It has been experiencing a significant warming since pre-industrial period. Projections pointed out to temperature increase between 2 and 4 °C by 2050. While, rainfall is anticipated to slightly decrease. There will be increase in sea level with increased sea surface temperatures and frequent intense storms that affecting coastal areas to the west. Regional level risks influencing Syria are water security as a result of reduced discharge in the Euphrates River.

- The United Arab Emirate (UAE): [*Average precipitation* = *94 mm/year, average temperature* = *27.2 °C, water availability* = *26 m*^{3}*/capita/year, water demand* = *504 m*^{3}*/capita/year*].

It is located on the southern part of the Arabian Gulf. Most population is located in the north-east parts, along the coastline such as Dubai and Abu Dhabi. UAE experiences a hot desert climate where rain falls intermittently as heavy downpours. Agricultural lands are rare due to water deficit and dominant dry lands. The country is known as an arid region with significant warming even before the industrial period. By 2050, temperature is expected to rise between 2 and 4 °C warmer than the existing climate. Like Saudi Arabia, projections on rainfall are undetermined whether to increase or decrease, but torrential rain is anticipated with more frequent cyclones. Sea levels will continue to rise, as well as sea surface temperatures will also increase. The regional level risks affecting UAE are heat stress and coastal impacts.

- Yemen: [*Average precipitation = 270 mm/year, average temperature = 23.4 °C, water availability = 125 m^3/capita/year, water demand = 120 m^3/capita/year*].

Yemen is located at the south-western part of the Arabian Peninsula and has a coastline along the Red Sea from the west and the Gulf of Aden from south. The country has a warm and desert climate, but affected by cyclones from the Pacific Ocean. The elevated terrain in the south-west witnesses highly variable microclimates due to orographic and coastal impacts, and this misleads the projections for this region uncertain. Yemen has already experienced a remarkable warming since pre-industrial period. By 2050, temperature in Yemen will increased between 2 and 4 °C than the current climate. However, the magnitude of temperature change is controlled by the geography, and cooler and wet localities may persist in some areas. Sea levels will rise; this, as well as the sea surface temperatures and this will be accompanied with and more intense storms impacting coastal zones. Regional level risks affecting Yemen are water stress, food security and the increased water temperatures in the Gulf of Oman affecting the ecosystems and fisheries.

5 Conclusion and Discussion

Between 45,000 and 60,000 people die each year due to natural disasters world-wide. Perhaps more than 95% of this number is due to hydroclimatic risks. It is estimated that between 35 and 40% of this number occurs in the Middle East Region. of course, this is a direct and immediate impact of these risks, but the prolonged impact can effect several times this number as for human life. In addition, many negative consequences are resulted from these risks including the major sectors, human health, and the social and economic status as a whole. The number of affect people in the developed countries is minimal if compared with low-income countries (and countries under conflicts) even if they both undergone the same risk magnitude, and this is attributed to the preparedness and implemented controls the act either in preventing the risk or reducing its extent and this reveals the feasibility of the precautionary measures. Moreover, most hydroclimatic risks are enhanced by the human interference which catalysts the occurrence of many risks, such as the construction in the pathways of watercourses, chaotic water abstraction from all water resources, excavation, wood logging and forest fires, etc.

Among these risks, drought is the most widespread, and it covers large areal extent, in almost all the Middle East countries and it might cover tens of thousands of hectares annually. Even though, it is a long-term physical process, yet it has been remarkable existing since 2000s and its impact has been recently reached and start exacerbating in many sectors notably water and the related agricultural purposes and food security. Usually, the significant role of human is accounted, notably drought is not only the decrease in precipitation over long time period but also the declined water supply. In other word, a country with sufficient water resources but poor water supply is considered as a country under drought risk. In addition, drought is always linked with desertification and land degradation which are attributed to the loss of natural resources and this implies the chain from reduced water availability, accompanied with loss of water and green cover and thus the abandoned lands.

Floods are another significant aspect of hydroclimatic risks which are commonly pronounced in the Middle East Region and it resulted in severe damages and loss of people over the last two decades. The paradox is that floods reoccurrence is mainly in dry regions like the Arabian Peninsula where climate change and its extreme processes became dominant; and thus, the spells of torrential rainfall has been lately exceeded the normal; besides, widespread human settlements constructed regard-less to any hydrological consideration along watercourses. Yet, prevention measures for floods are not efficient, notably various hydroclimatic models are applied with but they do not consider the natural characteristics of the area of interest. That mean the same hydroclimatic models is applied to regions with diverse natural components, and this often ends up with erroneous results. Unfortunately, floods cause people to day with a general estimation between 30 and 50 people annually in the Middle East Region.

Cyclones are another damaging risks, and their predication is always few days or even hours before they occur which sometime do not leave enough time to protect. They can be dry or wet cyclones that are derived from various direction but both cause unpredictable damages. Unlike other natural risks, these cyclones have least control by the topographic characteristics of land surface; especially they movement is usually at high altitudes. The existed cyclones in the Middle East Region have lately resulted in different aspects of interlinked risks such as floods, dust storms, snow avalanches, mass movements, etc.

Even though, wildfire is geographically occurred in the mountains regions, notably those in the Eastern Mediterranean, yet it is impact is always harmful, notably for the environment and the ecosystem as a whole. Thus thousands of hectares are lost annually by fire leaving vast burned regions, yet this also affects air pollution and the CO_2 emissions. Perhaps this contributes with the largest amount of GHGs emission released from the Middle East Region. in addition, this type of hydroclimatic risks increases desertification and land degradation. Unfortunately, prevention measures for wildfires are rare, and only pre-assessments are applied using satellite images and field surveys.

There are many other hydroclimatic risks discussed in this chapter and they have different degrees of impact and related repercussions on the social and economic status. It is obvious that the responding of the Middle East countries to hydroclimatic risks is unevenly treated, which is based mainly on the income level of these countries. This, and the respond of the developed countries to such risks, points out that the prevention of the existing risks can be managed and it is not tedious.

However, the low-income countries can upscale their situation be bridging with the international and regional initiatives and programmes (*e.g.*, UNCCD, UNDDR, WB), which are already existing, but must be consolidated and moved towards serious and effective implementations. Moreover, the adoption of models should be also oriented according to the natural setting and characteristics of each region; especially many satellite-based platforms are now available to provide valuable datasets which can support the Early Warning Systems. In this regard, it is not exaggeration to say that the changing date in climatic conditions and extremes is racing with the adaptive measures for risk prevention. It must be made clear that hydroclimatic risk reduction implies in a number of Sustainable Development Goals including SDG 3 (Good health and well-being), SDG6 (Clean water and sanitation), SDG8 (Descent work and economic growth), SGD11 (Sustainable cites and communities), SDG13 (Climate action) and SDGs 14 and 15 (Life below water and Life onland).

References

ACSAD (2008) ACSAD desertification bulletin-2008. Available at: https://www.researchgate.net/publication/267624860_ACSAD_Desertification_Bulletin_-_2008

Al Saud M (2015) Flood control management for the city and surroundings of Jeddah, Saudi Arabia. Springer Publisher, 169 pp. https://doi.org/10.1007/978-94-017-9661-3

Al Saud M (2020) Sustainable land management for NEOM region. Springer International Publishing, p 219. ISBN-13: 9783030576332

Al Saud M (2022) Applications of space techniques on the natural hazards in the MENA region. Springer Nature Publishing, 635 pp. https://doi.org/10.1007/978-3-030-88874-9

Aljenaid S (2024) Silent threat': Bahrain to build walls against rising sea levels. Personal contribution to The Straits Times magazine. Available at: https://www.straitstimes.com/world/middle-east/silent-threat-bahrain-to-build-walls-against-rising-sea

Cansler A, McKenzie D (2014) Climate, fire size, and biophysical setting control fire severity and spatial pattern in the northern Cascade Range, USA. Ecol Appl 24(5):1037–1056. https://doi.org/10.1890/13-1077.1

CBSE (Central Board of Secondary Education). (2006) Natural hazards and disaster management. A supplementary textbook in geography for Class XI, 51 pp

CMES (Centre for Advanced Middle East Studies) (2025) CMES regional outlook: dust storms in the Middle East—insights from satellite data and AI. Available at: https://www.cmes.lu.se/article/cmes-regional-outlook-dust-storms-middle-east-insights-satellite-data-and-ai

Dezfuli A (2020) Rare atmospheric river caused record floods across the Middle East. Bull Am Meteorol Soc. https://doi.org/10.1175/BAMS-D-19-0247.1

EcoMENA (2025) Desertification in MENA—causes and solutions. Available at: https://www.ecomena.org/desertification-mena/

Elsayed A, Nashwan M, Eltahan A, Shahid S (2024) Millions more Egyptians will be exposed to drought by 2100 under the goals of the Paris climate agreement. Int J Disast Risk Red 101(2024):104257. https://doi.org/10.1016/j.ijdrr.2024.104257

EM-DAT (2023) International disaster database. Université Catholique de Louvain, Brussels. Available at: https://www.emdat.be

Ezzeldin K, Adshead D, Smith P (2023) Adapting to a new climate in the MENA Region: An assessment of physical risk management and climate adaptation finance in the MENA region. UNEP Finance Initiative, 36 pp

Fawzy S, Osman AI, Doran J, Rooney DW (2020) Strategies for mitigation of climate change: a review. Environ Chem Lett 18:2069–2094

Hadley Center (2021) Climate risk report for the Middle East and North Africa (MENA) Region. Met Office- Hadley Center, ODI, UKaid, 90 pp

Jiang Z, Yang S, Liu Z (2022) Can ensemble machine learning be used to predict the groundwater level dynamics of farmland under future climate: a 10-year study on Huaibei Plain. Environ Sci Pollut Resour 29:44653–44667. https://doi.org/10.1007/s11356-022-18809-8

Joint Research Center (2024) Forest fires in Europe, Middle East and North Africa 2023, 205 pp. https://doi.org/10.2760/8027062

Kenawy A, Al-Awadhi T, Abdullah M (2025) A multidecadal assessment of drought intensification in the Middle East and North Africa: the role of global warming and rainfall deficit. In: Earth system and environment, 20 pp. https://doi.org/10.1007/s41748-025-00576-4

Khosravi Y, Ouarda (2025) Drought risks are projected to increase in the future in central and southern regions of the Middle East. Commun Earth Environ 6:384. https://doi.org/10.1038/s43247-025-02359-1

Lange M (2019) Mediterranean and the Middle East and North Africa Region and the water–energy nexus. Atmosphere 10:455. https://doi.org/10.3390/atmos10080455

MRC (Munich Reinsurance Company) (2013) Overall picture of natural catastrophes in 2013 dominated by weather extremes in Europe and Supertyphoon Haiyan, press release, Munich, January 7, 2014. Available at: http://www.munichre.com/en/media_relations/press_releases/2014/2014_01_07_press_release.aspx

ND-GAIN (2023) Notre Dame Global Adaptation Initiative index for country index. Available at: https://gain-new.crc.nd.edu/ranking

Ouarda T, Charron C, Niranjan Kumar K, Marpu P, Ghedira H, Molini A, Khayal I (2014) Evolution of the rainfall regime in the United Arab Emirates. J Hydrol 514(2014):258–270. https://doi.org/10.1016/j.jhydrol.2014.04.032

Peters K, Mayhew L, Slim H, Van Aalst M, Arrighi J (2019) Double vulnerability. The humanitarian implications of intersecting climate and conflict risk. ODI Working Paper 550

Raymond C, Horton RM, Zscheischler J et al (2020) Understanding and managing connected extreme events. Nat Clim Chang 10:611–621. https://doi.org/10.1038/s41558-020-0790-4

RICCAR (Regional Initiative for the Assessment of Climate Change Impacts on Water Resources and Socio-Economic Vulnerability in the Arab Region (2017) Arab climate change assessment report—main report. E/ESCWA/SDPD/2017/RICCAR/Report (ESCWA: 2017, Beirut)

Schaar J (2023) Addressing clime-related security risks in the Middle East and North Africa. UNDPPA (United Nations Department of Political and Peacebuilding Affairs), 37 pp

Shaban A (2020) Water resources of Lebanon. Springer Science Publisher, 229 pp. https://doi.org/10.1007/978-3-030-48717-1

Shomar B, Sankaran R (2024) Groundwater contamination in arid coastal areas: Qatar as a case study. Groundwater NGWA 62(6):847–859. https://doi.org/10.1111/gwat.13411

Stocker T (2013) Climate change 2013: the physical science basis: In: Working group I contribution to the Fifth assessment report of the intergovernmental panel on climate change. Cambridge University Press

UNEP (2024) Climate change and water-related disasters. Available at: https://www.unep.org/topics/fresh-water/disasters-and-climate-change/

UN-Geneva (2024) Latest news. Real cost of disasters is 10 times higher than previously thought, says UN. Available at: https://www.ungeneva.org/en/news-media/news/2025/05/106734/real-cost-disasters-10-times-higher-previously-thought-says-un

USAID & IWMI (2021) Reducing the impacts of drought in Jordan. How the MENA drought project is supporting Jordan to enhance its water and food security? Fact Sheet, 4 pp

WB (2018) World development indicators. World Bank, Washington, D.C. https://databank.worldbank.org

Winslow M, Thomas T (2016) Desertification in the Middle East and North Africa: warning signs for a global future. Regional aspects—desertification in the Middle East and North Africa. Agric Rural Dev 14(1):10–12

WMO (2024) State of the global climate 2024. Available at: https://wmo.int/publication-series/state-of-global-climate-2024worldbank.org/data/reports.aspx?source¼world-development-indicators.

Hydroclimate Vulnerability in the Middle East: A Qualitative Overview of Drought Exposure and Adaptive Capacity

Tayfun Kurt

1 Introduction

The Middle East, a region critically vulnerable to hydroclimatic crises, faces escalating threats from drought, freshwater scarcity, and ecological degradation, undermining sustainable development across diverse socio-political contexts (Tayfun 2025; Tinazzi 2024; Monshipouri and Mohajer 2024; Azizi and Leandro 2025; Al-Quraishi 2024; Darvishi Boloorani et al. 2024). These challenges are driven by the accelerating effects of climate change, including rising temperatures, declining precipitation, and increasing variability in seasonal water flows (IPCC 2022). Beyond climatic factors, water stress is deeply intertwined with socio-political structures, weak governance systems, and unsustainable water management practices, such as excessive groundwater extraction and poorly planned irrigation (El Idrissi et al. 2025; Fessahaye et al. 2025).

Droughts in the Middle East are rarely isolated events; they are chronic stressors exacerbated by human activities and limited regional cooperation, particularly in transboundary river basins like the Tigris-Euphrates. Rapid population growth, urban expansion, and geopolitical tensions further strain already scarce resources, undermining the adaptive capacity of states and communities (Azizi and Leandro 2025). To illustrate the scale of the crisis, Fig. 1 presents per capita renewable freshwater availability in selected Middle Eastern countries, measured against the Falkenmark Water Stress Index, which classifies water availability below 500 cubic meters per person per year as extreme scarcity—a level associated with severe constraints on food production, public health, and economic development.

Figure 1 highlights a stark reality: extreme water scarcity is a present condition across the Middle East, even in countries with relatively larger freshwater reserves like Iran and Türkiye. This widespread hydroclimatic stress underscores the urgency

T. Kurt (✉)
Independent Researcher, Istanbul, Türkiye
e-mail: tyfnkrt@hotmail.com

M. M. Al Saud (ed.), *Hydroclimate Changes: Implications of the Middle East Region*,
https://doi.org/10.1007/978-3-032-20053-2_3

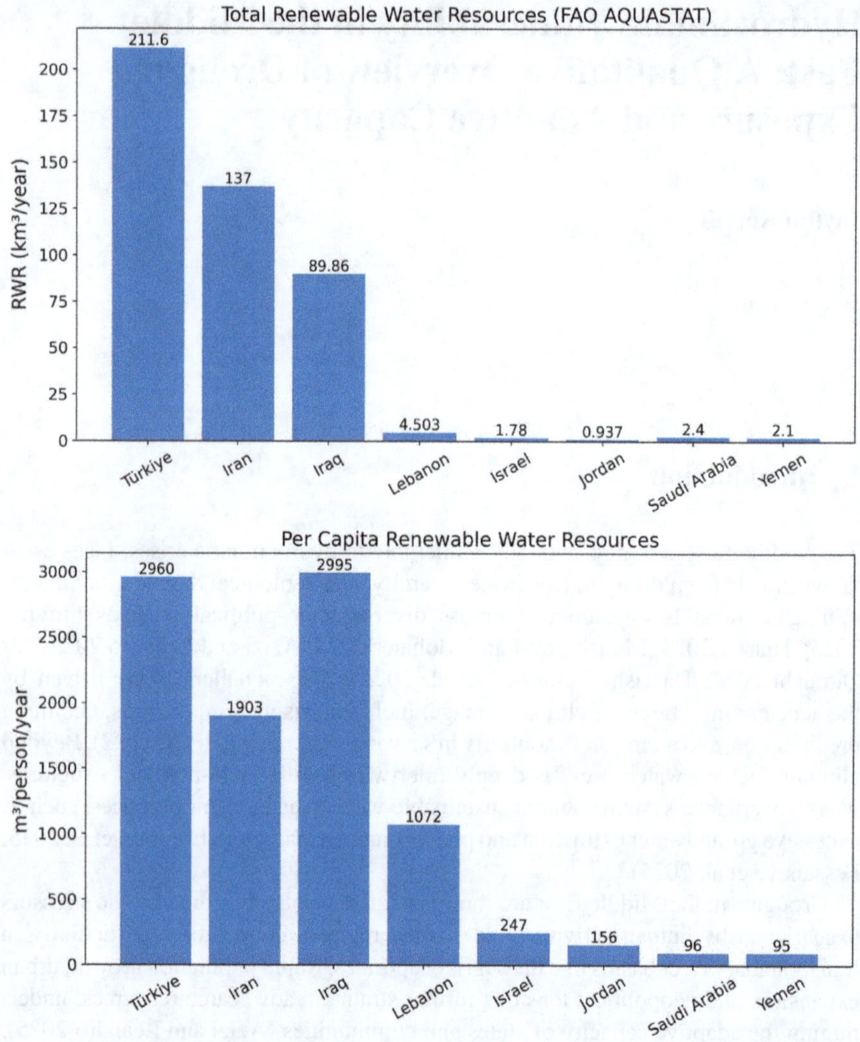

Fig. 1 Per Capita Renewable Water Resources and Water Scarcity Levels in Selected Middle Eastern Countries Adapted from FAO (2023a, b, c, d, e, f, g, h). *Note* Based on the Falkenmark Index: <500 m^3/person/year = Extreme Scarcity; 500–1000 = Water Scarcity; 1000–1700 = Water Stress; >1700 = Sufficient.

of addressing vulnerability in diverse ecological and political settings (FAO 2023a, b, c, d, e, f, g, h).

This chapter provides a qualitative synthesis of hydroclimatic vulnerability in the Middle East, focusing on drought exposure and adaptive capacity in countries such as Jordan, Iraq, Iran, Yemen, Lebanon, Syria, Saudi Arabia, and southeastern Türkiye. These countries were selected to reflect diverse ecological, socio-political, and governance contexts. Rather than relying on data-intensive modeling, often limited by inconsistent data and technical constraints in the region, the analysis draws on open-access academic literature, global datasets, and institutional reports. It examines how governance quality, socio-economic sensitivity, and infrastructure shape resilience, with a focus on low-cost, community-based innovations—such as rainwater harvesting in Jordan and traditional water-sharing systems in Yemen—and the critical role of cross-border cooperation.

While quantitative models dominate drought risk assessments, their limitations in data-scarce regions necessitate qualitative approaches to uncover context-specific vulnerabilities and adaptation strategies. Through country-specific case studies, this chapter explores localized responses and institutional challenges, arguing for a shift from crisis response to long-term resilience building. Ultimately, it aims to guide policymakers, researchers, and practitioners in developing inclusive, adaptive, and sustainable solutions to the region's escalating hydroclimatic challenges.

2 Methodology

This chapter adopts a qualitative, desk-based research design to assess hydroclimatic vulnerability in the Middle East, with a particular emphasis on drought exposure and adaptive capacity. Given the financial and institutional constraints that frequently limit access to primary data and large-scale fieldwork, the study relies exclusively on freely available secondary sources. This approach aligns with the chapter's aim to evaluate regional hydroclimatic risks and adaptation strategies while offering a cost-effective and replicable framework particularly useful for independent scholars and practitioners working under resource constraints (Change 2001).

The analytical framework is structured around three key dimensions—drought exposure, socioeconomic sensitivity, and adaptive capacity—consistent with widely used vulnerability concepts in the climate-impacts literature (e.g., the IPCC vulnerability framing and integrated vulnerability frameworks). This conceptual grounding follows the vulnerability taxonomy advanced by Turner et al. and the IPCC's treatment of exposure, sensitivity and adaptive capacity (Fig. 2).

Literature searches targeted peer-reviewed articles indexed in Scopus and Web of Science and were limited to studies published after 2015. Search terms included *"drought"*, *"water scarcity"*, *"hydroclimatic vulnerability"*, and *"adaptation"*. The choice of Scopus and Web of Science is motivated by their broad journal coverage and frequent use in systematic literature reviews (they were used as principal indexing sources for locating regionally relevant empirical and review papers). Complementary grey literature was collected from reputable technical sources such as FAO AQUASTAT country factsheets and major international organisations' technical reports (e.g., FAO, World Bank, UNDP) to ensure policy-relevant coverage.

Country selection used three criteria: (i) geographic representation across the Middle East, (ii) availability of qualitative and institutional information in public sources, and (iii) variation in water-stress levels and governance arrangements. Based on these criteria, the comparative analysis focuses on Jordan, Iraq, Saudi Arabia, and

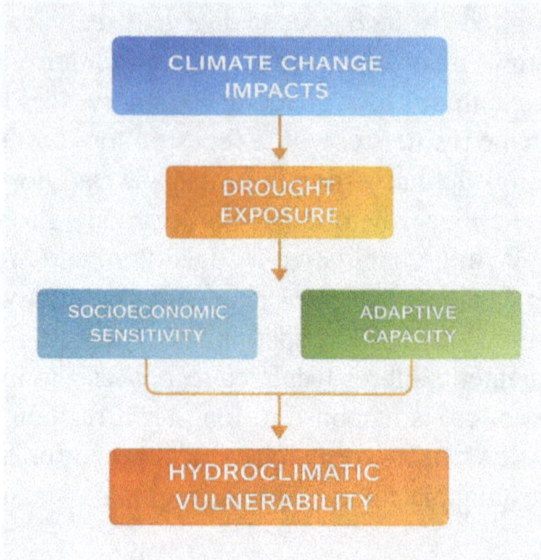

Fig. 2 Framework for assessing hydroclimatic vulnerability

southeastern Türkiye — cases chosen to capture a range of hydroclimatic regimes and institutional contexts. AQUASTAT country factsheets and national technical reports were used as baseline sources for each country's water-resource profile.

To operationalize the thematic comparative synthesis, documents were manually coded for recurring topics such as governance quality, water-management institutions, community resilience, infrastructural capacity, and documented adaptation measures. Manual coding was supplemented with iterative cross-checking to enhance reliability (double coding of a subset of sources and discussion of coding discrepancies). The thematic synthesis approach (line-by-line coding → descriptive themes → analytical themes) was applied to integrate diverse qualitative and policy materials while preserving links to the original texts.

Limitations: relying on secondary sources and desk-based synthesis implies trade-offs. Secondary data can be incomplete, may reflect the priorities of original authors, and often lack the spatial or temporal resolution required to resolve fine-scale hydrological dynamics (e.g., catchment-level recharge processes). Qualitative secondary analysis remains a defensible approach when conducted systematically, but findings should be interpreted with attention to evidentiary gaps and to the provenance of the underlying documents.

In sum, this methodology provides a rigorous, transparent, and replicable pathway for assessing hydroclimatic vulnerability under resource constraints. It combines established conceptual frameworks (IPCC; vulnerability literature), methodologically sound document analysis and thematic synthesis techniques, and verified open-access data sources (e.g., AQUASTAT) to deliver context-sensitive insights while clearly communicating limitations.

3 Regional Hydroclimate Vulnerability

The Middle East and North Africa (MENA) region is characterized by acute hydroclimatic vulnerability, driven primarily by persistent droughts and chronic water scarcity (El Kenawy 2024). The combination of natural aridity, climate change impacts, and socio-economic pressures creates a complex landscape of risks that varies significantly across countries and sub-regions.

Summary of Drought Risks and Water Scarcity Patterns

Drought events in the MENA region tend to be prolonged and recurrent, severely impacting water availability for domestic, agricultural, and industrial uses (Fig. 3). Seasonal variability in precipitation, combined with rising temperatures, intensifies evapotranspiration rates, further depleting surface and groundwater resources. Many countries in the region consistently register per capita renewable water resources well below the 500 m^3/year threshold defined by the Falkenmark Water Stress Index, placing them in the category of extreme water scarcity (Frenken et al. 2008; Fig. 4). For instance, Jordan, Yemen, and Israel face perennial water shortages aggravated by limited rainfall and over-extraction of groundwater. Meanwhile, countries like Iran and Türkiye, although relatively better endowed, experience significant spatial and temporal variability in water availability, rendering large populations vulnerable to episodic droughts.

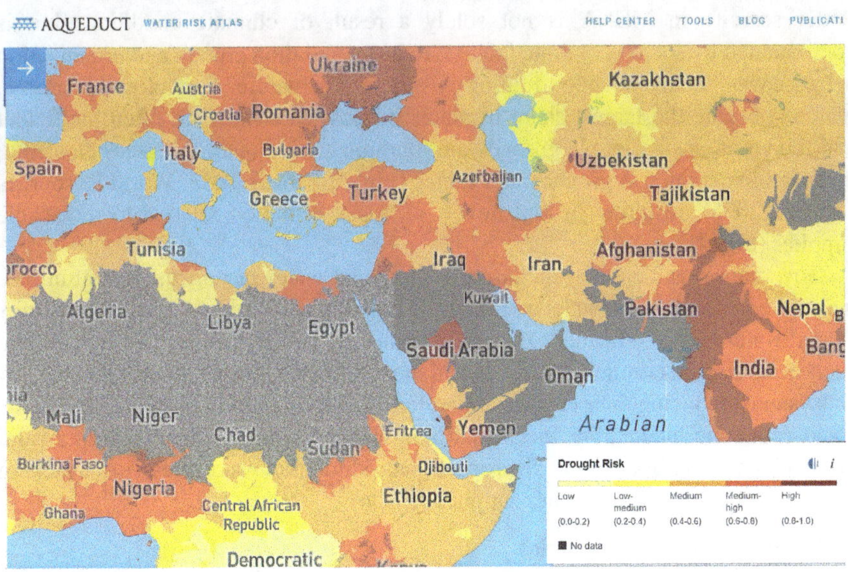

Fig. 3 Drought risk levels in the MENA region. *Source* World Resources Institute (2025), Aqueduct Water Risk Atlas

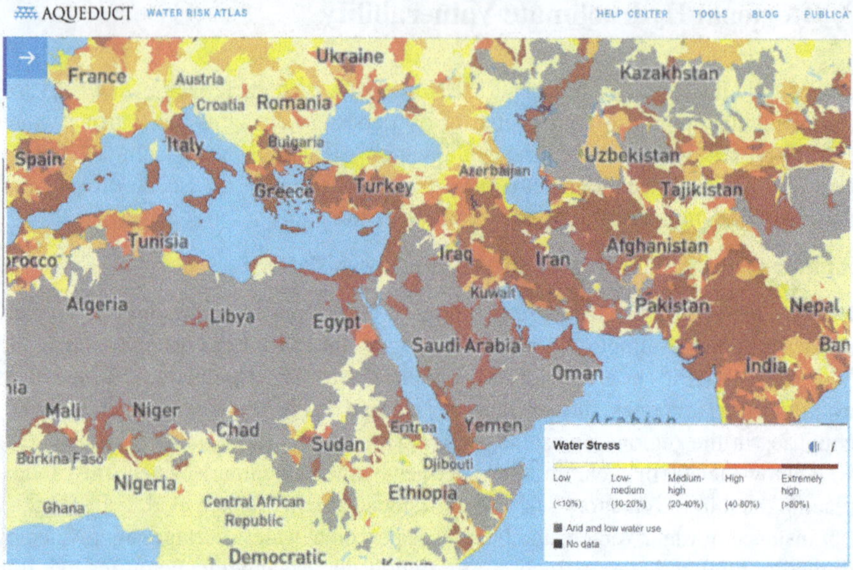

Fig. 4 Water stress levels in the MENA region based on baseline water depletion *Source: World Resources Institute* (2025), *Aqueduct Water Risk Atlas*

Socio-Economic and Environmental Factors Influencing Vulnerability

Water scarcity in MENA is not solely a result of climatic conditions but is compounded by socio-economic factors. Rapid population growth and urbanization increase water demand while straining existing infrastructure. Agricultural practices, which consume the largest share of water in most countries, are often inefficient and reliant on unsustainable groundwater pumping, which in many areas has led to measurable declines in groundwater tables (Fig. 5; Fanack 2025; World Resources Institute 2025), resulting in high levels of water depletion relative to renewable supplies (Fig. 4).

Governance challenges—including fragmented water management institutions, lack of enforcement of water regulations, and limited regional cooperation over shared water bodies—further exacerbate vulnerability. Environmental degradation, such as soil salinization and loss of wetlands, diminishes the natural resilience of ecosystems, reducing their capacity to buffer against drought impacts. Additionally, socio-economic inequalities mean that marginalized communities, including rural populations and refugees, bear disproportionate burdens of water stress, often lacking access to safe and reliable water supplies (Fig. 6).

Differences Between Countries

While the MENA region shares common hydroclimatic challenges, the degree and nature of vulnerability differ markedly between countries. Jordan, with its severe

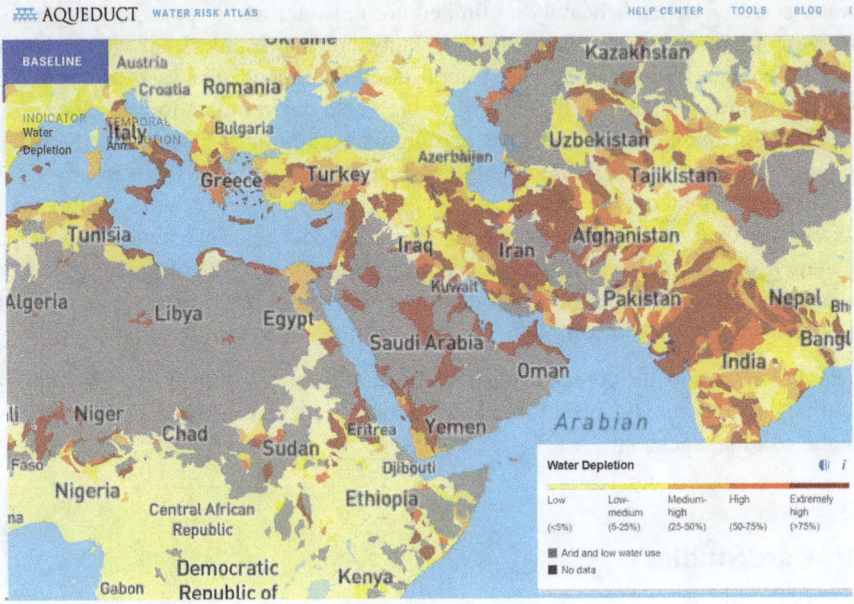

Fig. 5 Water depletion levels in the MENA region. *Source: World Resources Institute* (2025), *Aqueduct Water Risk Atlas*

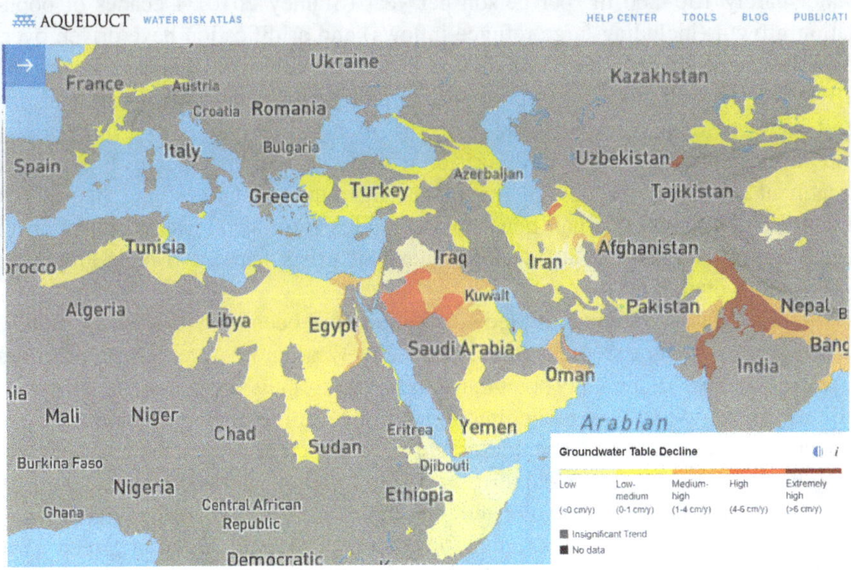

Fig. 6 Groundwater table decline levels in the MENA region. *Source: World Resources Institute* (2025), *Aqueduct Water Risk Atlas*

water scarcity, depends heavily on limited groundwater and imports, driving innovative community-based adaptation like rainwater harvesting. Iraq, with a larger but geopolitically contested river system, faces compounded risks from upstream water withdrawals and conflict. Saudi Arabia's arid environment forces reliance on desalination and non-renewable fossil groundwater, highlighting trade-offs between economic capacity and sustainability. Türkiye's greater freshwater resources are unevenly distributed, with eastern provinces experiencing higher vulnerability, and its control over the headwaters of key rivers introduces transboundary tensions. Lebanon, despite higher precipitation, confronts water management challenges exacerbated by political instability and infrastructure deficits. Yemen's ongoing conflict has devastated water infrastructure, pushing millions into acute water insecurity.

These country-specific contexts illustrate the multifaceted nature of regional vulnerability, underscoring the need for tailored, context-sensitive approaches to hydroclimatic risk management.

4 Case Studies

Jordan: Water Scarcity and Grassroots Adaptation

Jordan is among the world's most water-scarce countries, with average renewable water barely 150–200 m^3 per person per year (Stanley 2010). Decades of population growth (including large refugee inflows) and aridification have driven over-exploitation of groundwater, shrinking reservoirs and drying of aquifers (Stevanović 2018). As a result, surface water shortages and groundwater deficits are chronic, and even traditional rural wells and cisterns often fail (Priyan 2021). In response, the government and donors have invested in non-conventional sources: wastewater treatment and reuse, new desalination plants, and rainwater-harvesting schemes. For example, thousands of urban households practice rooftop rainwater collection or small cistern storage, and a growing number of urban gardens recycle greywater (Al-Addous et al. 2023).

In this context, community-led adaptation has become vital. Local NGOs, youth groups and grassroots organizations actively promote water-saving practices (Adkhamov 2023). One study notes that climate civil society organizations "empower communities by involving them directly in climate action, thereby building a more sustainable grassroots movement (Hussein 2025). Such efforts include village tree-planting (to protect watersheds), drip irrigation for small farms, and workshops on efficient water use. Many farmers and households have adopted drip irrigation and mulching to conserve moisture. International donors have funded rural "green jobs" projects that install solar-pump wells and rehabilitate communal irrigation canals. There are also modest examples of water vendors being replaced by cooperative water kiosks in refugee settlements, improving both access and quality.

Despite these advances, challenges remain. Water tariffs and inter-agency coordination in Jordan are weak, so informal connections (e.g. rooftop tanks and tanker vendors) fill gaps when piped supply fails. Research suggests that strengthening supportive policies and incentives is crucial: one review argues that "investing in wastewater treatment plants, rainwater harvesting, and optimizing virtual water utilization can help alleviate stress on surface and groundwater resources (Al-Addous et al. 2023). In practice, continuity of grassroots projects has improved via participatory planning and cooperation between municipalities and NGOs. But long-term resilience will depend on scaling up such bottom-up measures alongside improved national water management and strengthened community engagement (Hussein 2025).

Iraq: Institutional Challenges in Drought Management

Iraq's water situation has worsened under climate change and regional pressures. Roughly 70% of the country is arid; rising temperatures and declining rainfall have compounded severe droughts (Arkeh and Awadallah 2025) Agriculture (5% of GDP) is under acute stress: by some estimates, upstream dam construction in Türkiye, Iran and Syria has diverted as much as 60–70% of the Tigris–Euphrates flow. Winters are warming, summers lengthening, and reservoirs (like Mosul and Haditha) often fall to record lows. As a result, traditional farming and drinking-water supplies in central and southern Iraq have been exhausted in peak dry seasons (Arkeh and Awadallah 2025).

However, the overriding problems in Iraq are often institutional rather than purely technical. Analyses repeatedly point to fragmented governance, politicized water agencies and lack of coordination. According to experts, "Iraq's water crisis is…also the result of upstream restrictions and domestic neglect. Corruption and self-interest among Iraq's political elite weaken institutional capacity and create openings…for external actors" (Al-Shakeri 2025) In practice, multiple ministries (Water Resources, Agriculture, Environment, Oil) and regional governments (the Federal Ministry vs. the Kurdistan Regional Government) have overlapping mandates over water, with no unified management strategy (Tinti 2017). Long-term planning is weak: for example, the costly Basra desalination plant and new rainwater dams were only approved when shortages became extreme, and have since stalled amid shifting budgets (Al-Shakeri 2025). One analysis found that large projects often "stalled or faced long delays due to corruption, lack of policy continuity, and outright negligence.

The legacy of neglect is evident in decaying infrastructure: much of Iraq's canal network dates from the 1970s and operates at only ~ 60% efficiency, losing massive volumes to leakage and evaporation (Al-Shakeri 2025). In the political turmoil since 2010, roughly one-third of public water utility engineers and technicians have left or been sidelined, eroding operational capacity. Frequent power outages and fuel shortages also interrupt pumping. Against this backdrop, droughts have been managed haphazardly. A UNDP review stresses that effective drought management "requires an integrated approach, bringing together multiple stakeholders across several ministries" (RIMES 2022), yet coordination remains weak. There is currently no single national drought or water authority: responsibilities lie split among regional governments and temporary commissions.

In 2024–25, Iraq began drafting a National Drought Plan and upgrading its meteorological/drought monitoring systems. These efforts aim to link early warning with relief (for example, delivering subsidized fodder to herders in advance). Yet experts caution that without addressing the root governance failures—politicization of water decisions and endemic corruption—technical fixes alone will have limited effect (Al-Shakeri 2025).

Iran: Effects on Agriculture and Community Responses

Iran has suffered a severe, decades-long water crisis. Nationwide, climate change has reduced precipitation and boosted evaporation, causing surface water (rivers and wetlands) to shrink by roughly a third in recent decades. At the same time, over-pumping of groundwater has created a deficit of over 30 billion m^3 per year (Kowsar 2024). The result has been dramatic: traditional qanats (underground canals) have run dry, Lake Urmia and other wetlands have vanished, and even historically water-rich Khuzestan province now endures acute shortages. Per-capita water availability has fallen toward the "absolute scarcity" threshold of 500 m^3/year (Kowsar 2024). Iranians living standards have suffered; many farmers have been forced to abandon fields, accelerating rural-to-urban migration. One estimate is that around ten million people moved into city outskirts between 2013–2018 as agriculture became untenable (Kowsar 2024).

The agricultural sector (23% of employment) has been hit especially hard. Over 90% of Iran's water use goes to irrigation, mostly via inefficient surface flooding of fields (Country Reports 2019). Many irrigated areas now demand more water than nature provides. In response, farmers have drilled ever-deeper wells, tapping fossil aquifers to maintain production (Country Reports 2019). But this is unsustainable: the Central Plateau's groundwater levels are steadily dropping and land subsidence is reported in farming districts. Environmental studies note that without change, groundwater pumping will eventually yield minimal crop outputs and force economic collapse of rural communities.

According to (Country Reports 2019). Against these challenges, Iranian communities and authorities have taken some adaptive steps. The government has introduced subsidies and technical assistance for modern irrigation: for example, expanding drip irrigation and shifting some crop patterns to less water-intensive varieties. In recent years large-scale projects (like boosting solar-powered pumps and rainwater retention ponds) have been piloted in central provinces. Some regional authorities are even experimenting with cloud-seeding to induce rainfall in summer months, following neighboring countries' lead. Farmers benefiting from government drip-irrigation subsidies report overwhelmingly positive results, and most support expanding these programs (İpekçioğlu et al. 2022). On the community side, cooperatives for sharing water and equipment have formed in parts of Khuzestan and Kerman, improving access for smallholders during dry spells.

Nevertheless, societal stress is high. Iran witnessed mass protests in 2018–2021 in cities such as Isfahan and Khorramabad, driven in part by water shortages (Kowsar 2024). In rural Khuzestan in 2019–20, clashes over diverted rivers and dried-up lands led to violent confrontations with security forces. These events underscore that for many Iranians, water scarcity has become a political and social emergency. Analysts warn that if water mismanagement is not decisively reversed, further agricultural decline and unrest may follow. Long-term solutions will need to couple technical fixes (reuse treated wastewater, close wells, rehabilitate dams) with stronger local governance (giving water boards real authority) to restore community trust and viability(Country Reports 2019).

Yemen: Conflict and Water Vulnerability

Yemen has long been water-poor, but the civil war has made its water crisis catastrophic. Even before conflict, annual rainfall averaged only ~ 110 mm and per-capita availability was only ~ 150 m^3—among the lowest globally (Omar 2025). Farming depended almost entirely on groundwater, yet drilling has outpaced replenishment. Agriculture consumes over 90% of Yemen's scarce water, mostly via outdated flood irrigation that wastes water. These structural stresses were amplified by war. Since 2015, bombing and siege have demolished dams, pumping stations and pipelines across the country. Millions have been displaced, doubling or tripling demand in some relatively stable cities. With state water agencies effectively collapsed, officials can enforce neither allocations nor repairs. As one commentary notes, "the war has led to the destruction of water infrastructure, contamination of sources, and the displacement of millions, reducing access to safe water" (Omar 2025).

The human toll is enormous. A recent survey found that over 15.3 million Yemenis (roughly two-thirds of the population) lack enough safe water for basic needs (Human Rights Watch 2023). In besieged areas like Taizz, both government and Houthi forces have used water as a weapon—cutting off aquifers or blocking tankers. In Taizz city, for example, only 21 of 88 public wells are now functional, and people survive on scant network supply, sporadic rainwater collection, humanitarian deliveries or costly private trucks (Human Rights Watch 2023). Across rural districts (Amran, Saada, Abyan, etc.), families often rely on shallow wells and hand-collected rainwater. With aid agencies largely cut off, women and children must walk miles for a single jerrycan. Studies of Yemen's remote farming regions note that water scarcity—and conflict over remaining resources—has sharply reduced crop yields and forced many families to abandon farms (Omar 2025).

In the face of war, some bottom-up coping strategies have emerged. In oases and wadis across Tihama and Hadhramaut, villagers have refurbished traditional rainwater cisterns and built new sub-surface dams to catch flood flows. Local committees sometimes manage these collective resources with tribal rules to ration usage. In cities, residents increasingly harvest domestic wastewater for irrigation of kitchen gardens. Humanitarian groups have drilled boreholes and installed solar pumps in select camps. However, these piecemeal efforts are dwarfed by the crisis. Absent a peace and unified government to coordinate repairs, Yemen's water vulnerability remains extreme. Analysts emphasize that rebuilding Yemen's water security will require not just new pipes, but re-establishing governance—restarting water agencies, integrating displaced communities, and managing groundwater licenses—all of which have broken down during the conflict (Omar 2025).

Saudi Arabia: Large-Scale Water Projects and Challenges

Saudi Arabia has pursued massive engineering projects to secure water, but faces mounting challenges. The kingdom pumps heavily from deep (non-renewable) aquifers—meeting roughly 50–60% of its water needs– and in parallel has built the world's largest fleet of desalination plants (Suhail et al. 2024). By 2020 Saudi desalination capacity exceeded 2.2 billion m^3/year, about 14% of national supply. Hundreds of dams have also been constructed on scarce seasonal streams; one survey counted over 300 dams (mainly for flood control and reservoir storage) by 2009 (Suhail et al. 2024). These capital projects—along with extensive oil-for-water subsidies—have long kept urban taps running in desert cities.

Looking ahead, the authorities plan even larger infrastructure. Under Vision 2030, the goal is to meet roughly 90% of urban water demand by desalination (International Trade Administration 2024). This has led to an $80 billion national commitment to water projects. New initiatives include wastewater reuse (Saudi aims for 100% treatment and reuse of urban wastewater) and even cloud-seeding programs to stimulate rain. In 2023 Riyadh announced a Global Water Organization to spearhead international cooperation and invest in high-priority projects from large dams to advanced treatment plants desalination (International Trade Administration 2024).

Despite this scale, sustainability is an issue. Desalination is energy-intensive (current plants consume around 2.3 kWh per m^3 of water produced (Suhail et al. 2024) and produces brine waste. Overpumping of aquifers has lowered groundwater levels and increased salinity. Per capita water use is very high (around 360 L/day (International Trade Administration 2024) and 80% of irrigation is still via flood methods, with roughly half of that water wasted. Climate change may further stress supplies. Nonetheless, the kingdom's policies are shifting: new construction favors reverse-osmosis (RO) desalination (more efficient) (Suhail et al. 2024), and large solar farms are being paired with water treatment plants to cut carbon emissions. Private-sector partnerships (for example in the Red Sea Desalination project) are also on the rise. In the agricultural sector, limits on well drilling and incentives for efficient irrigation have curbed some groundwater overuse.

In summary, Saudi Arabia's strategy relies heavily on engineered solutions at grand scale. These have thus far staved off water shortages in cities, but continue to face economic and environmental constraints. Achieving long-term water security will require further balance—maximizing reuse and solar energy, continuing demand management, and protecting any remaining natural aquifer recharge zones. The planned mix of desalination (90%) with treated wastewater (including planned recycling of all urban effluent) reflects recognition of those limits (International Trade Administration 2024).

Lebanon: Urban Water Governance

Lebanon, though relatively water-rich by MENA standards, struggles with chronic urban water supply problems due to weak governance and infrastructure. Civil war legacies and political fragmentation mean that municipalities and state utilities are often unable to provide reliable service, especially in densely populated cities. Over the past decade, Beirut, Tripoli and many towns have typically received only a few hours of piped water per day, often on a multi-day rotating Schedule (Fayssal et al. 2024). Recent research documented a collapse in municipal water service amid the economic crisis: average weekly supply fell from about 49 h in 2019 to only 22 h by 2023. In compensation, private water vendors and trucks have become the norm: nationwide reliance on tanker water jumped from ~ 26% of water use in 2019 to 44% by 2023 (Fayssal et al. 2024).

Several factors underlie this: decades of under-investment have left pipe networks with many leaks, and power blackouts stall pumps in spring and summer. Corruption and politicization have also played a role: officials rarely meters demand or enforces billing, so the state utility (litani basin authority, plus municipal companies) runs large deficits. The influx of ~ 1.5 million Syrian refugees since 2011 (28% population increase) has further strained supply (Walnycki 2017). Refugee camps are often informal and fall outside utility coverage; their residents rely entirely on shared wells, tanks and NGO aid.

Governance of Lebanon's water sector is highly fragmented (Harb and Atallah 2015). There is no single national water plan; instead, multiple public institutions (water ministries, regional authorities, municipalities) and informal networks compete. In practice, many urban poor households pay private companies (or even local gunmen) for water. Humanitarian analyses note that "vulnerable residents rely on vendors, informal services and markets, when state provision…fails to meet their needs," with repeated crises and unplanned urbanization aggravating the problem. Traditional social ties often determine who gets water; for example, some informal settlements supply water only to those from the same community, heightening inequality (Bakchan et al. 2021).

Efforts are underway to improve governance through targeted projects. USAID's Lebanon Water Project (LWP) and other aid programs have been rehabilitating local treatment plants and distribution mains in select towns (Eid-Sabbagh et al. 2022). These programs emphasize *inclusive planning* (coordinating municipalities, NGOs and residents) and building capacity of small utilities (Walnycki 2017). The LWP report found that engaging local committees and informal vendors can incrementally improve access, even in crisis conditions. Another study highlights the positive impact of small investments: installing smart meters, fixing leaks, and solarizing pumps can nearly double continuous supply in some peri-urban districts.

Lebanese cities have begun using renewable energy to offset blackouts. Solar panels are increasingly installed on borehole pumps, reflecting lessons seen elsewhere in the region (El Helou et al. 2022). Regulations have been tightened on tanker operations, and media attention has made "water service" a municipal election issue. Nonetheless, larger challenges remain: Lebanon needs legal reform to separate politics from utility management, comprehensive billing to fund maintenance, and better data collection on water usage. The economic collapse since 2019 has only made these tasks harder. In short, Lebanon's urban water crisis is as much a governance and equity problem as a technical one. Sustaining any gains will require transparent institutions and investment in human capacity to run water services efficiently (Fayssal et al. 2024).

Syria: War Impacts on Water Infrastructure

Syria's decade-long conflict has devastated its water sector. Prior to 2011, the country relied on a mix of surface water (from seven main basins) and groundwater, managed through 166 dams and a network of irrigation Works (Abbara et al. 2021) Since then, nearly all parties to the war have **weaponized water**: sieges, dam-seizures and infrastructure attacks have been used to control or punish civilian areas. Early in the conflict, for instance, heavy fighting damaged Aleppo's main water pipeline in 2012, cutting supply to roughly 3 million people (Al Jazeera 2012). In 2019, airstrikes on Idlib destroyed eight local water facilities at once, leaving about 250,000 people without their usual sources (UNICEF 2019). Similar tactics were seen around Homs, Eastern Ghouta and Deir ez-Zor. Coalition airstrikes and ground assaults also shattered public sewage and pumping stations country-wide, often without immediate repair (Human Rights Watch 2020).

The result is a public health crisis. Widespread damage to water and sanitation networks has led to recurrent disease outbreaks. A recent study found "several attacks on water and related infrastructure" in Aleppo and Idlib between 2011–2019, noting that Syria's health surveillance recorded a significant rise in waterborne diarrheal illnesses coinciding with these attacks (Abbara et al. 2021). By 2018, some reports suggested that over 30–40% of Syria's trained water utility engineers had fled or been killed, and remaining facilities were often run by militias with little technical expertise (ICRC 2021). In besieged areas, civilians resort to unsafe alternatives: drinking from contaminated wells, collecting rainwater from rooftops, or relying on an unpredictable ration via damaged pipelines.

Looking ahead, reconstruction will be daunting. Rebuilding Syria's dams, pumping stations and pipe networks will require major investment once the fighting ends. Equally important is restoring institutions: the pre-war Syrian Public Company for Water and Wastewater (SPCWW) must be reestablished on a professional basis, and local water directorates repopulated with skilled managers. Some local groups (especially in Kurdish-administered northeast) have already started grassroots repairs: drilling communal wells with solar pumps, patching leaks in civilian pipelines, and reassembling desalination units near Mediterranean cities. These efforts show community resilience, but are small-scale. Syria's overall water security will only improve when a stable government (or governments) can enforce regulations, coordinate resource-sharing among communities, and rebuild the fractured infrastructure that war has shattered (Orto 2025; Schellens et al. 2024).

Southeastern Türkiye: Local Climatic Trends and Adaptation

The southeastern region of Türkiye (the upper reaches of the Tigris–Euphrates, including the GAP area) is already experiencing measurable climate shifts. Studies of historical data (1981–2022) confirm that precipitation has generally declined and drought conditions intensified after the mid-1990s (Kartal et al. 2024). Temperatures have risen in line with global warming, exacerbating evaporation and soil moisture loss. UN assessments note that as part of the southern Mediterranean belt, Türkiye shows a *decreasing* rainfall trend and *increasing* water stress in its eastern and south-eastern provinces (UNDP 2013). Some models even predict that by 2030 up to 20% of surface water in certain basins may be lost.

Local agriculture is adapting. The Southeast Anatolia Project (GAP)—Türkiye's massive dam-and-irrigation development—has already brought about 1.5 million ha under irrigation, which is over half of Türkiye's irrigated land. In this region, farmers have increasingly installed drip irrigation on cotton, wheat and pistachio orchards, often with government subsidy. A recent survey of farmers in Urfa and Mardin (two GAP provinces) found that those who received rural development grants for in-field drip systems overwhelmingly viewed the subsidies as successful, and strongly favored expanding them (İpekçioğlu et al. 2022). The same farmers reported bureaucratic hurdles in applying, but noted clear water savings and yield stability once systems were in place. In response to energy shortages, some irrigation pumps now use solar panels (Türkiye is piloting agrivoltaic projects combining shading and solar generation). At least one large project collects and stores floodwater in underground cisterns to recharge aquifers after wet seasons.

Communities have also shifted crop choices: in many areas, cotton has been partly replaced by less thirsty crops (rapeseed, safflower, or drought-tolerant wheat varieties), reflecting market changes and water availability. Local water cooperatives have begun coordinating sharing of irrigation schedules, reducing overuse. The government's "Rotation Systems" also periodically allocate water in turns among villages. On a planning level, authorities are installing more meteorological and streamflow gauges to improve early warning of drought, and forest managers are conserving headwater forests to stabilize runoff.

Nevertheless, challenges continue. Despite drip irrigation, total regional water use keeps rising with population and agricultural intensification. Because so much groundwater is still abstracted (e.g. via GAP deep wells), aquifers are slowly depleting. Experts warn that unless demand is curbed or rainfall recovers, water stress will worsen. They recommend continued investment in efficiency (e.g. fully lining irrigation canals, recycling municipal wastewater for peri-urban farms) and in diversifying rural livelihoods so that farming families are not entirely dependent on irrigation. Ongoing initiatives to improve climate resilience in the Southeast (including capacity building for local water managers) reflect this need. In summary, Southeastern Türkiye shows clear signs of regional drying; local adaptation has emphasized water-efficient agriculture and infrastructure (drip irrigation subsidies, water storage projects), but sustaining these gains will require systemic planning and support to cope with the anticipated "new normal" of heat and drought (Kartal et al. 2024).

5 Discussions

All case studies underscore a common reality: MENA's hydroclimatic stress is intensifying. Temperatures are rising and rainfall is declining across the region) compounding historic aridity (Thompson and Zakhirova 2021. For example, Southeastern Anatolia (Türkiye) has seen steadily decreasing precipitation and more extreme droughts since the mid-1990s (Kartal el al. 2024) while Syria endured an unprecedented 2006–2011 megadrought and now again faces record heat. MENA contains 15 of the world's 20 most water-scarce countries (Abu Hawash and Ghafar 2025) so even small climate shifts hit hard. Despite these similarities, contexts differ. Jordan—relatively stable and affluent—is investing in water infrastructure and even mandating rainwater capture (new buildings must have cisterns) (Hindiyeh et al. 2021). Iran, Türkiye and Iraq have large dam programs and aquifer development, whereas Yemen and Syria (war-torn) lack the capacity to maintain systems. Gulf states substitute desalination for scarce surface water. Across cases, weak governance or conflict often makes situations worse. In rural Iran, for instance, Moshizi et al. (2023) found that the "leading cause" of local water shortage was not climate per se but poor management and policy failures. Rural communities everywhere still rely on inherited practices—from Persian qanats to Yemen's *Haq al-Shirb* rotations—to stretch scarce water (Haddad 2025). But these traditions now coexist with modern stressors.

Traditional community systems still play a vital role. For example, Hama's ancient norias (large water wheels) continue to lift Orontes River water into medieval aqueducts (Figure), illustrating how local technology has long bolstered water supply (Haddad 2025). Across MENA, many villages maintain communal cisterns, tribal water rights, and timed rotational schemes. In Yemen the customary *Haq al-Shirb* ensures each farmer a fair turn on a stream, and across Iran villagers tend their qanat tunnels to tap groundwater (English 1998). These grassroots practices complement state policies. Jordan's 2016–2025 Water Strategy explicitly promotes rainwater harvesting at household and farm levels (Fakoussa and Kabis-Kechrid 2020), and even requires new homes to include storage tanks. NGOs and women's associations have helped spread efficiency techniques too—for instance, Jordan's Business and Professional Women's Organization trains local women to sell and demonstrate low-cost shower and faucet-saving devices (Roudi-Fahimi et al. 2000). Such community-led measures improve access and equity, but on their own they face limits. Even well-governed systems (like Iran's) can falter: downstream Iranian villages saw qanats running dry because upstream over-pumping and bureaucratic neglect were never checked Moshizi et al. (2023). Our cases show that without supportive policy and financing, local initiatives cannot compensate for degraded rivers or mass groundwater depletion.

National governance matters greatly. We found that many MENA states have climate or water plans on paper, but implementation often lags (Sowers et al. 2011) Jordan has issued NDCs and an Economic Modernization Vision that calls for water resilience, and has ministries focused on "green growth," yet analysts warn its climate governance must be strengthened across agencies and decentralized to municipalities (El-Sharif and Muasher 2024). In places like Iraq and Yemen, governance collapse has meant that adaptation happens piecemeal through humanitarian projects: for example, a UNDP initiative in Yemen is reintroducing traditional rainwater catchment (cisterns, terraces and even fog collectors) in vulnerable communities (UNPD 2013), but systemic enforcement of water law is impossible amid conflict. In Iraq's south and north, tribal councils and NGOs are now negotiating water allocations in lieu of a functional state regülatör (Löble 2024). This reflects a broader finding: Moshizi et al. (2023) and others emphasize that "poor water governance"—from lack of regulation to corruption—often underlies rural vulnerability more than climate aloneIn short, top-down policies must be matched by transparent institutions and stakeholder engagement. Where governance is inclusive, adaptation is more effective. Conversely, when governments are authoritarian or fractured, even generous international funds can be squandered on inappropriate big projects rather than local needs (Hamzawy and Arkeh 2025).

Low-cost adaptation measures (rainwater harvesting, small dams, share systems) have been shown to work locally, though their aggregate impact is modest. Rainwater capture, for example, is widely promoted: Jordan estimates that widespread household harvesting could yield ~ 7 million m^3/year and rural catchment ~ 15 million m^3 g (Al Karablieh and Salman 2016). In practice, modeling for Jordan's Lower Jordan Valley suggests urban RWH might supply ~ 20 × 10^6 m^3 in a normal year but only ~ 3 × 10^6 m^3 under drought (Hindiyeh et al. 2021). a substantial drop. Yemen's climate project explicitly trains farmers to rehabilitate cisterns and percolation ponds (Almeshreki et al. 2012). Elsewhere, countries like Morocco and Saudi Arabia are even building hundreds of small earth dams to capture wadi floods (Kantoush et al. 2022). These "soft" infrastructures boost local water tables and reduce erosion, and often cost far less than mega-dams. Similarly, traditional water-sharing customs remain crucial low-tech solutions: timed rotational rights (Yemen's *Haq al-Shirb*), communal watering committees, and inherited local rules help distribute scarce flow equitably (Haddad 2025). However, our review highlights that such approaches have limits in scale. They generally serve household and irrigation needs up to a point, but cannot meet rapidly growing urban or industrial demand. Therefore, while we find that community-led and low-cost solutions are effective as first lines of defense, they must be embedded in wider reforms (e.g. water pricing, infrastructure maintenance).

A final crosscutting theme is the critical need for regional cooperation. MENA's rivers and aquifers rarely respect political boundaries, so unilateral adaptation is risky. The Euphrates–Tigris basin (Türkiye, Syria, Iraq, with Iran on the Tigris) illustrates the point: Türkiye's massive dam-building under the GAP project has provoked repeated tensions, as Syria and Iraq demand more downstream flow and Türkiye resists in order to fill its reservoirs (Climate Diplomacy 2024). No binding treaty governs these waters; one analysis notes that decades of dispute have "prevented the three governments from effectively co-managing" the basin. Likewise, Jordan has long contested Syria's control of the Yarmouk River, a key Jordan tributary (Abu Hawash and Abdel Ghafar 2025). Experts argue that without institutionalized dialogue, droughts or dam releases upstream can spark severe crises. For example, an MEI report notes that regional leaders should adopt diplomatic "water diplomacy" and shared management to avert conflict. A few success stories exist (e.g. the Jordan–Saudi Arabia treaty on a shared aquifer), but they are the exception. Our cases reinforce that resilient adaptation in any one country is fragile if neighbors do not collaborate: water needs to be managed at the basin scale. In summary, effective hydroclimatic adaptation in MENA hinges on multilateral coordination—data-sharing, joint forecasting, and negotiated flow agreements—alongside national and local action.

6 Conclusion

This review of MENA case studies yields several clear findings. First, hydroclimatic vulnerability is acute everywhere in the region, driven by extreme aridity, population pressures, and rising temperatures. Second, simple measures—rainwater cisterns, small dams, fog collectors, equitable sharing customs—can significantly reduce hardship at the community level (United Nations Development Programme UNPD 2013). But these alone are **not** sufficient without supportive governance: fragmented or corrupt institutions often negate the benefits of local projects. Third, the most severely affected populations tend to be those with little political power (e.g. farmers, refugees, women), underscoring the need for inclusive approaches (Hamzawy and Arkeh 2025). Finally, our cases highlight that water scarcity intersects with conflict: managing scarcity poorly can amplify tensions (as in the Euphrates or Yemen), whereas cooperative management can build trust.

Policy recommendations: Given limited resources, we advocate prioritizing affordable, community-driven adaptation. In particular:

- **Empower local water committees and NGOs.** Build on successful community initiatives by funding participatory watershed planning and training local leaders. For example, Jordan's experience suggests that capacitating municipalities and civil society (including women's groups) to manage small schemes yields dividends (El-Sharif and Muasher 2024)
- **Scale up rainwater harvesting and efficiency.** Create subsidies or incentives for household cisterns, rooftop catchment, and drip irrigation. Enforce building codes for water storage (as Jordan has done (Hindiyeh et al. 2021). Introduce low-cost sensors or mobile apps to help farmers schedule irrigation.
- **Support traditional techniques.** Formalize and finance community water-sharing: e.g. legally recognize time-share agreements like *Haq al-Shirb*, hire local *mirabs* (watermasters) in rural Iran, and rehabilitate qanat canals and terraces. Partnerships with elders and tribes can help revive these practices in a modern contex (Haddad 2025).
- **Reform water governance.** Streamline existing laws and crack down on waste. For instance, Jordan's environmental ministry could be further strengthened to implement its climate laws. Decentralize decision-making so that local councils

tailor solutions to villagers' needs. Ensure that pricing and allocations reflect true scarcity, protecting basic needs first (echoing Islamic principles of fairness

- **Enhance regional cooperation.** Support basin-wide institutions or agreements (e.g. a formal Euphrates-Tigris commission). Share monitoring data across borders (e.g. river gauges, weather forecasts). Integrate water diplomacy into climate forums: as Abu Hawash and Abdel Ghafar (2025) argue, diplomacy and sustained dialogue are the only way to prevent "water wars" over shared rivers (Abu Hawash and Abdel Ghafar 2025).

Further research and data needs: There remain major knowledge gaps that should be closed. First, improved monitoring: researchers should leverage open-access hydro-climate data (satellite and ground) to assess trends. As (Khalifa and Brouziyne (2025) note, free remote-sensing products (e.g. CHIRPS, MODIS evapotranspiration, WRI Aqueduct) now allow consistent water assessments in data-poor MENA countries. Scholars must work to validate and calibrate these datasets regionally. Second, we urge more field-based qualitative work: household surveys, focus groups, and ethnographies are needed to capture how people perceive water risk and what coping strategies they actually use (see Arkeh and Awadallah 2025) on the value of "localizing authority" and engaging civil society. Third, interdisciplinary projects should link biophysical models with social data—for example, using GIS to map vulnerability while conducting in-depth interviews in hotspot villages. Finally, transparent data sharing between governments, NGOs and academics (e.g. publishing national water use and reservoir levels) would greatly improve adaptive planning. In conclusion, building MENA's resilience to hydroclimatic shocks will depend on blending traditional knowledge with modern tools, empowering communities, and strengthening cross-border collaboration. These concerted efforts—grounded in both quantitative data and qualitative understanding—are essential to safeguard water security for all in the region.

References

Abbara A, Zakieh O, Rayes D, Collin SM, Almhawish N, Sullivan R, … Karah N (2021) Weaponizing water as an instrument of war in Syria: Impact on diarrhoeal disease in Idlib and Aleppo governorates, 2011–2019. Int J Infect Dis 108:202–208. https://doi.org/10.1016/j.ijid.2021.05.030

Abu Hawash M, Abdel Ghafar A (2025) Water diplomacy: How to prevent water wars in the MENA region. Middle East Council on Global Affairs. Retrieved from: https://mecouncil.org/publication/water-diplomacy-how-to-prevent-water-wars-in-the-mena-region. Accessed 3 Sept 2025

Adkhamov U (2023) What are the roles of international water organizations in climate change adaptation? Master's thesis, Oregon State University. Available from: https://ir.library.oregonstate.edu/concern/graduate_thesis_or_dissertations/np193j48c. Accessed 3 Sept 2025

Al-Addous M, Bdour M, Alnaief M, Rabaiah S, Schweimanns N (2023) Water resources in Jordan: a review of current challenges and future opportunities. Water 15(21):3729. https://doi.org/10.3390/w15213729

Al Jazeera (2012) Syrian strikes on Aleppo kill dozens. Retrieved from: https://www.aljazeera.com/news/2012/9/9/syrian-strikes-on-aleppo-kill-dozens. Accessed 3 Sept 2025

Al Karablieh E, Salman A (2016) Water resources, use and management in Jordan: a focus on groundwater. International Water Management Institute (IWMI). Retrieved from: https://gw-mena.iwmi.org/wp-content/uploads/sites/3/2017/04/Rep.11-Water-resources-use-and-management-in-Jordan-a-focus-on-groundwater.pdf. Accessed 3 Sept 2025

Al-Quraishi AMF (2024) Geoinformatics approaches to climate change-induced soil degradation in the MENA Region: A review. In: Al-Quraishi A, Negm A, Benzougagh B (eds) Climate change and environmental degradation in the MENA Region. Handbook of environmental chemistry, vol 136. Springer, Cham. https://doi.org/10.1007/698_2024_1167

Al-Shakeri H (2025) Iraq's water crisis: Dammed by neighbours, failed by leaders [Expert comment]. Chatham House. Retrieved from: https://www.chathamhouse.org/2025/08/iraqs-water-crisis-dammed-neighbours-failed-leaders. Accessed 3 Sept 2025

Almeshreki DMH, Alsharjabi DKM, Bafadhl MOA, Almedheji MA, Al-Khulaidi DA, Alabiadh MKM, … Rahman Saleh Alazani MAS (2012) Combating land degradation in Yemen–a national report

Arkeh J, Awadallah N (2025) Climate change and state fragility in Iraq: Budgeting, governance, and the future of sustainability. Carnegie Endowment for International Peace. Retrieved from: https://carnegieendowment.org/research/2025/05/climate-change-and-state-fragility-in-iraq-budgeting-governance-and-the-future-of-sustainability?lang=en. Accessed 3 Sept 2025

Azizi MA, Leandro J (2025) Factors affecting transboundary water disputes: nile, Indus, and Euphrates-Tigris river basins. Water 17(4):525. https://doi.org/10.3390/w17040525

Bakchan A, Hacker ME, Faust KM (2021) Resilient water and wastewater infrastructure systems through integrated humanitarian-development processes: the case of Lebanon's protracted refugee crisis. Environ Sci Technol 55(9):6407–6420

Climate Diplomacy (2024) Türkiye, Syria and Iraq: Conflict over the Euphrates–Tigris. Climate Diplomacy Case Studies. Retrieved from: https://climate-diplomacy.org/case-studies/Türkiye-syria-and-iraq-conflict-over-euphrates-tigris. Accessed 3 Sept 2025

Country Reports (2019) Irrigation water management in Iran: Growing need for action. WaterHQ. Retrieved from https://waterhq.world/issue-sections/country-reports/iran/irrigation-water-management-in-iran-growing-need-for-action/

Change IPOC (2001) Climate change 2007: Impacts, adaptation and vulnerability. Genebra, Suíça

Darvishi Boloorani A et al (2024) Climate change, drought, and water scarcity in the MENA Region. In: Al-Quraishi A, Negm A, Benzougagh B (eds) Climate change and environmental degradation in the MENA Region. Handbook of Environmental Chemistry, vol 136. Springer, Cham. https://doi.org/10.1007/698_2024_1143

Eid-Sabbagh K, Roukoz S, Nassif MH, Velpuri N, Mateo-Sagasta J (2022) Analysis of water reuse potential for irrigation in Lebanon. IWMI Working Paper, vol. 181. International Water Management Institute

El Helou GC, Najjar A, Mardirossian S, Khallouf S, Chelala C, Chehab S, de Beyrouth ÉSDI (2022) 50% green electricity for Lebanon towards 2050. University of Beirut. (Unpublished report/PDF)

El Idrissi MC, Saber ER, El Mazi M (2025) Exploring farmers' practices and perceptions of groundwater extraction as a key component of integrated water management in the Dayet Aoua watershed, Middle Atlas Morocco. Environ Socio-Econ Stud 13(2):48–66

El Kenawy AMH, Torab MMM, Hereher M (eds) (2024) Hydroclimatic extremes in the Middle East and North Africa: assessment, attribution and socioeconomic impacts. Elsevier. https://doi.org/10.1016/C2020-0-01660-9

El-Sharif S, Muasher M (2024) Vulnerability and governance in the context of climate change in Jordan. Carnegie Endowment for International Peace. Retrieved from https://carnegieendowment.org/research/2024/05/jordan-climate-change-adaptation-commitments?lang=en

English P (1998) Qanats and lifeworlds in Iranian plateau villages. Transform Middle East Nat Environ Bull Series 103

Fakoussa D, Kabis-Kechrid LL (2020) Jordan's socio-economic woes and foreign policy: Employment, trade, and international cooperation—Policy briefs from the region and Europe

FAO (2023a) Water resources—Türkiye. AQUASTAT country factsheet. https://storage.googleapis.com/fao-aquastat.appspot.com/countries_regions/factsheets/water_resources/en/TUR-WRS.pdf. Accessed 3 Sept 2025

FAO (2023b) Water resources—Jordan. AQUASTAT country factsheet. https://storage.googleapis.com/fao-aquastat.appspot.com/countries_regions/factsheets/water_resources/en/JOR-WRS.pdf. Accessed 3 Sept 2025

FAO (2023c) Water resources—Saudi Arabia. AQUASTAT country factsheet. https://storage.googleapis.com/fao-aquastat.appspot.com/countries_regions/factsheets/water_resources/en/SAU-WRS.pdf. Accessed 3 Sept 2025

FAO (2023d) Water resources—Iraq. AQUASTAT country factsheet. https://storage.googleapis.com/fao-aquastat.appspot.com/countries_regions/factsheets/water_resources/en/IRQ-WRS.pdf. Accessed 3 Sept 2025

FAO (2023e) Water resources—Iran. AQUASTAT country factsheet. https://storage.googleapis.com/fao-aquastat.appspot.com/countries_regions/factsheets/water_resources/en/IRN-WRS.pdf. Accessed 3 Sept 2025

FAO (2023f) Water resources—Lebanon. AQUASTAT country factsheet. https://storage.googleapis.com/fao-aquastat.appspot.com/countries_regions/factsheets/water_resources/en/LBN-WRS.pdf. Accessed 3 Sept 2025

FAO (2023g) Water resources—Yemen. AQUASTAT country factsheet. https://storage.googleapis.com/fao-aquastat.appspot.com/countries_regions/factsheets/water_resources/en/YEM-WRS.pdf. Accessed 3 Sept 2025

FAO (2023h) Water resources—Israel. AQUASTAT country factsheet. https://storage.googleapis.com/fao-aquastat.appspot.com/countries_regions/factsheets/water_resources/en/ISR-WRS.pdf. Accessed 3 Sept 2025

Frenken K (Ed.) (2008) Irrigation in the Middle East: Region in figures, AQUASTAT survey. FAO Land and Water Division

Fanack W (2025) Groundwater of the MENA region. Retrieved May 9, 2025, from: https://water.fanack.com/publications/groundwater-of-the-mena-region/. Accessed 3 Sept 2025

Fayssal N, Al Tawil L, Danageuzian H, Sabat M, ElMerehbi H, Darazi D, … Romanos J (2024) Navigating the water–energy nexus amidst the Lebanese economic crisis. J Water Clim Change 73(5):917–929. https://doi.org/10.2166/aqua.2024.329

Fessahaye AK, Xie Y, Hu Y, Dai G, Zheng H, Teng W (2025) Groundwater over-extraction: Comprehensive review of socio-economic impacts and pathways to sustainable management. Euro J Theoretic Appl Sci 3(2):190–211. https://doi.org/10.59324/ejtas.2025.3(2).15

Haddad M (2025) Approaches used by Muslims to allocate and distribute water among users and sectors throughout history. Muslim Heritage. Retrieved from: https://muslimheritage.com/allocate-and-distribute-water. Accessed 3 Sept 2025

Hamzawy A, Arkeh J (2025) Climate resilience in the MENA region: Navigating challenges, empowering communities, and transforming governance. Carnegie Endowment for International Peace. Retrieved from: https://carnegie-production-assets.s3.amazonaws.com/static/files/MEP_Climate-Resilience-2025.pdf. Accessed 3 Sept 2025.

Harb M, Atallah S (2015) Lebanon: a fragmented and incomplete decentralization. In: Local governments and public goods: Assessing decentralization in the Arab world, pp. 187.

Hindiyeh MY, Matouq M, Eslamian S (2021) Rainwater harvesting policy issues in the MENA region: Lessons learned, challenges, and sustainable recommendations. In: Eslamian S (ed) Handbook of water harvesting and conservation: basic concepts and fundamentals. Wiley, pp 457–473. https://doi.org/10.1002/9781119478911.ch30

Hussein H (2025) Toward inclusive climate action in Jordan. Arab Reform Initiative. Retrieved from: https://s3.eu-central-1.amazonaws.com/storage.arab-reform.net/ari/2025/04/25112328/ENGLISH-2025-EPP-Toward-Inclusive-Climate-Action-in-Jordan-1.pdf. Accessed 3 Sept 2025

Human Rights Watch (2020) Targeting life in Idlib: Syrian and Russian strikes on civilian infrastructure. Human Rights Watch. Retrieved from https://www.hrw.org/report/2020/10/15/targeting-life-idlib/syrian-and-russian-strikes-civilian-infrastructure

Human Rights Watch (2023) Yemen: Warring parties deepen water crisis. Human Rights Watch. Retrieved from https://www.hrw.org/news/2023/12/11/yemen-warring-parties-deepen-water-crisis

İpekçioğlu Ş, Aydın B, Anlağan Taş M, … Bayraktar MS (2022) Producers' approach to drip irrigation subsidies in Türkiye: The case of Southeastern Anatolia Region. ADYUTAYAM Dergisi 10(2):92–107

ICRC (International Committee of the Red Cross) (2021) Syria water crisis: Up to 40% less drinking water after 10 years of war. Retrieved from https://www.icrc.org/en/document/syria-water-crisis-after-10-years-war. Accessed 3 Sept 2025

International Trade Administration (2024) Saudi Arabia country commercial guide: Water. U.S. Department of Commerce. Retrieved from: https://www.trade.gov/country-commercial-guides/saudi-arabia-water. Accessed 3 Sept 2025

IPCC (2022) Climate Change 2022: Impacts, Adaptation and Vulnerability. Contribution of Working Group II to the Sixth Assessment Report of the Intergovernmental Panel on Climate Change. Cambridge University Press. https://www.ipcc.ch/report/ar6/wg2/downloads/report/IPCC_AR6_WGII_FullReport.pdf

Kantoush SA, Saber M, Abdel-Fattah M, Sumi T (2022) Integrated strategies for the management of wadi flash floods in the Middle East and North Africa (MENA) Arid Zones: The ISFF Project. In: Wadi Flash Floods, pp 3–34

Kartal V, Yavuz VS, Ariman S, Kaya K, Alkanjo S, Simsek O (2024) Climate change trends in the Southeastern Anatolia region of Türkiye: precipitation and drought. J Water Clim Change 15(12):5893–5919

Khalifa M, Al-Zu'bi M, Brouziyne Y (2025) Pathway from water-conflict to water-peace in the Middle East and North Africa. Water Pract Technol 20(3):671–684

Kowsar N (2024) A thirsty reality: Iran's dire water situation. Atlantic Council, 22 January 2024. Retrieved from: https://www.atlanticcouncil.org/blogs/iransource/iran-water-environment-us-policy/. Accessed 3 Sept 2025

Löble N (2024) Strengthening enhanced climate and community resilience in Iraq. Berghof Foundation. Retrieved from: https://berghof-foundation.org/work/projects/strengthening-enhanced-climate-and-community-resilience-in-iraq. Accessed 3 Sept 2025

Monshipouri M, Mohajer R (2024) Climate change and the potential for conflict in the Middle East: emerging threats to security and human rights. The Maghreb Rev 49(1):14–41

Moshizi ZM, Yousefi A, Amini AM et al (2023) Rural vulnerability to water scarcity in Iran: an integrative methodology for evaluating exposure, sensitivity and adaptive capacity. GeoJournal 88:2121–2136. https://doi.org/10.1007/s10708-022-10726-0

Omar A (2025) Struggling over every drop: Yemen's crisis of aridity and political collapse [Blog post]. Sada, Carnegie Endowment for International Peace. 29 April 2025. Retrieved from: https://carnegieendowment.org/sada/2025/04/struggling-over-every-drop-yemens-crisis-of-aridity-and-political-collapse?lang=en. Accessed 3 Sept 2025

Orto N (2025) Syria's water crisis: Assessing the intersection of climate change and geopolitical interests. Fanack Water. 12 February 2025. Retrieved from: https://water.fanack.com/publications/syrias-water-crisis-assessing-the-intersection-of-climate-change-and-geopolitical-interests/. Accessed 3 Sept 2025

Priyan K (2021) Issues and challenges of groundwater and surface water management in semi-arid regions. In: Pande CB, Moharir KN (eds) groundwater resources development and planning in the semi-arid region. Springer, Cham. https://doi.org/10.1007/978-3-030-68124-1_1

RIMES (Regional Integrated Multi-Hazard Early Warning System) (2022) Institutional survey & assessment report on drought early warning system in Iraq. Pathumthani, Thailand: United Nations Development Programme. Retrieved from: https://www.undp.org/sites/g/files/zskgke326/files/2024-06/institutional_survey_assessment_report_on_drought_early_warning_system_in_iraq.pdf. Accessed 3 Sept 2025

Roudi-Fahimi F, Creel L, de Souza RM (2000) Finding the balance: Population and water scarcity in the Middle East and North Africa. Population Reference Bureau. 17 July 2000. Retrieved from: https://www.prb.org/resources/finding-the-balance-population-and-water-scarcity-in-the-middle-east-and-north-africa. Accessed 3 Sept 2025

Schellens M, Ozdemir A, Greenwood M, Zwijnenburg W (2024) War and water security challenges along the Orontes River in Syria. PAX for Peace. April 2024. Retrieved from: https://paxforpeace.nl/wp-content/uploads/sites/2/2024/11/PAX_report_Thirst-for-Peace_2024.pdf. Accessed 3 Sept 2025

Sowers J, Vengosh A, Weinthal E (2011) Climate change, water resources, and the politics of adaptation in the Middle East and North Africa. Clim Change 104(3):599–627

Stanley B (2010) Case Study: Jordan–Dealing with the Water Deficit in Jordan. Recycling household water to irrigate home gardens makes good environmental and financial sense. IDRC–International Development Research Centre. Retrieved from: https://www.idrc.ca/en/research-in-action/case-study-jordan-dealing-water-deficit-jordan. Accessed 3 Sept 2025

Stevanović Z (2018) Karst aquifers in the arid world of Africa and the Middle East: Sustainability or humanity? Karst Water Environment: Advances in Research, Management and Policy. Springer, Cham, pp 1–43

Suhail M, Faraj TK, Ahmad W, Ravshanov AX, Khan MN (2024) Issues of water resources in Saudi Arabia: Past, present, and future. Sustainability 16(10):4189. https://doi.org/10.3390/su16104189

Thompson WR, Zakhirova L (2021) Climate change in the Middle East and North Africa: 15,000 years of crises, setbacks, and adaptation. Routledge

Tayfun K (2025) A new integrated gis-based approach to drought susceptibility mapping: case study of the medina region, Saudi Arabia. Oceanography & Fisheries Open Access J 18(5):555996. https://doi.org/10.19080/OFOAJ.2025.18.555996

Tinazzi I (2024) Water scarcity, migrations and climate change: an assessment of their nexus. Dissertation, Ca' Foscari University of Venice, Academic Year 2022/2023.

Tinti A (2017) Water resources management in the Kurdistan region of Iraq. IRIS Policy Report. Institute of Regional and International Studies (IRIS), The American University of Iraq, Sulaimani. Retrieved from: https://auis.edu.krd/iris/sites/default/files/Water%20Policy%20Report%20IRIS_FINAL%20ES.pdf. Accessed 3 Sept 2025

UNICEF (2019) Eight water facilities come under attack in Idlib, northwest Syria, affecting water for approximately 250,000 people. 20 July 2019. Retrieved from: https://www.unicef.org/mena/press-releases/eight-water-facilities-come-under-attack-idlib-northwest-syria-affecting-water. Accessed 3 Sept 2025

UNDP (2013) Adaptation to climate change through integrated water harvesting technologies in Yemen. UNDP Climate Change Adaptation. Retrieved from: https://www.adaptation-undp.org/projects/ldcf-yemen. Accessed 3 Sept 2025

Walnycki A (2017) Making Lebanon's water flow: delivering better basic urban services. International Institute for Environment and Development (IIED). November 2017. Retrieved from: https://www.iied.org/sites/default/files/pdfs/migrate/17442IIED.pdf. Accessed 3 Sept 2025

World Resources Institute (2025) Aqueduct Water Risk Atlas. Retrieved September 3, 2025, from: https://www.wri.org/applications/aqueduct/water-risk-atlas/#/?dvanced=false&basemap=terrain&indicator=drr_cat&lat=24.383658896597737&lng=47.02148437500001&mapMode=view&month=1&opacity=0.5&ponderation=DEF&predefined=false&projection=absolute&scenario=optimistic&scope=baseline&threshold&timeScale=annual&year=baseline&zoom=3. Accessed 3 Sept 2025

The Growing Threat of Extreme Weather Events in the Middle East

Mohammed Mahmoud

1 Introduction

The Middle East, long defined by its arid climate and scarce water resources, is increasingly confronting the growing threat of extreme weather events that challenge both human and environmental resilience. Once considered rare, episodes of severe heatwaves, flash floods, dust storms, and unseasonal heavy rainfall are now occurring with greater frequency and intensity, driven in large part by climate change. These shifts pose mounting risks to public health, infrastructure, and economic stability, emphasizing the urgent need to better understand their drivers and consequences in one of the world's most vulnerable regions.

At the same time, the region stands at the intersection of multiple, interconnected climate challenges, where rising temperatures and shifting weather patterns are giving rise to unprecedented phenomena. In recent years, the Middle East has witnessed the emergence of tropical storms and even rare cyclonic activity, exemplified by the unprecedented landfall of Cyclone Shaheen in Oman. Similarly, the rise of Medicanes such as Storm Daniel highlights how the warming Mediterranean is becoming a new hotspot for destructive storm systems. These evolving threats not only place immense strain on ecosystems and societies, but also fuel misinformation and conspiracy theories, as extreme events are frequently misattributed to human interventions like cloud seeding. Together, these dynamics reveal how climate change is reshaping the region's weather extremes and how societies interpret and respond to these escalating risks.

M. Mahmoud (✉)
United Nations University Institute of Water, Environment, and Health, Richmond Hill, ON, Canada
e-mail: mohammed.mahmoud@unu.edu

M. M. Al Saud (ed.), *Hydroclimate Changes: Implications of the Middle East Region*,
https://doi.org/10.1007/978-3-032-20053-2_4

2 The Intersection of Climate Change, Warming, and Extreme Weather

Over the past decade, climate change has had increasingly severe impacts across the Middle East, with extreme heat and warming emerging as the most prominent threats. Already experiencing higher temperatures than most of the world, the region has seen daily highs exceed historical averages, with numerous countries and cities breaking records during recent summer seasons. Consequently, the Middle East has experienced some of the world's most extreme warming events, underscoring the region's growing vulnerability to heat extremes. In the last couple of years alone, several major cities in the Middle East have experienced record-breaking heat. In Mecca, Saudi Arabia, temperatures during the 2024 Hajj season reached 51.8 °C, the highest on record for the city, leading to over a thousand heat-related deaths among pilgrims (Time 2023). In May 2025, Dubai and Abu Dhabi recorded 51.6 °C, setting a new national May record for the UAE (Reuters 2025). Soon after, in August 2025, Amman, Jordan, endured its hottest conditions in history, with the city reaching 43.5 °C and experiencing its warmest nights ever documented (Guardian 2025).

Extreme heat in the Middle East has significant implications for public health, safety, and development. For public health, frequent peak temperatures above 40 °C increase the risk of heat-related illnesses, including heatstroke and heat exhaustion, even indoors, placing heavy demands on power systems to maintain effective cooling. In terms of safety, sustained high temperatures accelerate the degradation of critical infrastructure such as roads, bridges, buildings, and utility facilities, requiring updated maintenance protocols and, in some cases, cooling measures to preserve structural integrity. From a development perspective, extreme heat exacerbates water scarcity, challenging the region's ability to support population growth and economic activity (Mahmoud 2021b).

This elevated warming trend in the region extends beyond land, affecting surrounding oceans and seas, where rising sea surface temperatures are contributing to more extreme weather events. Warmer waters fuel severe storms, including thunderstorms and cyclones, which, despite bringing much-needed rainfall to an otherwise arid region, often result in catastrophic impacts. Intense precipitation, powerful winds, and subsequent flooding from these storms pose significant risks to communities, infrastructure, and overall regional resilience, highlighting the escalating vulnerability of Middle Eastern countries to climate-driven hazards (Mahmoud 2024).

3 Dust Storms

Dust storms are a growing environmental hazard across the Middle East, with increasing frequency, intensity, and impacts. In countries with sparse vegetation, such as Kuwait, they occur more than three months a year, while Bahrain, Qatar, and the UAE experience lower but still significant exposure. Saudi Arabia is affected across nearly the entire country, driven by dry al-Bawareh winds from the eastern Mediterranean and storms originating as far away as Libya. These storms have wide-ranging impacts on the environment and society. They increase air pollution, and cause damage to agriculture, infrastructure, and essential services such as transportation, water, and energy systems. Human activities, including land degradation, deforestation, desertification, poorly managed irrigation, and drained wetlands, often worsen their severity (Keynoush 2022).

An example of dust storm severity in the Middle East occurred in April 2022, when a series of severe dust storms swept across Iraq, Syria, and Turkey, causing widespread health, infrastructure, and societal impacts. In Iraq, particularly in Baghdad, thousands of people were hospitalized during the initial dust storm, with more people affected during a subsequent storm later in the month (Al Jazeera 2022). In Syria, hospitals reported increased cases of respiratory distress, while schools, offices, and transportation networks were also impacted, interrupting economic and daily activities (Gulf Cultural Club 2022). In Turkey, the storms led to significant spikes in particulate matter, especially along the western coast, worsening air quality and posing serious risks to individuals with pre-existing respiratory conditions (Zeydan and Yildrim 2024).

Addressing the challenges of dust storms in the Middle East will require strategies that go beyond immediate response and focus on long-term prevention and adaptation. Efforts to restore degraded lands, improve soil management, and protect natural vegetation can reduce dust generation at its source, while investments in urban planning and infrastructure can help communities withstand exposure when storms occur. Regional cooperation is also critical, as dust storms often cross national borders, affecting air quality, health, and economic activity across multiple countries. By combining environmental management, scientific monitoring, and community engagement, countries in the Middle East can begin to reduce the human, economic, and ecological toll of these persistent storms and build resilience against a hazard that is likely to grow in intensity under a warming climate.

4 Tropical Storms

Tropical cyclones commonly form in the north Indian Ocean, particularly over the southeastern Arabian Sea, where warm sea surface temperatures near the equator and high atmospheric moisture create ideal conditions for their development, much like in the Bay of Bengal to India's east. Nearly half of the cyclones that form in the Arabian Sea never make landfall, and those that do often weaken significantly, especially when striking the Arabian Peninsula, where they are typically downgraded to depressions. Some systems move westward toward the Horn of Africa, but the cooler waters of the western Arabian Sea further diminish their strength. Historically, Oman and Yemen have borne the greatest impact from these storms due to their extensive Arabian Sea coastlines. Though few cyclones strike them at full intensity, those that do reach land still cause widespread flooding, destruction, and loss of life.

Recent studies indicate a significant increase in the frequency and intensity of tropical cyclones in the Arabian Sea, particularly over the past two decades. Between 2001 and 2019, the number of very severe cyclones (Category 3 and above) rose by approximately 150%, compared to the period from 1982 to 2000 (Deshpande et al. 2012). This trend is attributed to rising sea surface temperatures and increased atmospheric moisture availability, which provide more energy for cyclones to develop and intensify. Additionally, the duration of these storms has increased, with very severe cyclones lasting significantly longer than in previous decades. As a result, countries along the Arabian Peninsula, particularly Oman, Yemen, and parts of southern Iran, are experiencing more frequent and stronger storms that make landfall.

This trend has heightened the risk of severe flooding, infrastructure damage, and economic losses across the region. Recent storms in the Middle East further illustrate the scale of extreme rainfall now occurring. Cyclone Shaheen in Oman (2021), Storm Daniel in Libya (2023), and the April 2024 Arabian Peninsula storm in the United Arab Emirates each produced extraordinary daily rainfall totals ranging from more than 200 mm to over 450 mm, highlighting the growing potential for single storms to deliver volumes that far exceed historical norms (Fig. 1).

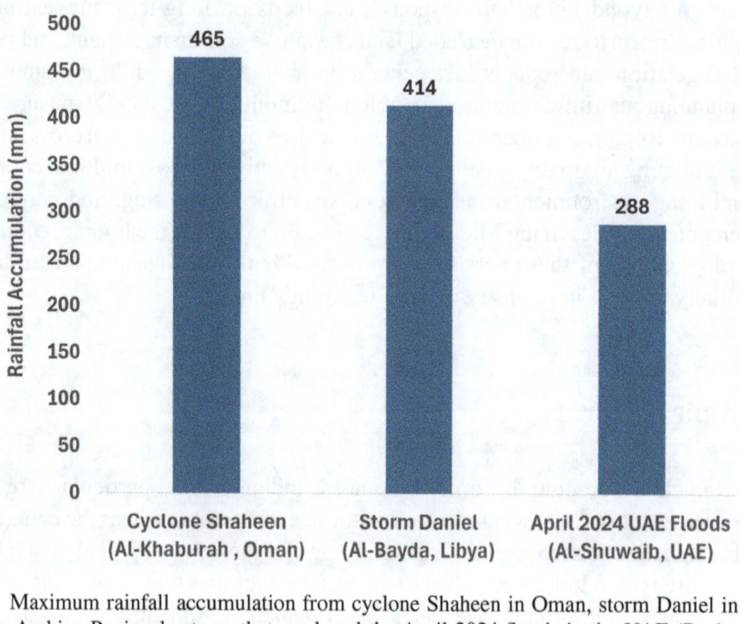

Fig. 1 Maximum rainfall accumulation from cyclone Shaheen in Oman, storm Daniel in Libya, and the Arabian Peninsula storm that produced the April 2024 floods in the UAE (Derived from Oman Meteorology 2025; WMO 2023; and Khaleej Times 2024)

5 The Unusual Case of Cyclone Shaheen

On October 3, 2021, Cyclone Shaheen made landfall near Muscat, Oman, after crossing the Gulf of Oman from the Arabian Sea. Classified by the India Meteorological Department as a severe cyclonic storm, it struck with sustained winds of about 70 miles per hour. The cyclone unleashed torrential rains that inundated Oman's valleys, while powerful winds generated storm surges along the coast, damaging infrastructure, destroying homes, and displacing communities. At least 14 people in Oman lost their lives, and the storm's effects extended beyond its landfall, bringing heavy rainfall and strong winds to the United Arab Emirates and causing structural damage, injuries, and fatalities in Iran.

Cyclone Shaheen stands out as an unusual and troubling event in the modern record of tropical cyclones originating in the North Indian Ocean, raising questions about whether climate change is reshaping the behavior and impacts of such storms in the Middle East. Unlike most cyclones that weaken and dissipate before reaching the Arabian Peninsula, Shaheen emerged from the remnants of Cyclone Gulab, which had formed in the Bay of Bengal, crossed India while weakening to a depression, and then unexpectedly reintensified upon reaching the Arabian Sea. This regeneration produced a much stronger system, with Shaheen gaining the classification of a very severe cyclonic storm as it entered the Gulf of Oman, ultimately striking Oman with greater force than the original storm from which it developed. Such persistence over land and reemergence at sea is exceedingly rare, with only two similar historical cases recorded in 1921 and 1959, neither of which culminated in landfall at such intensity. Equally significant is Cyclone Shaheen's trajectory. Very few Arabian Sea cyclones push into the Gulf of Oman, and even fewer make landfall on the Arabian Peninsula as powerful storms. In fact, only two comparable events are on record, the first in 1890 and the second being Shaheen itself more than 130 years later, making it one of the most anomalous cyclones in the region's documented history.

Similarly, Cyclone Gonu offers a striking example of the severe impacts that can occur when a tropical cyclone maintains its full strength at landfall. In June 2007, Gonu formed over the Arabian Sea and reached the highest possible classification of a super cyclonic storm before striking the southeastern tip of Oman as an extremely severe cyclonic storm. With sustained winds exceeding 100 miles per hour, the cyclone brought torrential rainfall, widespread inland flooding, and massive coastal waves, causing extensive damage to buildings and infrastructure, crippling power supplies, and displacing communities. The resulting economic losses in Oman were estimated at $4 billion, and at least 50 lives were lost, making Cyclone Gonu the most destructive natural disaster in Oman's modern history and a sobering illustration of the potential severity of Arabian Sea cyclones fueled by the amplification of modern climate change (Mahmoud 2021a).

6 Rise of the Medicanes: Storm Daniel

Rising global sea surface temperatures are reshaping the geography of tropical storms, extending their reach into regions historically less prone to cyclones. While tropical cyclones have long formed in the warm waters of the Indian Ocean and the Caribbean Sea, warming oceans are increasingly producing extreme weather in areas previously considered atypical for such events. The Pacific Ocean, for example, has experienced rare hurricanes that previously would have been unlikely to develop. Similarly, the Mediterranean Sea has historically seen only occasional and weak cyclones, known as medicanes, but climate change-driven warming is creating conditions conducive to more frequent and intense storms in this region, signaling a potential shift in the patterns and risks of extreme weather.

Medicanes have a documented history of forming over the Mediterranean Sea, though they are relatively rare compared with tropical cyclones in other regions. Studies have documented several notable events, showing that these storms tend to form in the fall months when sea surface temperatures are relatively warm. They typically affect countries along the Mediterranean coastline, such as Turkey and Libya, bringing heavy rainfall, strong winds, and coastal flooding. While generally smaller and less intense than tropical cyclones elsewhere, medicanes have occasionally caused significant local damage.

A stark example of the emerging risk of Medicanes occurred in September 2023 with Storm Daniel, which struck eastern Libya. Originating as a medicane over the Mediterranean and moving southward from Greece, the storm unleashed torrential rainfall on the coastal city of Derna. The extreme precipitation, combined with Libya's aging and poorly maintained infrastructure, triggered the catastrophic collapse of two upstream dams, Derna Dam and Mansour Dam. The sudden release of massive volumes of water overwhelmed the city, washing away entire neighborhoods and compounding the flooding already underway. The human toll has been devastating, with over 10,000 confirmed deaths and thousands more missing, highlighting the deadly consequences of extreme weather events intensified by climate change and the vulnerabilities of urban areas to such unprecedented disasters (Altaeb 2024).

7 Extreme Weather and Conspiracy Theories

In mid-April 2024, an intense storm swept across the Arabian Peninsula, unleashing nearly three years' worth of rain in just a few days and causing widespread devastation. Over 250 mm of rainfall were reported in parts of the UAE and Oman within 24 h, triggering severe flooding across the UAE, northern Oman, Bahrain, and parts of Saudi Arabia. The storm inundated streets, damaged infrastructure, disrupted transportation and utilities, and destroyed homes and businesses. The human toll was significant, with around 200 deaths reported across the UAE, Oman, Iran, Pakistan, and Afghanistan, the majority in the latter two countries. Agricultural lands were flooded, critical services were interrupted, and communities across the region faced extensive displacement and property loss.

The storm revealed the varying vulnerabilities of communities across the Arabian Peninsula. In Dubai, critical infrastructure such as roads, bridges, and drainage systems were overwhelmed, forcing the closure of schools and workplaces and stranding thousands of residents. In Oman, flash floods and landslides isolated entire villages, making rescue operations challenging and highlighting gaps in emergency preparedness. Even in Bahrain, where flooding was less extensive, waterlogged streets and disrupted utilities emphasized how urban areas with dense populations and limited drainage are particularly susceptible. These impacts illustrate that while the storm was a single meteorological event, its consequences were shaped by regional infrastructure, urban planning, and disaster response capacities, highlighting the importance of resilience measures across the region.

Following the storm, conspiracy theories quickly emerged on social media, blaming cloud seeding for the unprecedented rainfall. These claims suggested human intervention had unnaturally amplified the storm, sparking widespread speculation and fear. In reality, the UAE's National Centre of Meteorology had conducted no seeding operations targeting this storm, and cloud seeding itself can only enhance rainfall by ten to thirty percent under specific conditions, far below the volume observed during this event. Cloud seeding cannot increase the total atmospheric moisture, which is the key factor in extreme precipitation, and operations are carefully planned and only carried out when necessary, which was not the case given the storm's scale. Instead, analyses indicate that the intense rainfall was driven by natural factors, including El Niño conditions and climate warming of 1.2 °C, which together increased rainfall intensity by an estimated ten to forty percent. The storm demonstrates how natural atmospheric dynamics, amplified by climate change, can produce extreme weather events without human intervention, despite the spread of disinformation (Morfin 2024).

8 Conclusion

The evidence from recent years makes clear that the Middle East is entering a new era of climate vulnerability, where extreme heat, dust storms, tropical cyclones, and even rare Mediterranean hurricanes converge to create unprecedented risks for societies and ecosystems. Events such as Cyclone Shaheen, Storm Daniel, and record-breaking heatwaves demonstrate how rising temperatures and warming seas are not only intensifying existing hazards but also introducing new and unfamiliar ones to the region. These extremes threaten public health, infrastructure, and economic stability, while the spread of conspiracy theories in their aftermath highlights the additional challenge of misinformation in shaping public perception and policy. Taken together, the growing frequency and severity of these events signal that climate change is no longer a distant concern but a present and escalating crisis in the Middle East.

Coastal areas in the Middle East represent a particularly high-risk frontier in the region's exposure to extreme weather. Low-lying cities and critical infrastructure along the Arabian Gulf, the Red Sea, and the Mediterranean are increasingly threatened by storm surges, flash floods, and coastal erosion, all exacerbated by rising sea levels. The concentration of ports, desalination plants, and densely populated urban centers along these coasts means that even a single severe event can produce disproportionate economic and societal impacts. Recent cyclones and extreme storms have demonstrated how quickly coastal communities can be overwhelmed, revealing gaps in preparedness and natural coastal defenses.

Dealing with the reality of extreme weather in the Middle East requires a coordinated approach that strengthens resilience across multiple fronts. Heat action plans, improved urban design, and expanded cooling access can help reduce the dangers of extreme heat, while land restoration and better water management are key to limiting dust storms. Investments in flood-resilient infrastructure, early warning systems, and stronger emergency response capacities are critical for reducing the impact of cyclones and severe storms. At the same time, transparent communication and public education are needed to counter misinformation and build trust in science-based climate strategies. Ultimately, addressing these challenges will require urgent investment in adaptation, resilience, and public awareness to safeguard communities against the growing threat of extreme weather events.

References

Al Jazeera (2022) Thousands sickened, one dead as latest severe sandstorm hits Iraq. Available at: https://www.aljazeera.com/news/2022/5/5/thousands-sick-as-latest-severe-sandstorm-sweeps-across-iraq

Altaleab M (2024) Post-disaster reconstruction: tackling water security in Derna after storm Daniel. Middle East Institute. Available at: https://www.mei.edu/publications/post-disaster-reconstruction-tackling-water-security-derna-after-storm-daniel

Deshpande M, Singh V, Ganadhi K et al (2012) Changing status of tropical cyclones over the north Indian Ocean. Clim Dyn 57:3545–3567 (2021). https://doi.org/10.1007/s00382-021-05880-z

Guardian (2025) Hellish: heatwave brings hottest nights on record to the Middle East. Available at: https://www.theguardian.com/world/2025/aug/15/hellish-heatwave-brings-hottest-nights-on-record-to-the-middle-east

Gulf Cultural Club (2022) Apocalyptic skies': the dust storms devastating Gulf states and Syria. Available at: https://gcclub.org/2022/06/03/apocalyptic-skies-the-dust-storms-devastating-gulf-states-and-syria/

Keynoush B (2022) Severe sand and dust storms are an underrated risk in the Gulf region, despite mitigation measures. Available at: https://www.mei.edu/publications/severe-sand-and-dust-storms-are-underrated-risk-gulf-region-despite-mitigation

Khaleej Times (2024) UAE witnesses record-breaking rains, highest in 75 years. Available at: https://www.khaleejtimes.com/uae/uae-witnesses-record-breaking-rains-highest-in-75-years

Mahmoud M (2021a) Cyclone Shaheen: a reminder of the Arabian Peninsula's vulnerability to extreme weather events. Middle East Institute. Available at: https://mei.edu/publications/cyclone-shaheen-reminder-arabian-peninsulas-vulnerability-extreme-weather-events

Mahmoud M (2021b) Extreme heat: the urgent climate impact. Middle East Institute. Available at: https://mei.edu/publications/extreme-heat-urgent-climate-impact

Mahmoud M (2024) The looming climate and water crisis in the middle east and North Africa. Carnegie Endowment for International Peace. Available at: https://carnegieendowment.org/research/2024/04/the-looming-climate-and-water-crisis-in-the-middle-east-and-north-africa?lang=en

Morfin O (2024) Debunking the role of cloud seeding in the April Arabian Gulf floods. Middle East Institute. Available at: https://mei.edu/publications/debunking-role-cloud-seeding-april-arabian-gulf-floods

Oman Meteorology (2025) Tropical Cyclone SHAHEEN 1–4 October 2021. Al Arabiya. Available at: https://met.caa.gov.om/en/

Reuters (2025) UAE summer temperatures surge near record highs after hottest ever spring. Available at: https://www.channelnewsasia.com/world/uae-united-arab-emirates-heat-hottest-518-5276246

Time (2023) How extreme heat is proving a major challenge for Hajj. Available at: https://time.com/6291161/extreme-heat-hajj-climate/

WMO (2023) Storm Daniel leads to extreme rain and floods in Mediterranean, heavy loss of life in Libya. Available at: https://wmo.int/media/news/storm-daniel-leads-extreme-rain-and-floods-mediterranean-heavy-loss-of-life-libya

Zeydan Ö, Yıldırım Y (2024) Impact of dust transport events on PM_{10} concentrations on the western coasts of Turkey: a case of April 2022. Air Qual Atmos Health 17:799–811. https://doi.org/10.1007/s11869-023-01481-3

Spatiotemporal Dynamics and Intensification of Drought in the Levant: Vegetation Response Across the Middle East

Shifa Mathbout, George Boustras, and Pierantonios Papazoglou

1 Introduction

The Middle East (ME) is increasingly recognized as one of the world's most climate-vulnerable regions, emerging as a global hotspot for extreme drought events (Elkhalfi et al. 2025; Reis et al. 2025). Despite encompassing a wide range of climatic zones, from humid coastal areas to hyper-arid deserts, the region is undergoing consistent climatic shifts, characterized by rising temperatures and declining precipitation (Kiro et al. 2020; Abou Samra and Ali 2025; Vicente-Serrano et al. 2025). These long-term changes not only disrupt seasonal weather patterns but also exacerbate chronic water scarcity, accelerate soil moisture depletion, and impose unprecedented stress on agricultural productivity and ecological systems (Schillinger et al. 2022; Sun et al. 2024; Roy et al. 2025). Recent scientific studies document a significant increase in the frequency, duration, and severity of droughts across the region (Mathbout et al. 2018; Zeleňáková et al. 2022; Al Zubi et al. 2024). Unlike previous isolated or short-lived dry periods, contemporary droughts tend to be prolonged and extreme, driven by persistent rainfall deficits coupled with rising atmospheric evaporative demand (AED), which intensifies surface drying (Gebrechorkos et al. 2025; Ye et al. 2025). Soil moisture depletion further reduces evaporative cooling, leading to elevated surface temperatures and extended drought durations (Mukherjee et al. 2023; Wang et al. 2025). These synergistic effects produce environmental stress that exceeds the sum of heat and drought impacts occurring independently, threatening food security, ecosystem resilience, and sustainable water resource management throughout the ME. The Eastern Mediterranean-Levant subregion, comprising Israel, Jordan,

S. Mathbout (✉) · G. Boustras · P. Papazoglou
Center of Excellence in Risk and Decision Sciences (CERIDES), European University, Nicosia, Cyprus
e-mail: s.mathbout@research.ac.euc.cy

S. Mathbout
Climate Change and Landscape Ecology Group, University of Barcelona, 08001 Barcelona, Spain

M. M. Al Saud (ed.), *Hydroclimate Changes: Implications of the Middle East Region*,
https://doi.org/10.1007/978-3-032-20053-2_5

Lebanon, Palestine, and Syria, is particularly vulnerable due to its dense population and limited freshwater resources (Kiro et al. 2020). This area ranks among the most sensitive globally to rising temperatures and projected climate impacts (Kelley et al. 2012, 2015; Langgut et al. 2021). Both observational data and climate models reveal a drying trend during the winter wet season, largely driven by increased greenhouse gas concentrations (Mathbout and Martin-Vide Lopez-Bustins 2023), signalling heightened risks for water scarcity and drought-related challenges across the Levant. In arid and semi-arid areas, vegetation exhibits heterogeneous response patterns: although drought events typically suppress photosynthetic activity and net primary productivity, the magnitude and spatial imprint of these effects are modulated by species-specific physiological tolerance, anthropogenic land-use regimes, and hydrological management practices (Zhang et al. 2023; Liu et al. 2025). The ongoing drying trend in the Levant, like other semi-arid regions, is accompanied by more severe flooding events (Armon et al. 2019), often linked to a higher frequency of tropical moisture intrusions (de Vries et al. 2013; Neugebauer et al. 2015). In Syria, over a decade of ongoing conflict has caused extensive damage to water infrastructure, disrupted agricultural production, and weakened institutions tasked with managing environmental risks (Eklund et al. 2022; Mathbout et al. 2025). As a result, large segments of the population remain highly vulnerable even to moderate drought events. Similarly, in the State of Palestine, persistent political tensions and limited access to land and water resources have obstructed effective climate adaptation efforts (Tubi and Feitelson 2016; Otekunrin 2025). Compounding these challenges, outdated infrastructure and unequal resource distribution further amplify the socio-economic impacts of drought (Gleick 2014; Dinc and Eklund 2023; Eboreime et al. 2025). With population growth and urbanization driving increasing demand for water and food, the intersection of climate extremes and political instability creates a multifaceted and deeply interconnected risk landscape (Javan et al. 2024; Bani Mfarrej 2025). In this context, drought acts as more than a natural hazard, it multiplies vulnerabilities, exacerbating humanitarian crises and destabilizing already fragile social and environmental systems (Marzi et al. 2021; Eklund et al. 2024). Tackling these complex challenges necessitates integrated, cross-sectoral approaches that address both environmental stressors and socio-political realities (Ward et al. 2020; Kabir et al. 2024).

2 Study Area and Materials

2.1 Regional Context of the Levant: Climate Patterns and Implications for Vegetation Dynamics

The Levant, located along the eastern Mediterranean coast (Por 1975), extends roughly 800 km north to south and 150 km east to west. Bounded by the Mediterranean Sea to the west and the Arabo-Syrian Desert to the east, the region stretches from the Orontes River in the north to the Isthmus of Suez in the south. It is characterised by four primary north–south geomorphological zones: the coastal plain, the western (cis-rift) mountain ranges, the rift valley, and the eastern (trans-rift) mountains, gradually transitioning into desert landscapes eastward. The region's climate is strongly influenced by its varied topography, including coastal plains, mountain ranges such as the Lebanon and Anti-Lebanon, and interior plateaus. Orographic uplift causes moist Mediterranean air to ascend, producing higher precipitation on windward slopes and arid conditions on leeward sides due to rain shadow effects. Annual precipitation across the Levant exhibits high spatial variability, ranging from only a few millimetres in arid desert regions to over 1200 mm in the western coastal zones and southeastern mountain ranges of Syria. This pronounced variability in rainfall is coupled with a wide temperature range, from approximately 5 °C in highland areas to 25 °C in warmer lowland regions, which drives elevated levels of atmospheric demand. As a result, ET_0 values can exceed 2500 mm annually, particularly in eastern Syria and the southern and eastern parts of Jordan (Fig. 1). The proximity to the Mediterranean Sea exerts a maritime moderating effect, resulting in a typical Mediterranean climate marked by wet, mild winters and hot, dry summers. Situated between approximately 29° and 35°N latitude, the Levant undergoes pronounced seasonal changes in solar radiation, driving temperature variability.

The Levant exhibits a distinct spatial correlation between elevation and vegetation cover, governed by topographic variability and climatic gradients. Lowland zones, including coastal and riverine areas, are dominated by artificial surfaces and dense vegetation due to favourable hydrological conditions and intensive land use. In contrast, mid- to high-elevation zones, particularly along the orographic belts of Lebanon and western Syria, are characterised by tree cover and shrublands supported by increased precipitation and moderated temperatures.

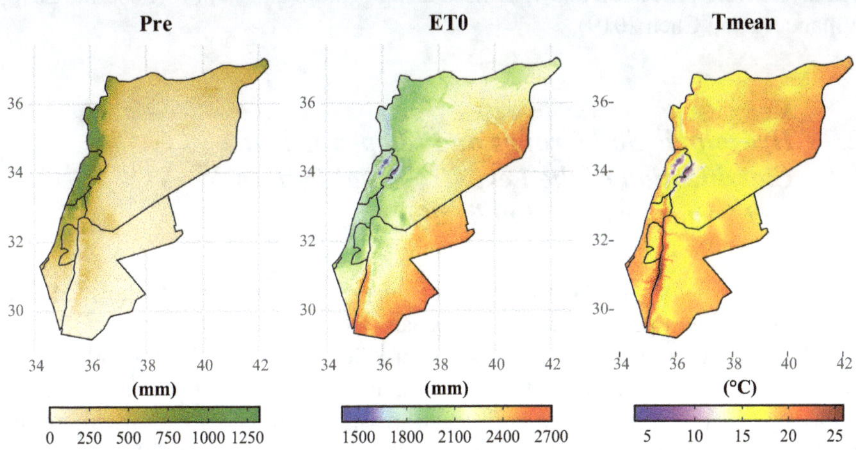

Fig. 1 Historical patterns of total annual precipitation (Pre), reference evapotranspiration (ET_0), and mean annual temperature (Tmean) across the Levant, 1958–2024

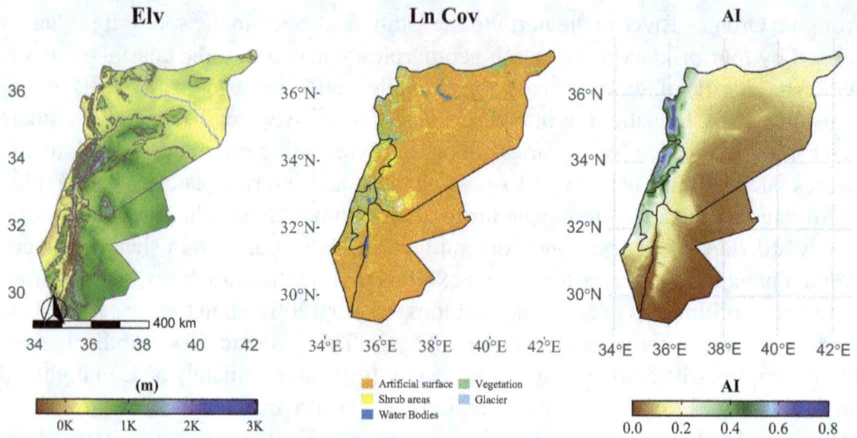

Fig. 2 Hypsometric Map (Elv), land cover classification (Ln Cov), and the The UNESCO aridity index (AI) Across the Levant (1958–2024)

At higher altitudes, vegetation becomes sparse, transitioning to shrub cover and barren land under harsher environmental constraints (Fig. 2). In this study, we mapped the spatial distribution of vegetation cover and topographic elevation across the Levant, focusing on Israel, Jordan, Lebanon, Syria, and Palestine. Elevation data were obtained from the USGS Elevation Point Query Service via the *elevatr* package in R (Hollister et al. 2023), providing a high-resolution digital elevation model. Vegetation cover was derived from classified land cover rasters into key ecological categories. Administrative boundaries and river networks were sourced from Natural Earth datasets (Natural Earth, 2023; https://www.naturalearthdata.com/). All geospatial layers were processed and visualised using *ggplot2* and *sf* in R (Pebesma 2018; Villanueva and Chen 2019).

2.2 *Hydrologic Soil Groups and Köppen Climate Classification of the Levant: Spatial Patterns and Environmental Implications*

The Levant's soils are classified into distinct Hydrologic Soil Groups (HSGs) according to runoff potential and texture characteristics (Table 1).

A large portion of the region falls within HSG-B and HSG-C, both associated with moderate runoff potential. HSG-B soils, moderately sandy (50–90% sand, 10–20% clay), are prevalent in semi-arid steppe areas, supporting rainfed agriculture and moderate vegetation cover. HSG-C soils, with higher clay content (20–40% clay, < 50% sand), are widespread in inland zones, retaining more water but exhibiting slower drainage and greater vulnerability to drought stress. HSG-D soils, characterised by high runoff potential (>40% clay, < 50% sand), are found in parts of the northeastern

Table 1 Classification of Hydrologic Soil Groups (HSGs) and their associated runoff potential, determined according to soil composition

HSG	Description
A	Low runoff potential (>90% sand and < 10% clay)
B	Moderately low runoff potential (50–90% sand and 10–20% clay)
C	Moderately high runoff potential (<50% sand and 20–40% clay)
D	High runoff potential (<50% sand and > 40% clay)
A/D	High runoff potential unless it drained (>90% sand and < 10% clay)
B/D	High runoff potential unless drained (50–90% sand and 10–20% clay)
C/D	High runoff potential unless drained (<50% sand and 20–40% clay)
D/D	High runoff potential unless it drained (<50% sand and > 40% clay)

interior, the west and northwest of Syria, the southwest and parts of northwest Jordan, and along certain floodplains. These soils often require artificial drainage for effective agricultural use (Fig. 3). Coastal plains, dominated by fertile alluvial soils, exhibit low to moderate runoff potential and support diverse cropping systems due to their favourable sand–clay balance.

The HSG classification follows USDA texture-based guidelines, and the HYSOGs250m dataset provides a gridded 250 m resolution characterisation of runoff

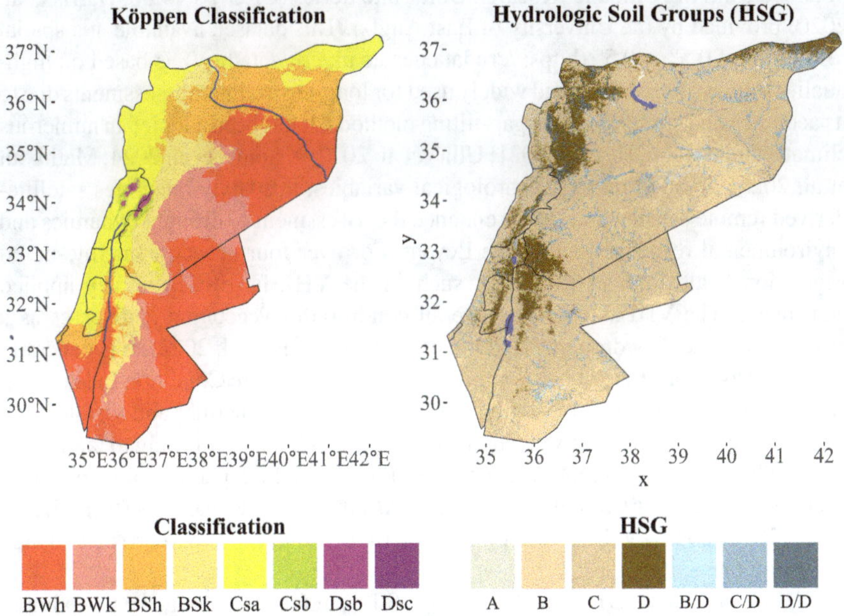

Fig. 3 Spatial distribution of Köppen climate classification and Hydrologic Soil Groups (HSG) at 250 m resolution in the Levant

potential, making it suitable for applications from regional to global scales. Complementing this, the Köppen climate classification in the Levant (Fig. 3) reveals a striking climatic diversity, rom Mediterranean conditions with hot, dry summers and mild, wet winters along the coasts, to arid and semi-arid climates dominating the inland plateaus and desert fringes. Mountain ranges exhibit temperate climates with cooler temperatures and higher precipitation, while transitional zones show mixed climatic influences. Widely applied in ecological and climatological research, the Köppen system provides a valuable framework for interpreting regional climate variability and its interaction with soil characteristics (Peel et al. 2007; Malamos et al. 2025).

3 Methods

3.1 Comprehensive Process for Gathering and Preparing Data

This study combines ground-based meteorological data with satellite-derived remote sensing products to investigate drought dynamics and land surface responses across the Levant. Monthly gridded data on precipitation (Pre), average temperature (Tmean), and potential evapotranspiration (PET) for the period 1958–2024 were obtained from the Climatic Research Unit Time Series (CRU TS v4.06; Harris et al. 2020) provided by the University of East Anglia. This dataset, available at a spatial resolution of 0.5° × 0.5° (https://crudata.uea.ac.uk/cru/data/hrg/), is based on high-quality station observations and widely used for long-term climate assessments due to its accuracy and comprehensive gap-filling methods. It has been adopted in numerous climate studies (e.g., He et al. 2021; Ullah et al. 2022; Xiaomei et al. 2024; Mathbout et al. 2025). In addition to meteorological variables, this study integrates satellite-derived remote sensing datasets to enhance the assessment of drought dynamics and environmental responses across the Levant. For over four decades, satellite-based vegetation health indices (Table 2), such as the VHI, have been widely applied worldwide. The VHI assesses the overall condition of vegetation and serves as a key tool for detecting drought (Karnieli et al. 2010; Zeng et al. 2022). It is generally computed as the average of two components: the Temperature Condition Index (TCI) and the Vegetation Condition Index (VCI), derived from thermal and visible band data, respectively. The relative weights of VCI and TCI vary according to land cover type, reflecting the different influences of moisture and temperature on vegetation cycles (Bento et al. 2020; Zeng et al. 2023). As outlined by Kogan (2000) and Bento et al. (2018, 2020), VCI for each pixel and time within a given year is calculated following these principles:

Where, $NDVI_{Max}$, BT_{Max} and $NDVI_{Min}$, BT_{Min} refer to the long-term maximum and minimum averages of NDVI and BT, respectively and all categories of Vegetation Health Indices are presented in Table 3.

Table 2 List of vegetation and agricultural drought indices used in this study

Index	Equation		References
NDVI	$NDVI = \frac{NIR-Red}{NIR+Red}$	(1)	Gao 1996
VCI	$VCI = \frac{NDVI-NDVI_{Min}}{NDVI_{Max}-NDVI_{Min}}$	(2)	Kogan1995,1997; Bento et al. 2018, 2020
TCI	$TCI = \frac{BT_{Max}-BT}{BT_{Max}-BT_{Min}}$	(3)	Kogan1995,1997,2000
VHI	$VHI = \alpha XVCI + +(1-\alpha)XTCI$	(4)	Kogan 1997; Kogan et al. 2004

Table 3 Classification of drought and vegetation health indices (VHI, VCI, TCI, BT, and NDVI)

VHI, VCI, TCI, BT	Class	NDVI	Class
< 10	Extremely drought	< 0.0	Non-vegetative surfaces (water, snow)
10–20	Severely drought	0.0–0.2	Barren or very sparse vegetation
20–30	Moderate drought	0.2–0.6	Sparse or moderate health
30–40	Mild drought	0.6–0.9	Dense, healthy vegetation
> 40	No drought		

The left two columns present drought severity categories derived from satellite-based vegetation indices (VHI, VCI, TCI, and BT), whereas the right two columns display NDVI intervals used to evaluate vegetation cover.

NDVI data from NOAA's Center for Satellite Applications and Research (STAR) (https://www.star.nesdis.noaa.gov/smcd/emb/vci/VH/vh_ftp.php) provide 4 km spatial resolution and weekly coverage from 1981–2024. Commonly used in drought monitoring, these data support analysis of vegetation condition and agricultural water stress. VHI data from the same source combine vegetation condition with temperature-based stress, enabling earlier drought detection and distinguishing heat from moisture stress across the Levant's diverse climates (Ghobadi and Badehian 2025). NDVI and VHI were aggregated into annual averages using the arithmetic mean, with background values (−9999) removed via spatial masking to ensure clean datasets for further analysis. To further capture land surface characteristics, the study also employs land cover classifications from the Global Land Cover 2000 (GLC 2000) dataset, developed using the FAO's Land Cover Classification System (LCCS) (Bartholomé and Belward 2004). The LCCS framework combines visual interpretation with expert knowledge to classify land surfaces based on observable biophysical features, allowing for consistency across diverse ecological and socioeconomic contexts (Arafat et al. 2014). Together, these remote sensing datasets support a spatially explicit and ecologically informed framework for analysing the spatiotemporal progression of drought events and their impacts across the Levant's heterogeneous agroecological zones (Arafat et al. 2014). The Köppen-Geiger climate classification data was sourced from the openly available global climate dataset provided by

Beck et al. (2018), accessible at (https://koeppen-geiger.vu-wien.ac.at/) and widely used for climate zone mapping and analysis.

3.2 Selection of Meteorological Drought Indices (MetDIs) Used in the Levant

In arid and semi-arid regions such as the Levant, accurate drought assessment requires integrating diverse and complementary indices. This study employs both meteorological and agricultural drought metrics to capture the full spectrum of drought impacts. For meteorological drought, we applied the SPEI (Vicente-Serrano et al. 2010) at 3-, 6-, and 12-month timescales. These durations reflect short-term anomalies (SPEI-3), seasonal drought conditions (SPEI-6), and long-term trends (SPEI-12), all of which are essential for water resource management and agricultural planning in the Levant. To complement the SPEI and incorporate soil moisture dynamics, we used the sc-PDSI (Wells et al. 2004), which adjusts the original PDSI formulation for regional climatic characteristics, thereby enhancing its reliability across diverse environments. The sc-PDSI was selected because it overcomes key limitations of the original Palmer Drought Severity Index (PDSI; Palmer 1965) by introducing an automated calibration process that adapts the index to local climate conditions (Wells et al. 2004). While the traditional PDSI uses fixed empirical constants derived from U.S. climate data, the sc-PDSI adjusts these parameters based on the statistical characteristics of the study region, improving its comparability and accuracy across diverse climatic zones such as the Levant. In this study, drought indices were first calculated separately for each country in the Levant reflecting the fact that each nation operates under distinct water management strategies, adaptation policies, and socio-environmental contexts. This country-by-country approach preserves important spatial variability arising from the region's diverse Köppen climate zones, which range from arid desert to Mediterranean climates with markedly different precipitation regimes and seasonal patterns. After individual national assessments were completed, the resulting datasets were assembled into a combined raster covering the whole Levant, enabling a unified visual representation while retaining the underlying country-specific drought characteristics. The classifications for the MetD categories are provided in Table 4.

The Modified Mann–Kendall test (Hamed and Rao 1998) was employed to detect trends in the MetDIs while accounting for potential autocorrelation in the time series

Table 4 Categorization of drought conditions based on SPI, SPEI, PDSI, and sc-PDSI

sc-PDSI	SPEI	Class
− 2.0 to − 3.0	− 1.0 to − 1.5	Moderately dry
− 3.0 to − 4.0	− 1.5 to − 2.0	Severely dry
≤ − 4.0	≤ − 2.0	Extremely dry

Table 5 The Kappa statistics (K) values are presented along with their corresponding agreement strength levels

Kappa statistic	Strength of agreement
< 0	Poor
0.00–0.20	Slight
0.21–0.40	Fair
0.41–0.60	Moderate
0.61–0.80	Substantial
0.81–1.00	Almost perfect

data. This approach provides a more reliable assessment of significant temporal changes by reducing the influence of serial dependence. Pearson correlation and Cohen's Kappa (Cohen 1960, 1968) were used to assess relationships and agreement between vegetation health and drought indices. Pearson correlation quantified the strength of linear relationships, while Cohen's Kappa evaluated categorical agreement between indices based on monthly severity classes. Continuous drought values were converted into ordinal categories (e.g., 0 = no drought, 1 = moderate, 2 = severe, 3 = extreme) to calculate weighted Kappa, which ranges from − 1 (less than chance agreement) to 1 (perfect agreement), with 0 indicating chance-level agreement (Table 5) Calculations were performed using the **irr** package in R (Gamer et al. 2022).

3.3 Assessment of CRU TS Gridded Dataset: Statistical Validation and Accuracy Evaluation

The accuracy of MetDIs in the Levant depends heavily on the quality of the meteorological parameters used in their calculation. The CRU TS dataset, widely applied and validated in previous research (Felegari et al. 2022; Liu et al. 2023), was evaluated against in situ measurements from 2637 (Fig. 4) meteorological stations across the region. The analysis covered 54 years (1970–2024) and assessed CRU TS performance over two 20-year intervals, 1970–1996 and 1997–2024, to examine potential variations in data reliability across time. These intervals were selected for their alignment with climatological norms, enabling a structured division of the long-term series into comparable segments.

Observed station data underwent rigorous quality control, including the removal of outliers beyond ± 3.5 standard deviations, elimination of duplicates, and homogeneity testing via the Standard Normal Homogeneity Test (SNHT). CRU data were interpolated to exact station coordinates using nearest-neighbour remapping. Performance evaluation employed multiple statistical measures, (RMSE, Coefficient of Correlation (CC), and Nash–Sutcliffe Efficiency (NSE), where lower RMSE and

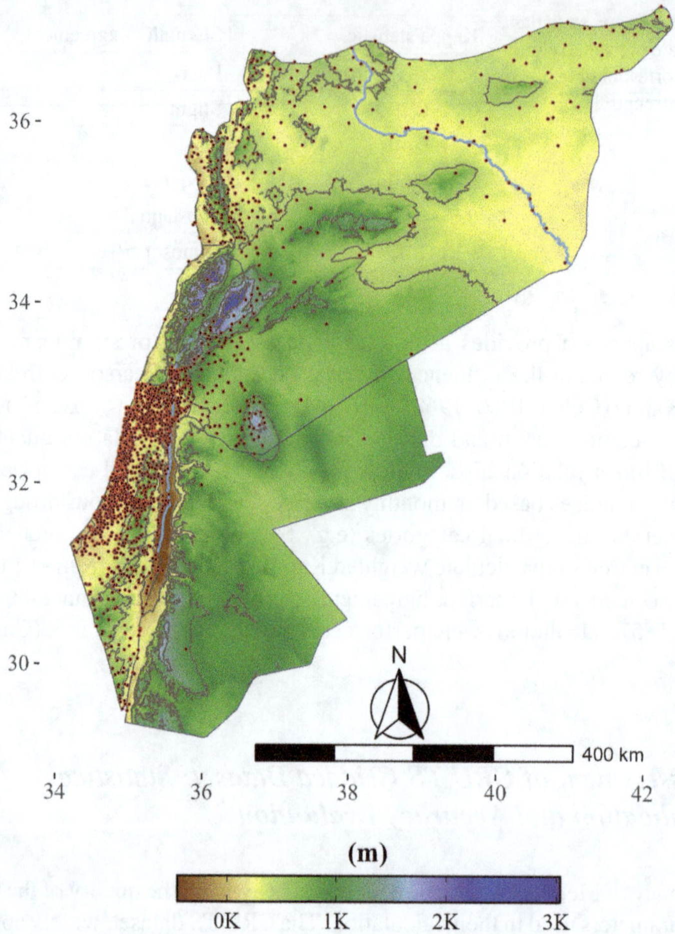

Fig. 4 Hypsometric map of the levant and spatial distribution of meteorological stations used to validate CRU data

higher CC and NSE values indicated stronger agreement. Pearson's correlation coefficient was also applied to assess the linear relationship between CRU and observed records.

4 Results

4.1 Evaluation of CRU TS Dataset Performance Against Observed Data in the Levant

The data extracted from the CRU database demonstrate strong correlation coefficients (CCs) when compared to observed temperature and precipitation data in the Levant (Fig. 5). Notably, CCs consistently exceed 0.80 for precipitation during the rainy season and surpass 0.90 for temperature across all seasons, indicating robust agreement. Furthermore, RMSE values remain low, below 1.5 for temperature and under 3.0 for precipitation, underscoring the accuracy of the CRU data in representing observed climatic conditions, particularly at multi-year averaged monthly time scales. The CRU dataset exhibits a slight overestimation of historical precipitation averages in January and December over the Levant; however, monthly climatology plots confirm that most of the precipitation occurs between October and May, with peaks in these months.

Despite this minor bias, CRU precipitation closely aligns with observed climatological averages. Additionally, monthly mean air temperature analyses reveal that the CRU data accurately capture the average annual temperature variability, mirroring observed monthly fluctuations and key warming and cooling trends across all seasons. This consistency reflects the dataset's ability to reliably reproduce significant climatic events in the region. The Taylor diagram synthesizes these performance metrics by simultaneously depicting the correlation, RMSE, and standard deviation between observed and modelled datasets in a single visualization. Correlation values ranging from 0.85 to 0.92 and RMSE values between 0.9 and 2.4 for precipitation and between

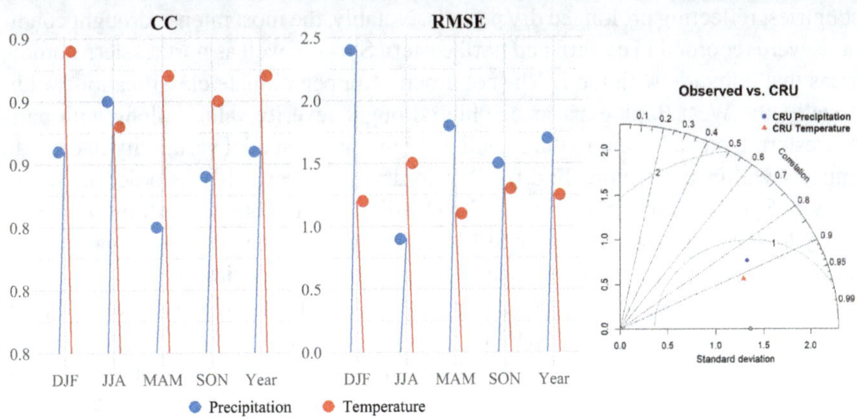

Fig. 5 Descriptive statistics comparing monthly temperature and precipitation from the CRU TS dataset with observed data averaged over 1970–1996 and 1997–2024

1.1 and 1.5 for temperature demonstrate strong model fidelity. Points nearer the reference observation on the diagram indicate better model performance with high correlation and low error, confirming that the CRU dataset effectively captures the seasonal and annual climatic variability in the Levant region. This integrated approach facilitates a comprehensive assessment of data quality and agreement, making the Taylor diagram a valuable tool for evaluating the reliability of climate datasets.

4.2 Comprehensive Assessment of the Levant's Historical Meteorological Droughts (1958–2024) Using Multiple MetDIs at the Regional Scale

Drought conditions were assessed separately for Syria, Jordan, Lebanon, Israel, and Palestine to capture country-specific climatic variability, then compiled into a combined raster for a unified regional view. Combining drought indices across Syria, Jordan, Lebanon, Israel, and Palestine into a single regional assessment would smooth important spatial variability due to the presence of multiple Köppen climate classifications within the area. These range from arid desert to Mediterranean climates, each with distinct precipitation regimes and seasonal variability, resulting in varied drought patterns and impacts. Therefore, analysing drought at the country level allows for more precise characterisation of drought dynamics and provides actionable information tailored to national water resource management and policy needs, avoiding the loss of critical details that arise from averaging across climatically diverse regions. In this study, drought conditions are considered based on the criteria outlined in Table 4.

DEs are defined as occurring when these conditions persist for at least two consecutive months. The highest drought severities were observed for SPEI-12 across all countries, reflecting prolonged dry periods. Notably, the most intense drought conditions were recorded in eastern and northeastern Syria as well as northeastern Jordan, areas that coincide with the BWh (hot desert) Köppen climate classification. Additionally, the West Bank experienced high drought severity values, along with parts of western Israel adjacent to the coastline, highlighting spatial variability in drought impacts within the region (Fig. 6). Severe droughts were also recorded in southwestern Syria, a region classified as Csa (Mediterranean climate), exhibiting particularly high drought severities for SPEI-12. These prolonged drought conditions in the Mediterranean climate zone can lead to significant environmental and socioeconomic impacts, including reduced agricultural productivity, water shortages, and increased vulnerability of ecosystems.

By aggregating the years 2020 to 2024 into the 2010s decade, the analysis reveals a pronounced increase in the frequency of dry events (DEs) during the 2010s compared to the 1990s and 2000s (Fig. 7). According to SPEI-3, the highest frequency, reaching up to 1.5 events per decade, was observed in western Lebanon, southwestern Syria, Jordan, and Israel. Similarly, DEs derived from SPEI-6 and SPEI-12 indices also exhibited elevated frequencies in the 2010s relative to the preceding two decades,

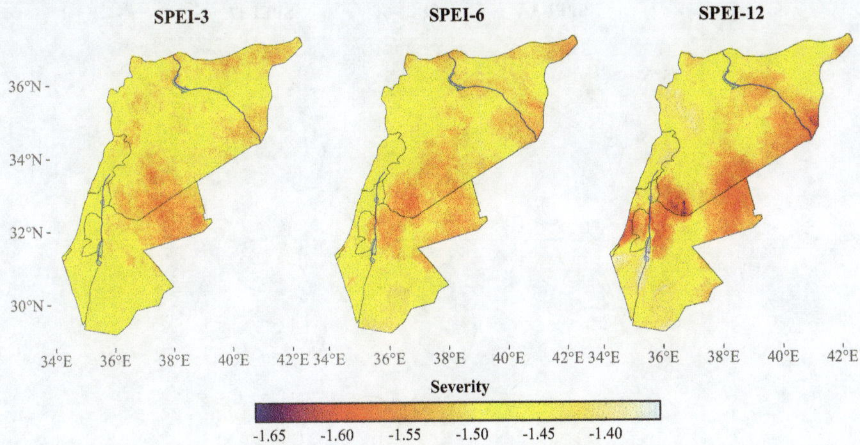

Fig. 6 Spatial patterns of short-, medium-, and long-term drought severities across Syria, Jordan, Lebanon, Israel, and Palestine using SPEI index

particularly across the Levant. Notably, southern Jordan and eastern Israel, especially near the Jordanian border, recorded frequencies exceeding 1.25 events per decade. Most of the Syrian territory experienced DE frequencies ranging from 0.75 to 1.2 events in the 2010s, compared to 0.70 to 1 event per decade during the 1990s and 2000s. These findings underscore a significant escalation in the occurrence of dry events over recent decades throughout the Levant, signalling increasing drought stress in this climatically sensitive region.

The drought severity patterns derived from the sc-PDSI exhibit a strong resemblance to those obtained from the SPEI-12, indicating that both indices effectively capture long-term drought variability (Fig. 8). However, the sc-PDSI consistently indicates greater drought severity than SPEI-12, particularly in the northern and northeastern regions of Syria. This area has experienced successive severe DEs over the past decade, as reported by Mathbout et al. (2018), and Mathbout et al. (2025). The persistence of drought in this region is primarily driven by prolonged rainfall deficits coupled with rising temperatures, which together enhance evapotranspiration rates and consequently diminish soil moisture availability.

The heightened sensitivity of sc-PDSI to drought severity arises from its mechanistic foundation as a soil moisture balance model. By integrating precipitation, temperature, and local soil water-holding capacity, the sc-PDSI simulates the cumulative impact of sustained moisture deficits on soil and surface water resources. This approach enables it to capture more intense and prolonged drought conditions compared to SPEI-12, which is based on the statistical standardization of climatic water balance over a 12-month timescale. Moreover, the Levant region has experienced an average of more than 0.5 drought events per year, with the highest frequency, up to 0.9 drought events annually, recorded in eastern and northeastern Syria, western Israel, western Jordan, and southwestern Jordan. These results highlight the heightened vulnerability of these areas to recurrent drought stress, a situation

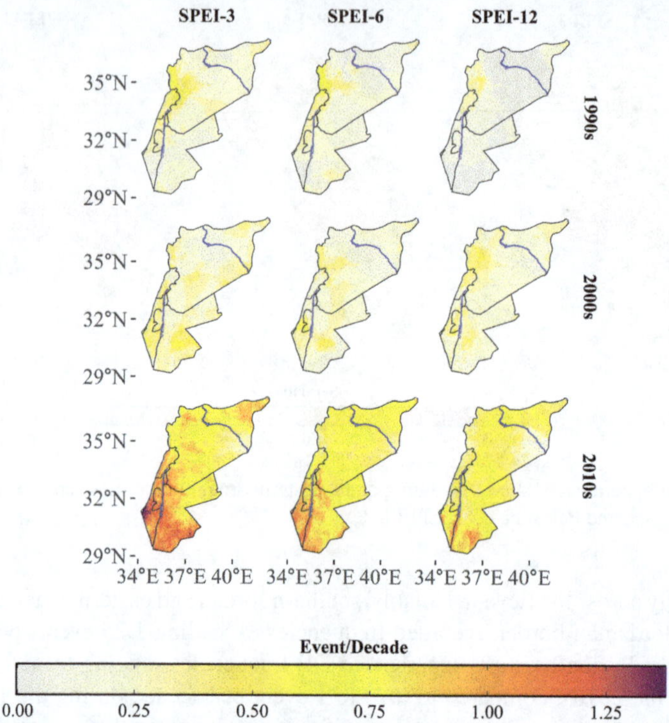

Fig. 7 Spatial distribution and decadal variability of DE frequencies according to (SPEI-3, SPEI-6, SPEI-12) in the Levant from the 1990s to the 2010s

likely aggravated by climate change and ongoing regional water resource challenges. The temporal trends of drought indices SPEI-3, SPEI-6, SPEI-12, and sc-PDSI across the Levant region reveal significant negative trends in all indices for all countries, indicating worsening drought conditions over time. Statistically significant change points for the SPEI indices occur consistently after the year 2000, while the sc-PDSI shows a notable shift after 2004 (Fig. 9).

Historically, the Levant experienced long and severe droughts during the late 1950s and 1960s, as evidenced by all MetDIs analysed. More recently, the year 2021 recorded the lowest drought index values during spring, highlighting an extreme drought event. The ISPA (Israel and Palestinian Authorities) endured prolonged drought conditions from 2016 through 2018. Severe to extreme drought episodes were also documented from November 1998 to November 2001, between 2005 and 2012, and from 2021 continuing through 2024. These findings emphasize the increasing frequency and severity of prolonged droughts impacting the Levant, with important implications for water resource management and regional climate resilience. These results corroborate the spatial patterns observed in the 2010s, which revealed high severity and increased frequency DEs across the Levant region. The identified

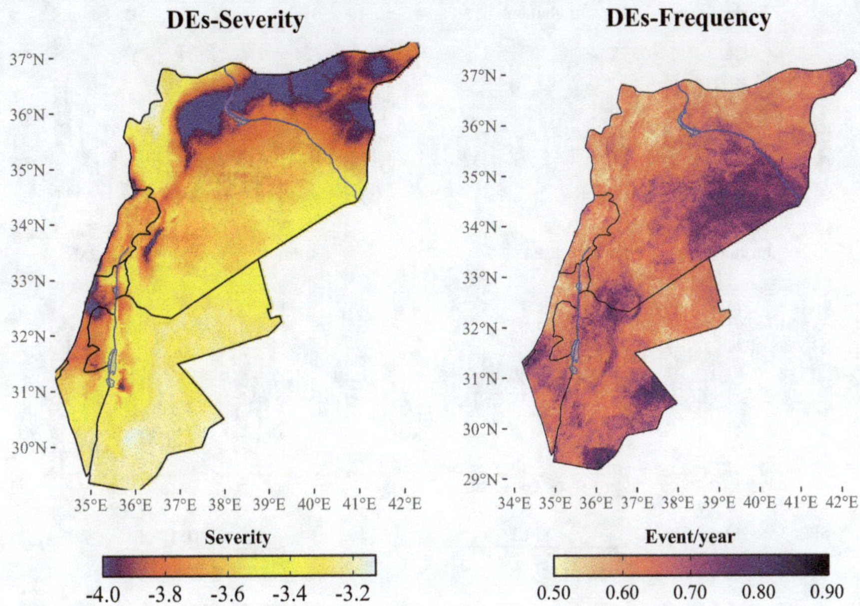

Fig. 8 Spatial distribution of drought severity patterns (left panel) and average annual number of drought events per grid cell from 1958 to 2023 (right panel) based on the sc-PDSI

change points and significant negative trends align with documented intensifications in drought conditions during this period, reinforcing the evidence of escalating drought stress across all the Levant countries. This further supports the ME as a hotspot area for drought vulnerability and climatic extremes.

4.3 Evaluating Drought Dynamics and Vegetation Stress in the Levant Using Integrated NDVI and VHI Analysis

To assess drought impacts on vegetation health and ecosystem dynamics across the Levant between 1980 and 2024, we used both the NDVI and the VHI as complementary remote sensing indicators. Weekly NDVI composites from weeks 10 to 40 (Fig. 10) reveal clear seasonal patterns of vegetation growth and decline, strongly modulated by precipitation, temperature, soil properties, and Köppen–Geiger climate zones. Maximum greenness occurred between weeks 11 and 17, with NDVI exceeding 0.48 in week 14 along western and southwestern Syria, the coastal mountains of Lebanon, and northwestern Israel. These Mediterranean and warm-temperate zones receive abundant winter–spring rainfall, and their fine-textured soils (<50% sand, > 40% clay) provide high root-zone water retention despite limited infiltration, supporting vigorous early-season vegetation. In contrast, southern Syria,

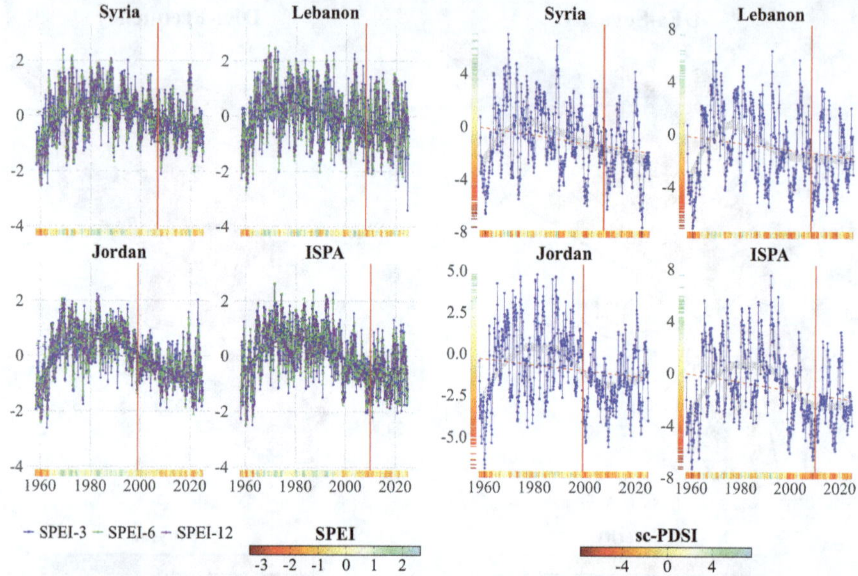

Fig. 9 Temporal trends and change points of drought indices: SPEI-3, SPEI-6, and SPEI-12 are shown in the left panel, and sc-PDSI in the right panel, across the Levant countries. The grey shaded areas represent the loess-smoothed confidence intervals around the trends, indicating variability in the temporal drought patterns. Change points are marked by vertical red lines, highlighting significant shifts in drought behaviour across Syria, Lebanon, Jordan, and ISPA countries

Jordan, and the interior plateau, dominated by desert (BWh/BWk) and steppe (BSh/BSk) climates with coarse or shallow soils of low moisture-holding capacity, rarely show NDVI values above 0.10, even during peak precipitation periods. Northern and northeastern Syria displays intermediate behaviour, maintaining NDVI of 0.30–0.40 through week 17 before declining below 0.20 by week 20 as rising temperatures and reduced rainfall induce moisture stress. This spatiotemporal variability aligns with the mean NDVI distribution across 1981–2021, segmented by winter (weeks 9–10, late February to early March), spring (weeks 12–22, mid-March to late May), summer (week 25, mid-June), and autumn (week 41, early October), highlighting how climate regime and soil type jointly control vegetation productivity and drought sensitivity in the region.

To complement the NDVI-based assessment of vegetation dynamics, we analysed the VHI during weeks 10 to 14, focusing on the most humid and driest years identified through MetDIs. This period corresponds to late winter to early spring, when vegetation in the Levant initiates green-up following winter precipitation and exhibits its peak response to rainfall, particularly in areas with the highest NDVI values, exceeding 0.43. Assessing VHI during this early-season growth phase enables evaluation of the combined effects of moisture and thermal stress on vegetation health, providing a comprehensive understanding of drought impacts and differentiating

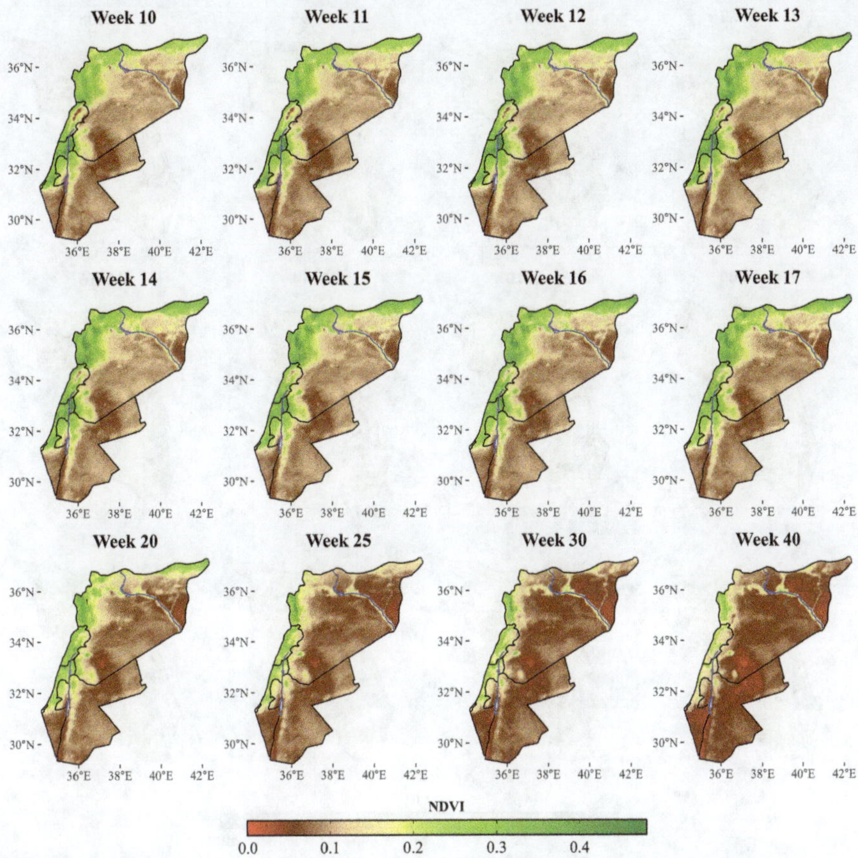

Fig. 10 Spatiotemporal distribution of mean NDVI across the Levant from 1981 to 2024, shown for weeks 10 to 17 (late winter to mid-spring), week 20 (early summer onset), week 25 (mid-summer), week 30 (late summer), and week 40 (autumn), illustrating seasonal vegetation dynamics across different climatic zones

regions where vegetation remained resilient from areas where low rainfall or elevated temperatures induced early stress (Fig. 11).

During the driest years identified in this study across the Levant, 2008, 2010, 2014, 2016, 2018, and 2021, the most severe vegetation impacts were observed in eastern, central, and northeastern Syria, where VHI values dropped below 20, indicating extreme vegetation stress. In 2014, 2018, and 2021, extensive areas of Jordan were also affected, with similarly low VHI values, reflecting strong drought-induced stress. These declines in VHI were closely associated with low TCI values (Fig. 12), often falling below 35, showing that elevated temperatures exacerbated moisture deficits and intensified vegetation stress. Particularly in 2008, very low TCI values were recorded across the Levant, with northeastern Syria falling below 20, and large portions of western Syria, Lebanon, and Jordan registering TCI values

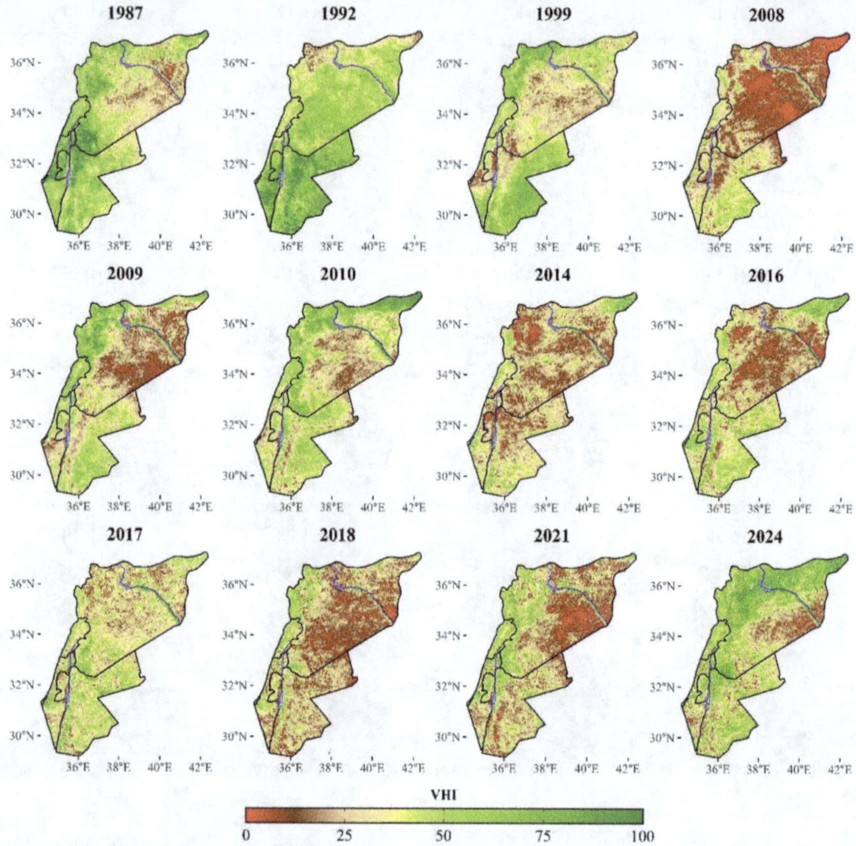

Fig. 11 Spatiotemporal distribution of the VHI across the Levant during the most humid and driest years identified by the MetDIs, highlighting vegetation responses to drought extremes

below 25. By 2021, the majority of Syria experienced similarly low TCI values, further emphasising the widespread vegetation vulnerability during these extreme drought events. The spatial concurrence of low VHI and TCI highlights the combined influence of thermal and hydrological stress on vegetation during the driest years across the Levant. These results reveal that vegetation in regions with high water-holding soils and Mediterranean climate influence, such as western Syria and coastal Lebanon, can partially buffer drought impacts, maintaining higher NDVI and VHI values, whereas arid and semi-arid areas exhibit rapid declines in vegetation health during drought events. The integration of VHI with NDVI and MetDIs thus highlights the vulnerability and resilience of vegetation across the Levant, illustrating how extreme and severe droughts translate into observable ecological stress.

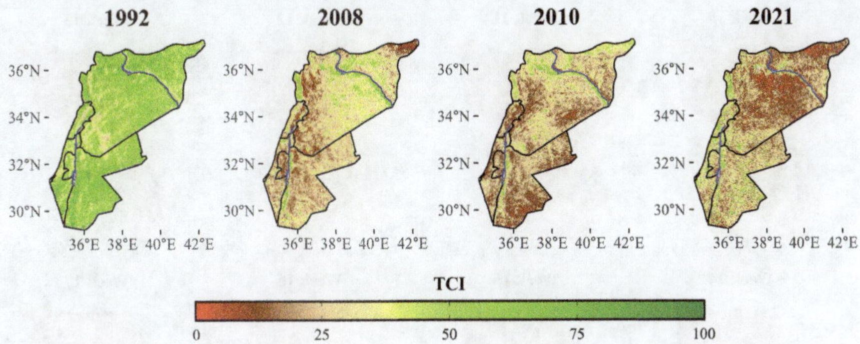

Fig. 12 Spatial distribution of TCI across the Levant during selected dry and humid years identified in this study

4.4 Multi-year Hotspots of Extreme Vegetation Stress Across the Levant (1981–2024)

Persistent hotspots of severe vegetation stress across multiple years reveal areas most vulnerable to drought. Using VHI raster data, we focused on extreme conditions (VHI < 20) that indicate critical stress. To capture the most relevant period for vegetation growth, we mapped weekly VHI during the growing season (weeks 10 to 40) and aggregated these layers across years, counting the frequency of extreme events at each pixel (Fig. 13). This approach generates maps that depict how often severe drought occurs, revealing spatial patterns of persistent dryness and pinpointing regions with chronic vegetation vulnerability. The persistence mapping of extreme VHI values (<20) in this study aligns with widely recognised ecological drought thresholds, where VHI below 20 is indicative of severe to extreme vegetation stress and potential long-term ecosystem degradation (e.g., Kogan 1995; Bento et al. 2018; Zeng et al. 2022). Referencing these thresholds underscores the ecological significance of the identified hotspots and supports their interpretation as priority areas for targeted drought adaptation measures. Analysis of VHI extreme events across multiple weeks shows clear spatiotemporal patterns of severe vegetation stress across the Levant. Between weeks 10 and 17 (approximately early March to late April), all Levant countries exhibited high frequencies of extreme stress, with particularly pronounced hotspots in northern and western Jordan, Israel/Palestine, and southern Lebanon.

These weeks coincide with the early to mid-growing season, a critical period for vegetation development, and align with the highest NDVI values recorded in the region, indicating that vegetation is actively growing and therefore most sensitive to water stress during these periods. During week 30 (~late July), additional hotspots emerged, particularly in Syria's central-eastern regions, northeastern areas, and even in the humid western and coastal zones, where some pixels experienced extreme stress in more than 22 years over the 1981–2024 period. This indicates that these areas had already been exposed to prolonged periods of extreme drought prior to this

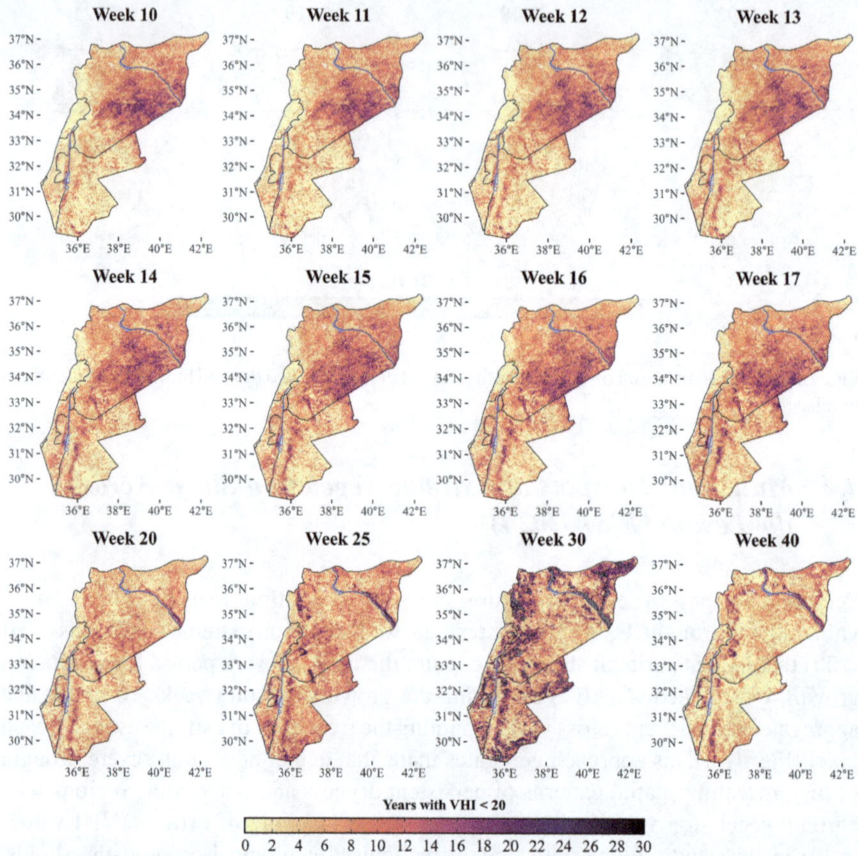

Fig. 13 Frequency of severe vegetation stress hotspots (VHI < 20) across the Levant from 1980 to 2024, aggregated over multiple weeks during the growing season. Areas with higher frequencies indicate persistent ecological drought conditions and chronic vegetation vulnerability

week. Other countries in the Levant also experienced high frequencies of extreme events during this period, highlighting a region-wide vulnerability in mid-summer. Conversely, week 40 (early October) showed the lowest frequency of extreme stress, reflecting the dry summer season and reduced vegetation water demand in most areas. Over the 1981–2024 period, the central-eastern parts of Syria experienced the highest persistence of severe stress, with some pixels showing extreme events in over 25 years, representing approximately 62.5% of the 40-year period, emphasizing the chronic nature of vegetation vulnerability. The northeastern parts experienced between 8 and 15 years of severe stress, indicating recurrent but less persistent drought conditions compared to the central-eastern regions. These results show that persistent VHI hotspots coincide with periods of maximum vegetation activity, making them particularly informative for understanding drought impacts. Overall, these findings demonstrate that certain regions are repeatedly exposed to critical water stress, highlighting the importance of identifying persistent hotspots for targeted adaptation and mitigation strategies in areas most vulnerable to drought.

4.5 Assessing Concordance of Drought and Vegetation Health Indicators: A Statistical Approach

The relationship between the VHI and MetDIs was examined using Pearson correlation against sc-PDSI and SPEI at 12-, 6-, and 3-month accumulation periods. Correlations were positive and statistically significant for all combinations except between VHI and sc-PDSI in Jordan, indicating that vegetation generally responded to drought variations detected by these indices (Fig. 14). Short- and mid-term indicators (SPEI-3 and SPEI-6) showed stronger alignment with vegetation health than longer-term measures, confirming that vegetation responds most sensitively to short-term moisture variability. The ISPA region consistently exhibited the strongest relationships, reflecting a clear vegetation response to changing moisture conditions. In contrast, Jordan showed weaker relationships, likely due to irrigation or land-management practices that buffer vegetation against drought. Syria, Lebanon, and Israel displayed intermediate to high correlations, with no single index performing uniformly poorly. These findings emphasize that vegetation health is more closely linked to short- to mid-term drought conditions, while longer-term indices remain informative but less responsive.

In addition to linear correlations, Cohen's Kappa statistics were used to assess the categorical agreement between vegetation health conditions and drought classifications derived from the same indices. Pearson correlation measures how strongly two variables change together as continuous values, whereas Cohen's Kappa assesses how often both indicators assign conditions to the same category (e.g., "normal" or "drought") beyond what would occur by chance. Using these measures together offers a clearer and more reliable view of how vegetation health corresponds to different drought intensities across the region (Fig. 15). Results from Kappa analysis indicate that VHI aligns only weakly to moderately with individual drought indices, with the strongest agreement observed at shorter accumulation periods. Among the vegetation–drought comparisons, SPEI-3 achieved the highest Kappa value (around 0.31), followed by SPEI-6 (0.13), SPEI-12 (0.11), and sc-PDSI (0.06), confirming that vegetation responds more readily to short-term moisture variations. In contrast, agreement between the drought indices themselves is substantially stronger. For example, SPEI-12 and sc-PDSI show very high agreement (Kappa ~ 0.66), while SPEI-6 and SPEI-12 (0.46) and SPEI-6 and sc-PDSI (0.25) also exhibit closer alignment than VHI has with any single drought index. These findings indicate that, although meteorological indices classify drought in broadly similar ways, vegetation health does not always mirror these classifications directly, especially at longer accumulation periods where the signal becomes weaker.

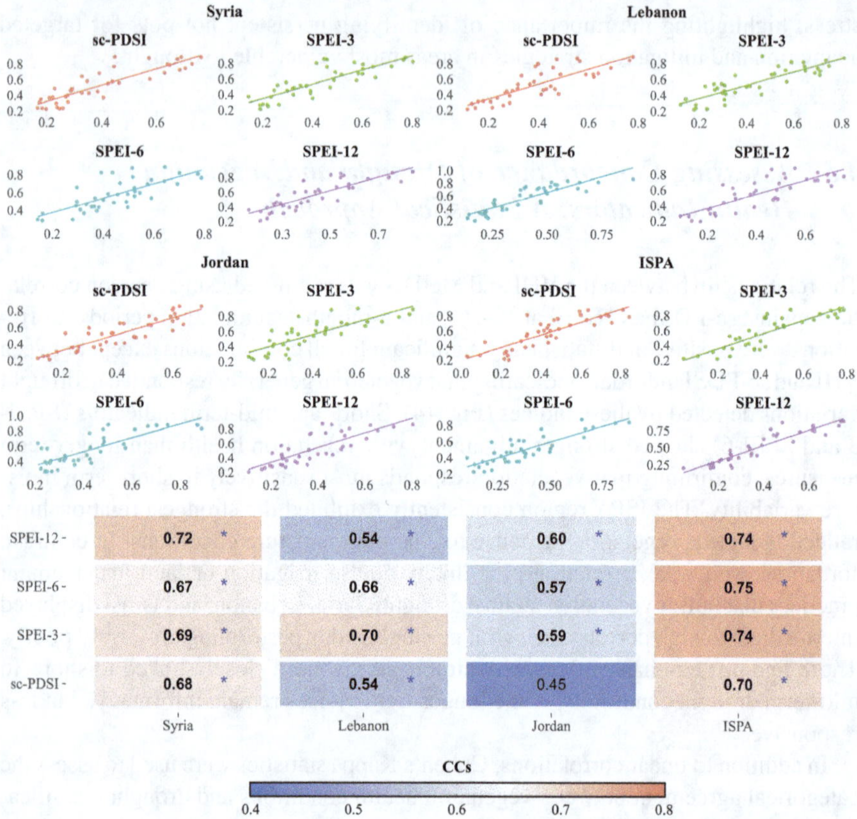

Fig. 14 Relationships between the VHI and MetDIs (sc-PDSI, SPEI-12, SPEI-6, SPEI-3) across the study region. The upper panel shows scatter plots of VHI versus each drought index for Syria, Lebanon, Jordan, Israel, and the ISPA region, illustrating country-level correlation patterns. The lower panel presents a heatmap of Pearson correlation coefficients, with asterisks marking statistically significant relationships at the 0.01 level

Studying agreement is crucial because correlation alone may overstate or understate functional relationships. Two indices can be highly correlated yet disagree on classification at specific times or locations, which is critical for operational drought monitoring and decision-making. By combining correlation and agreement metrics, we gain a clearer picture of how closely vegetation signals track actual drought conditions, and whether different indices can be used interchangeably or should be applied in combination. This dual approach enhances the reliability of drought assessment and supports more informed resource management strategies.

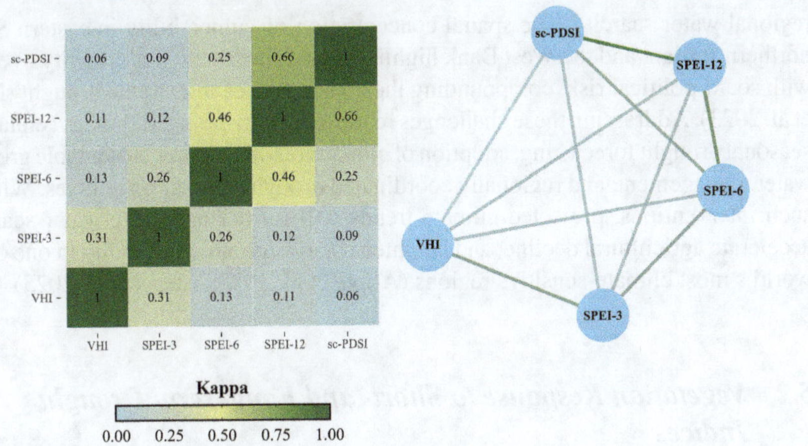

Fig. 15 Cohen's Kappa agreement among the VHI and MetDIs (SPEI-3, SPEI-6, SPEI-12, sc-PDSI). The left panel presents a heatmap of Kappa values, with numerical values displayed in each cell for clarity. Colours range from red (low agreement) through light blue (moderate agreement) to dark green (high agreement), allowing visual comparison of classification consistency. The right panel shows a network diagram where edge thickness and colour intensity represent the strength of agreement (Cohen's Kappa) between index pairs, thicker and darker green edges indicate stronger agreement, while thinner red edges indicate weaker agreement

5 Discussion and Conclusion

5.1 *Drought Intensification and Regional Variability*

This study provides robust evidence of a marked intensification of drought conditions across the Levant since the early 2000s, consistent with regional assessments reporting more frequent and severe dry events (Kiro et al. 2020; Zittis et al. 2022; Essa et al. 2023; Garrido-Perez et al. 2024). Both SPEI and sc-PDSI exhibit statistically significant negative trends with distinct change points in the early twenty-first century, reflecting the combined impacts of declining precipitation and rising atmospheric evaporative demand (Mathbout and Martin-Vide Lopez-Bustins 2023; Öz et al. 2024; Mukherjee et al. 2023). Spatial analyses reveal drought hotspots concentrated in eastern and northeastern Syria, northern and western Jordan, and parts of the West Bank, where arid to semi-arid climates coincide with fragile water resources and limited adaptive capacity. In contrast, Mediterranean coastal zones demonstrate partial resilience, supported by orographic precipitation and relatively higher soil moisture availability. These spatial contrasts underscore the influence of Köppen climate regimes, topography, and groundwater reserves in shaping drought severity and persistence, aligning with recent projections that identify the Middle East as a climate change hotspot (Zittis et al. 2022; Lazoglou et al. 2024; Christou et al. 2025). This integrated analysis further confirms that short-term drought events exert the most immediate impacts on vegetation health, while persistent long-term deficits intensify regional water scarcity. The spatial concentration of vulnerability in eastern Syria, northern Jordan, and the West Bank highlights how environmental fragility overlaps with socio-political risk, compounding the consequences of extreme drought (Neira et al. 2023). Addressing these challenges requires a multi-pronged strategy: enhanced seasonal drought forecasting, adoption of climate-resilient crops, sustainable groundwater management, and regionally coordinated drought adaptation policies. Without such interventions, projected climate trends will further aggravate water scarcity, accelerate agricultural decline, and heighten socio-economic instability in one of the world's most climate-sensitive regions (Afuye et al. 2025; Waqas et al. 2025).

5.2 *Vegetation Response to Short- and Long-Term Drought Indices*

The combined use of NDVI, VHI, and MetDIs demonstrates that vegetation health in the Levant responds most strongly to short- and medium-term moisture deficits (SPEI-3, SPEI-6), consistent with findings from other Mediterranean semi-arid systems (Bento et al. 2018; Won et al. 2025). This sensitivity reflects rapid physiological responses of vegetation to water stress, such as stomatal closure, reduced photosynthetic activity, and leaf shedding, which directly diminish canopy greenness and thermal stability detectable by remote sensing (Flexas and Medrano 2022). In contrast, longer-term indices (SPEI-12, sc-PDSI) exhibit weaker correspondence with vegetation stress, reflecting the cumulative yet less immediate nature of prolonged drought impacts, which typically involve slower processes like root depletion, carbon reserve exhaustion, and eventual canopy dieback (karla et al. 2023; Khan et al. 2025; Rahman et al. 2025) These weaker correlations are particularly evident in irrigated or managed landscapes such as Jordan, where groundwater abstraction, irrigation, and land-management interventions buffer vegetation from climatic water deficits (Cook-Patton et al. 2021; Abu-Hamatteh et al. 2025). Persistent hotspots of extreme vegetation stress in central-eastern Syria and northern Jordan reflect chronic water scarcity, compounded by land degradation, groundwater depletion, and socio-political instability (Eklund et al. 2022). The ISPA region overall shows the strongest vegetation–drought coupling, while Cohen's Kappa analysis indicates that vegetation health indices classify drought conditions only weakly to moderately in agreement with meteorological indices, especially at longer accumulation periods. Short-term measures (SPEI-3) achieve the highest categorical consistency, whereas meteorological indices themselves exhibit much stronger mutual agreement (e.g., Kappa > 0.6 between sc-PDSI and SPEI-12), underscoring their shared sensitivity to long-term hydro-climatic deficits. Overall, these findings highlight that correlation alone may overestimate the operational interchangeability of vegetation and climate indices (Vergni et al. 2021; Mathbout et al. 2025). Moreover, CMIP6 projections for the Mediterranean and Middle East indicate a likely intensification of drought conditions (Ibarra-Berastegui et al. 2025; Khosravi et al. 2025), raising urgent needs for proactive adaptation planning to safeguard ecosystems, livelihoods, and food security.

The combined use of NDVI, VHI, and MetDIs demonstrates that vegetation health in the Levant responds most strongly to short- and medium-term moisture deficits (SPEI-3, SPEI-6), consistent with findings from other Mediterranean semi-arid systems (Bento et al. 2018; Won et al. 2025). In contrast, longer-term indices (SPEI-12, sc-PDSI) exhibit weaker correspondence with vegetation stress, reflecting the cumulative yet less immediate nature of prolonged drought impacts. These weaker correlations are particularly evident in irrigated or managed landscapes such as Jordan, where groundwater abstraction, irrigation, and land-management interventions buffer vegetation from climatic water deficits (Cook-Patton et al. 2021; Abu-Hamatteh et al. 2025). Persistent hotspots of extreme vegetation stress in central-eastern Syria and northern Jordan reflect chronic water scarcity, compounded by land degradation, groundwater depletion, and socio-political instability (Eklund et al. 2022). The ISPA region overall shows the strongest vegetation–drought coupling, while Cohen's Kappa analysis indicates that vegetation health indices classify drought conditions only weakly to moderately in agreement with meteorological indices, especially at longer accumulation periods. Short-term measures (SPEI-3) achieve the highest categorical consistency, whereas meteorological indices themselves exhibit much stronger mutual agreement (e.g., Kappa > 0.6 between sc-PDSI and SPEI-12), underscoring their shared sensitivity to long-term hydro-climatic deficits. Overall, these findings highlight that correlation alone may overestimate the operational interchangeability of vegetation and climate indices (Vergni et al. 2021; Mathbout et al. 2025). Moreover, CMIP6 projections for the Mediterranean and Middle East indicate.

a likely intensification of drought conditions (Ibarra-Berastegui et al. 2025; Khosravi et al. 2025), raising urgent needs for proactive adaptation planning to safeguard ecosystems, livelihoods, and food security.

5.3 *Methodological Limitations and Uncertainties*

While the CRU TS dataset correlates well with in situ observations, it slightly overestimates precipitation in some months, potentially influencing SPEI values (Mathbout and Martin-Vide Lopez-Bustins 2023; Elagib et al. 2024; Yin et al. 2025). The coarse spatial resolution (0.5°) may obscure microclimatic variability, particularly in Lebanon's mountains and coastal plains. Satellite-derived VHI assumes stable sensor calibration over four decades; however, changes in instrumentation and atmospheric conditions may introduce radiometric biases (Cimini et al. 2024; Liu et al. 2024). In addition, anthropogenic land-use changes, such as agricultural intensification, irrigation expansion, and urban encroachment, can modify vegetation signals independently of meteorological drought conditions. For example, irrigated croplands may mask precipitation deficits in vegetation indices, while conflict-driven land abandonment may produce false drought "signatures" in regions where vegetation recovery occurs spontaneously (Eklund et al. 2022). Socio-environmental datasets, including the FAO's Global Land Cover SHARE database and the World Bank's Land Use Indicators, could help disentangle these human-driven effects from climate variability. Future studies should integrate such ancillary datasets to better identify where vegetation changes arise from human management versus climatic stress. Moreover, vegetation indices may be decoupled from meteorological drought by shifts in cropping patterns or seasonal planting calendars that are not captured by coarse-resolution satellite products (Burka et al. 2024). Statistical methods used here primarily capture linear associations; however, the interactions between climate, vegetation, and human activities are often non-linear and involve lagged feedback. Future work should incorporate machine learning models or coupled climate–vegetation simulations to capture these complexities (Maity et al. 2024; Cross et al. 2025).

References

Abu Samra RMA, Ali RR (2025) The Arab world at a crossroads: assessing future risks under changing climate. Environ Sci Eur 37:130. https://doi.org/10.1186/s12302-025-01162-1

Abu-Hamatteh Z, Al-Shawabkeh AF, Abu-Hammad N (2025) Climate change in Jordan: challenges and mitigation strategies. Open Environ Res J 18. https://doi.org/10.2174/0118742130359972250520065707

Afuye GA, Majahana JM, Kalumba AM, Zhou L, Mazinyo SP (2025) Drought disasters, vulnerability, severity, preparedness and response to the water sector: a comprehensive model approach for water governance and policy in South Africa. Progr Disast Sci 26:100417. https://doi.org/10.1016/j.pdisas.2025.100417

Al Zubi R, Gill TE, Lee JA, Ardon-Dryer K, Kandakji T (2024) Spatiotemporal analysis of sand and dust emission point sources detected from satellite imagery in Syria, Jordan, and Iraq. Sci Total Environ 949:175157. https://doi.org/10.1016/j.scitotenv.2024.175157

Arafat SM, Saleh NS, Aboelghar M et al (2014) Mapping of North Sinai land cover according to FAO-LCCS. Egypt J Rem Sens Space Sci 17(1):29–39. https://doi.org/10.1016/j.ejrs.2014.03.002

Armon M, Morin E, Enzel Y (2019) Overview of modern atmospheric patterns controlling rainfall and floods into the Dead Sea: Implications for the lake's sedimentology and paleohydrology. Quatern Sci Rev 216:58–73. https://doi.org/10.1016/j.quascirev.2019.06.005

Bani Mfarrej MF (2025) Exploring the nexus between climate change, water scarcity, and security dynamics in the Middle East and North Africa. Next Res 2(1):100168. https://doi.org/10.1016/j.nexres.2025.100168

Bartholomé E, Belward AS (2004) GLC2000: A new approach to global land cover mapping from Earth observation data. Int J Rem Sens 25(9):1959–1977. https://doi.org/10.1080/01431160412331291297

Beck HE, Zimmermann NE, McVicar TR, Vergopolan N, Berg A, Wood EF (2018) Present and future Köppen-Geiger climate classification maps at 1-km resolution. Scientific Data 5:180214. https://doi.org/10.1038/sdata.2018.214

Bento VA, Gouveia CM, DaCamara CC, Trigo IF (2018) A climatological assessment of drought impact on vegetation health index. Agric Meteorol 259:286–295. https://doi.org/10.1016/j.agrformet.2018.05.014

Bento VA, Gouveia CM, DaCamara CC, Libonati R, Trigo IF (2020) The roles of NDVI and land surface temperature when using the vegetation health index over dry regions. Global Planet Change 190:103198. https://doi.org/10.1016/j.gloplacha.2020.103198

Burka, A, Biazin B, Bewket W (2024) Spatial drought occurrences and distribution using VCI, TCI, VHI, and Google Earth Engine in Bilate River Watershed, Rift Valley of Ethiopia. J Maps. https://doi.org/10.1080/19475705.2024.2377672

Christou ED, Dailianis T, Chatzinikolaou E, Karachle PK, Damalas D, Tsagarakis K, Strogyloudi E, Karris G, Zervoudaki S, Giannakourou A, Salomidi M, Reizopoulou S (2025) Climate change in the "vulnerable" Eastern Mediterranean and adjacent areas: a literature review of ecological impacts and threats. Mar Environ Res 211:107390. https://doi.org/10.1016/j.marenvres.2025.107390

Cimini D, Barlakas V, Carminati F, De Angelis F, Di Paola F, Fassò A, Gallucci D, Gentile S, Hewison T, Larosa S, Madonna F, Mattioli V, Montopoli M, Romano F, Rosoldi M, Viggiano M, Von Engeln A, Ricciardelli E (2024) Title missing. Bull Atmospher Sci Technol 5:9. https://doi.org/10.1007/s42865-024-00073-y

Cohen J (1960) A coefficient of agreement for nominal scales. Educ Psychol Measur 20(1):7–46. https://doi.org/10.1177/001316446002000104

Cohen J (1968) Weighted kappa: Nominal scale agreement with provision for scaled disagreement or partial credit. Psychol Bull 70(4):213–220. https://doi.org/10.1037/h0026256

Cook-Patton SC, Drever CR, Griscom BW, Hamrick K, Hardman H, Kroeger T, Pacheco P, Raghav S, Stevenson M, Webb C, Yeo S, Ellis PW (2021) Protect, manage and then restore lands for climate mitigation. Nat Clim Chang 11:1027–1034. https://doi.org/10.1038/s41558-021-01198-0

Cross JF, Owen JS Jr, Shreckhise JH, Fields JS, Nackley L, Altland JE, Drewry DT (2025) Parsimonious models of root zone temperature in soilless substrates through ensemble machine learning. Smart Agricult Technol 12:101289. https://doi.org/10.1016/j.atech.2025.101289

de Vries H, Woollings T, Anstey J et al (2013) Atmospheric blocking and its relation to jet changes in a future climate. Clim Dyn 41:2643–2654. https://doi.org/10.1007/s00382-013-1699-7

Dinc P, Eklund L (2023) Syrian farmers in the midst of drought and conflict: the causes, patterns, and aftermath of land abandonment and migration. Clim Dev 15(5):349–362. https://doi.org/10.1080/17565529.2023.2223600

Eboreime E, Anjorin O, Obi-Jeff C, Ojo TM, Hertelendy A (2025) From drought to displacement: assessing the impacts of climate change on conflict and forced migration in West Africa's Sahel Region. J Clim Chang Health 23:100448. https://doi.org/10.1016/j.joclim.2025.100448

Eklund L, Mohr B, Dinc P (2024) Cropland abandonment in the context of drought, economic restructuring, and migration in northeast Syria. Environ Res Lett 19(1):014077. https://doi.org/10.1088/1748-9326/ad1723

Eklund L, Theisen OM, Baumann M et al (2022) Societal drought vulnerability and the Syrian climate-conflict nexus are better explained by agriculture than meteorology. Commun Earth Environ 3:85. https://doi.org/10.1038/s43247-022-00405-w

Elagib NA, Ali MMA, Schneider K (2024) Evaluation and bias correction of CRU TS4.05 potential evapotranspiration across vast environments with limited data. Atmospher Res 299:107194. https://doi.org/10.1016/j.atmosres.2023.107194

Elkhalfi O, Chaabita R, Ghoujdam M, Zehraoui K, El Alaoui H, Belhaj I (2025) The impact of climate change on food security in the Middle East and North Africa: challenges and adaptation strategies. J Agricult Food Res 21:101963. https://doi.org/10.1016/j.jafr.2025.101963

Essa YH, Hirschi M, Thiery W et al (2023) Drought characteristics in the Mediterranean under future climate change. npj Clim. Atmos Sci 6:133. https://doi.org/10.1038/s41612-023-00458-4

Felegari S, Sharifi A, Moravej K, Golchin A, Tariq A (2022) Investigation of the relationship between NDVI index, soil moisture, and precipitation data using satellite images. Sustain Agric Syst Technol 314–325. https://doi.org/10.1002/9781119808565.ch15

Flexas J, Medrano H (2002) Drought-inhibition of photosynthesis in C3 plants: stomatal and non-stomatal limitations revisited. Ann Bot 89(2):183–189. https://doi.org/10.1093/aob/mcf027

Gamer M, Lemon J, Fellows I, Singh P (2022) Various coefficients of interrater reliability and agreement (R package version 0.84.1). CRAN. https://CRAN.R-project.org/package=irr

Gao B-C (1996) NDWI—a normalized difference water index for remote sensing of vegetation liquid water from space. Rem Sens Environ 58(3):257–266. https://doi.org/10.1016/S0034-4257(96)00067-3

Garrido-Perez JM, Vicente-Serrano SM, Barriopedro D, García-Herrera R, Trigo R, Beguería S (2024) Examining the outstanding Euro-Mediterranean drought of 2021–2022 and its historical context. J Hydrol 630:130653. https://doi.org/10.1016/j.jhydrol.2024.130653

Gebrechorkos SH, Sheffield J, Vicente-Serrano SM, Funk C, Miralles DG, Peng J, Dyer E, Talib J, Beck HE, Singer MB, Dadson SJ (2025) Warming accelerates global drought severity. Nature 642:628–635. https://doi.org/10.1038/s41586-025-09047-2

Ghobadi M, Badehian Z (2025) Assessment of agricultural drought severity using multi-temporal remote sensing data in Lorestan region. Sci Rep 15:18528. https://doi.org/10.1038/s41598-025-03087-4

Gleick PH (2014) Water, drought, climate change, and conflict in Syria. Weather Clim Soc 6:331–340. https://doi.org/10.1175/WCAS-D-13-00059.1

Hamed KH, Rao AR (1998) A modified Mann-Kendall trend test for autocorrelated data. J Hydrol 204(1–4):182–196. https://doi.org/10.1016/S0022-1694(97)00125-X

Harris I, Osborn TJ, Jones P, Lister D (2020) Version 4 of the CRU TS monthly high-resolution gridded multivariate climate dataset. Sci Data 7:109. https://doi.org/10.1038/s41597-020-0453-3

He QN, Ju WM, Dai SP, He W, Song L, Wang SH, Li X, Li G (2021) Drought risk of global terrestrial gross primary productivity in recent 40 years detected by a remote sensing-driven process model. J Geophys Res 126(6):G02004. https://doi.org/10.1029/2020JG005944

Hollister J, Shah T, Wheeler B (2023) Elevatr: access elevation data from various APIs. R package version 0.4.2. https://CRAN.R-project.org/package=elevatr

Ibarra-Berastegui G, Sáenz J, Ulazia A, González-Rojí SJ, Esnaola G (2025) CMIP6 projections of wave climate change in the Mediterranean Sea by the end of the twenty-first century. Ocean Eng 335:121704. https://doi.org/10.1016/j.oceaneng.2025.121704

Javan K, Altaee A, BaniHashemi S, Darestani M, Zhou J, Pignatta G (2024) A review of interconnected challenges in the water–energy–food nexus: urban pollution perspective towards sustainable development. Sci Total Environ 912:169319. https://doi.org/10.1016/j.scitotenv.2023.169319

Kabir K, de Vries Robbe S, Godinho C (2024) Climate change mitigation policies in agriculture: an overview of sociopolitical barriers. Wiley Interdisciplin Rev Clim Change 15(6):e916. https://doi.org/10.1002/wcc.916

Kalra A, Goel S, Elias AA (2023) Understanding role of roots in plant response to drought: way forward to climate-resilient crops. Plant Genome 16(4):e20395. https://doi.org/10.1002/tpg2.20395

Karnieli A, Agam N, Pinker RT, Anderson M, Imhoff ML, Gutman GG, Panov N, Goldberg A (2010) Use of NDVI and land surface temperature for drought assessment: merits and limitations. J Clim 23(3):618–633. https://doi.org/10.1175/2009JCLI2900.1

Kelley CP, Mohtadi S, Cane MA, Kushnir Y (2015) Climate change in the fertile crescent and implications of the recent Syrian drought. Proc Natl Acad Sci 112(11):3241–3246. https://doi.org/10.1073/pnas.1421533112

Kelley C, Ting M, Seager R et al (2012) The relative contributions of radiative forcing and internal climate variability to the late 20th century winter drying of the Mediterranean region. Clim Dyn 38:2001–2015. https://doi.org/10.1007/s00382-011-1221-z

Khan AA, Wang Y-F, Akbar R, Alhoqail WA (2025) Mechanistic insights and future perspectives of drought stress management in staple crops. Front Plant Sci 16:1547452. https://doi.org/10.3389/fpls.2025.1547452

Khosravi Y, Ouarda TB, Homayouni S (2025) Developing an ensemble machine learning framework for enhanced climate projections using CMIP6 data in the Middle East. npj Clim Atmospher Sci 8:174. https://doi.org/10.1038/s41612-025-01033-9

Kiro Y, Goldstein SL, Kushnir Y, Olson JM, Bolge L, Lazar B, Stein M (2020) Droughts, flooding events, and shifts in water sources and seasonality characterize last interglacial Levant climate. Quat Sci Rev 248:106546. https://doi.org/10.1016/j.quascirev.2020.106546

Kogan FN (1995) Application of vegetation index and brightness temperature for drought detection. Adv Space Res 15(11):91–100. https://doi.org/10.1016/0273-1177(95)00079-T

Kogan FN (1997) Global drought watch from space. J Clim 10(6):621–636. https://doi.org/10.1175/1520-0477(1997)078%3c0621:GDWFS%3e2.0.CO;2

Kogan FN (2000) Satellite-observed sensitivity of world land ecosystems to El Niño/La Niña. Rem Sens Environ 74(3):445–462. https://doi.org/10.1016/S0034-4257(00)00137-1

Kogan F, Stark R, Gitelson A, Jargalsaikhan L, Dugrajav C, Tsooj S (2004) Derivation of pasture biomass in Mongolia from AVHRR-based vegetation health indices. Int J Rem Sens 25(14):2889–2896. https://doi.org/10.1080/01431160410001697619

Langgut D, Cheddadi R, Sharon G (2021) Climate and environmental reconstruction of the Epipaleolithic Mediterranean Levant (22.0–11.9 ka cal. BP). Quatern Sci Rev 270:107170. https://doi.org/10.1016/j.quascirev.2021.107170

Lazoglou G, Papadopoulos-Zachos A, Georgiades P, Zittis G, Velikou K, Manios EM, Anagnostopoulo C (2024) Identification of climate change hotspots in the Mediterranean. Sci Rep 14:29817. https://doi.org/10.1038/s41598-024-80139-1

Liu H, Liu Y, Chen Y et al (2023) Dynamics of global dryland vegetation were more sensitive to soil moisture: evidence from multiple vegetation indices. Agric For Meteorol 331:109327. https://doi.org/10.1016/j.agrformet.2023.109327

Liu L, Bliankinshtein N, Huang Y, Gyakum JR, Gabriel PM, Xu S, Wolde M (2024) Radiative closure tests of collocated hyperspectral microwave and infrared radiometers. Atmospher Measure Techn 17:2219–2233. https://doi.org/10.5194/amt-17-2219-2024

Liu Z, Lin H, Li H, Li M, Zhou P, Wang Z, Niu J (2025) Response mechanisms of vegetation productivity to water variability in arid and semi-arid areas of China: a decoupling analysis of soil moisture and precipitation. Atmosphere 16(8):933. https://doi.org/10.3390/atmos16080933

Maity R, Srivastava A, Sarkar S, Khan MI (2024) Revolutionizing the future of hydrological science: impact of machine learning and deep learning amidst emerging explainable AI and transfer learning. Appl Comput Geosci 24:100206. https://doi.org/10.1016/j.acags.2024.100206

Malamos N, Tegos A, Bourantas G, Chalvantzis C, Koutsoyiannis D (2025) Global reference evapotranspiration clustering and its relation to the Köppen-Geiger climate classification. J Hydrol 660A:133342. https://doi.org/10.1016/j.jhydrol.2025.133342

Marzi S, Mysiak J, Essenfelder AH, Pal JS, Vernaccini L, Mistry MN, Alfieri L, Poljansek K, Marin-Ferrer M, Vousdoukas M (2021) Assessing future vulnerability and risk of humanitarian crises using climate change and population projections within the INFORM framework. Glob Environ Chang 71:102393. https://doi.org/10.1016/j.gloenvcha.2021.102393

Mathbout S, Boustras G, Papazoglou P, Martin Vide J, Raai F (2025) Integrating climate indices and land use practices for comprehensive drought monitoring in Syria: impacts and implications. Environ Sustain Indicat 26:100631. https://doi.org/10.1016/j.indic.2025.100631

Mathbout S, Lopez-Bustins JA, Martin-Vide J et al (2018) Spatial and temporal analysis of drought variability at several time scales in Syria during 1961–2012. Atmos Res 200(1):153–168. https://doi.org/10.1016/j.atmosres.2017.09.016

Mathbout S, Martin-Vide A, Lopez-Bustins JA (2023) Drought characteristics projections based on CMIP6 cli-mate change scenarios in Syria. J Hydrol Reg Stud 50:101581. https://doi.org/10.1016/j.ejrh.2023.101581

Mukherjee S, Mishra AK, Zscheischler J, Entekhabi D (2023) Interaction between dry and hot extremes at a global scale using a cascade modeling framework. Nat Commun 14(1):277. https://doi.org/10.1038/s41467-022-35748-7

Neira M, Erguler K, Ahmady-Birgani H, Al-Hmoud ND, Fears R, Gogos C, Hobbhahn N, Koliou M, Kostrikis LG, Lelieveld J, Majeed A, Paz S, Rudich Y, Saad-Hussein A, Shaheen M, Tobias A, Christophides G (2023) Climate change and human health in the Eastern Mediterranean and Middle East: literature review, research priorities and policy suggestions. Environ Res 216:114537. https://doi.org/10.1016/j.envres.2022.114537

Neugebauer I, Brauer A et al (2015). Evidences for centennial dry periods at ~3300 and ~2800 cal. yr BP from micro-facies analyses of the Dead Sea sediments. The Holocene 25(8). https://doi.org/10.1177/0959683615584208

Otekunrin OA (2025) A critical assessment of the interplay of conflict, hunger, poverty, and food insecurity in Africa. Food Hum 4:100544. https://doi.org/10.1016/j.foohum.2025.100544

Öz FY, Özelkan E, Tatlı H (2024) Comparative analysis of SPI, SPEI, and RDI indices for assessing spatio-temporal variation of drought in Türkiye. Earth Sci Inform 17:4473–4505. https://doi.org/10.1007/s12145-024-01401-8

Palmer WC (1965) Meteorological drought (Research Paper No. 45). U.S. Weather Bureau, Washington, DC.

Pebesma E (2018) Simple features for R: Standardized support for spatial vector data. The R J 10(1):439–446. https://doi.org/10.32614/RJ-2018-009

Peel MC, Finlayson BL, McMahon TA (2007) Updated world map of the Köppen-Geiger climate classification. Hydrol Earth Syst Sci 11:1633–1644. https://doi.org/10.5194/hess-11-1633-2007

Por F (1975) An outline of the zoogeography of the Levant. Zool Scr 4(1):5–20. https://doi.org/10.1111/j.1463-6409.1975.tb00713.x

Rahman G, Jung M-K, Kim T-W, Kwon H-H (2025) Drought impact, vulnerability, risk assessment, management and mitigation under climate change: a comprehensive review. KSCE J Civ Eng 29(1):100120. https://doi.org/10.1016/j.kscej.2024.100120

Reis J, Öztürk Ş, Ayta S, Tülek Z, Siva A, Can G, Spencer PS (2025) Health challenges of climate change in the Eastern Mediterranean and Middle East region, with a focus on Türkiye: an environmental neurological and brain health perspective. J Neurol Sci 470:123423. https://doi.org/10.1016/j.jns.2025.123423

Roy D, Gillespie SA, Hossain MS (2025) Revisiting the drought–food insecurity nexus: a social–ecological systems perspective. Glob Food Secur 46:100874. https://doi.org/10.1016/j.gfs.2025.100874

Schillinger J, Özerol G, Heldeweg M (2022) A social-ecological systems perspective on the impacts of armed conflict on water resources management: case studies from the Middle East. Geoforum 133:101–116. https://doi.org/10.1016/j.geoforum.2022.05.001

Sun J, Zhang Q, Liu X, Sun J, Chen L, Wu Y, Hu B, Zhang G (2024) Flash droughts in a hotspot region: spatiotemporal patterns, possible climatic drivings and ecological impacts. Weather Clim Extrem 45:100700. https://doi.org/10.1016/j.wace.2024.100700

Tubi A, Feitelson E (2016) Drought and cooperation in a conflict prone area: Bedouin herders and Jewish farmers in Israel's northern Negev, 1957–1963. Polit Geogr 51:30–42. https://doi.org/10.1016/j.polgeo.2015.11.009

Ullah S, You Q, Sachindra DA, Nowosad M, Ullah W, Bhatti AS, Jin Z, Ali A (2022) Spatiotemporal changes in global aridity in terms of multiple aridity indices: an assessment based on the CRU data. Atmos Res 268:105998. https://doi.org/10.1016/j.atmosres.2021.105998

Vergni L, Todisco F, Di Lena B (2021) Evaluation of the similarity between drought indices by correlation analysis and Cohen's Kappa test in a Mediterranean area. Nat Hazards 108:2187–2209. https://doi.org/10.1007/s11069-021-04775-w

Vicente-Serrano SM, Begueria S, Lopez-Moreno JI et al (2010) A new global 0.58 gridded dataset (1901–2006) of a multiscalar drought index: comparison with current drought index datasets based on the Palmer drought se-verity index. J Hydrometeorol 11:1033–1043. https://doi.org/10.1175/2010JHM1224.1

Vicente-Serrano SM, Tramblay Y, Reig F et al (2025) High temporal variability not trend dominates Mediterra-nean precipitation. Nature 639:658–666. https://doi.org/10.1038/s41586-024-08576-6

Villanueva RAM, Chen ZJ (2019) ggplot2: Elegant graphics for data analysis (2nd ed.). Measur Interdisciplin Res Perspect 17(3):160–167. https://doi.org/10.1080/15366367.2019.1565254

Wang M, Wang Y, Liu X, Hou W, Wang J, Li S, Zhao L, Hu Z (2025) Vapor pressure deficit dominates vegetation productivity during compound drought and heatwave events in China's arid and semi-arid regions: evidence from multiple vegetation parameters. Eco Inform 88:103144. https://doi.org/10.1016/j.ecoinf.2023.103144

Waqas M, Naseem A, Wannasingha Humphries U, Hlaing PT, Shoaib M, Hashim S (2025) A comprehensive review of the impacts of climate change on agriculture in Thailand. Farm Syst 3(1):100114. https://doi.org/10.1016/j.farsys.2024.100114

Ward, P. J., de Ruiter, M. C., Mård, J., Schröter, K., Van Loon, A., Veldkamp, T., von Uexkull, N., Wanders, N., AghaKouchak, A., Arnbjerg-Nielsen, K., Capewell, L., Llasat, M. C., Day, R., Dewals, B., Di Baldassarre, G., Huning, L. S., Kreibich, H., Mazzoleni, M., Savelli, E., Teutschbein, C., … Wens, M. (2020). The need to integrate flood and drought disaster risk reduction strategies. Water Secur 11:100070. https://doi.org/10.1016/j.wasec.2020.100070

Wells N, Goddard S, Hayes MJ (2004) A self-calibrating palmer drought severity index. J Clim 17(12):2335–2351. https://doi.org/10.1175/1520-0442(2004)017%3c2335:ASPDSI%3e2.0.CO;2

Won J, Lee J, Kim S (2025) Vegetation drought condition index for probabilistic monitoring and forecasting of vegetation drought. Weather Clim Extrem 49:100786. https://doi.org/10.1016/j.wace.2025.100786

Xiaomei P, Binwu Z, Wenshu W, Ding A, Shengchun X (2024) Different drought indices showed different variations and applicability to dendrochronological studies. Dendrochronologia 84:126179. https://doi.org/10.1016/j.dendro.2024.126179

Ye L, Cheng L, Zhang L, Lei X, Liu P (2025) Evolution of long-term global drought during past 70 years based on estimated evaporation using the generalized complementary relationship. J Hydrol 650:132532. https://doi.org/10.1016/j.jhydrol.2024.132532

Yin X, Zhang Z, Lin Z, Yin J (2025) Assessment of rainfall frequencies from global precipitation datasets. Atmosphere 16:66. https://doi.org/10.3390/atmos16010066

Zeleňáková M, Abd-Elhamid HF, Krajníková K, Smetanková J, Purcz P, Alkhalaf I (2022) Spatial and temporal variability of rainfall trends in response to climate change—a case study: Syria. Water 14:1670. https://doi.org/10.3390/w14101670

Zhang Y, Liu X, Jiao W, Wu X, Zeng X, Zhao L, Wang L, Guo J, Xing X, Hong Y (2023) Spatial heterogeneity of vegetation resilience changes to different drought types. Earth's Fut 11:e2022EF003108. https://doi.org/10.1029/2022EF003108

Zeng J, Zhang R, Qu Y, Bento VA, Zhou T, Lin Y, Wu X, Qi J, Shui W, Wang Q (2022) Improving the drought monitoring capability of VHI at the global scale via ensemble indices for various vegetation types from 2001 to 2018. Weather Clim Extrem 35:100412. https://doi.org/10.1016/j.wace.2022.100412

Zeng J, Zhou T, Qu Y, Zhang R, Bento VA, Lin Y, Wu X, Qi J, Shui W, Wang Q (2023) An improved global vegetation health index dataset in detecting vegetation drought. Sci Data 10:338. https://doi.org/10.1038/s41597-023-02255-3

Zittis G, Almazroui M, Alpert P, Ciais P, Cramer W, Dahdal Y, Fnais M, Francis D, Hadjinicolaou P, Howari F, Jrrar A, Kaskaoutis DG, Kulmala M, Lazoglou G, Mihalopoulos N, Lin X et al (2022) Climate change and weather extremes in the Eastern Mediterranean and Middle East. Rev Geophys 60(3):e2021RG000762. https://doi.org/10.1029/2021RG000762

Local Climate Zone Mapping in MENA: A Streamlined Remote Sensing and GIS Method with Hyperspectral Imagery

Mohammad Younes, Alberto Vavassori, and Maria Antonia Brovelli

1 Introduction

Over the past two centuries, global industrialisation has driven urban development at an increasingly rapid rate. Approximately 55% of the world's population currently lives in cities, and this percentage is projected to increase to 68% by 2050 (United Nations 2019). Economic activities concentrated within urban regions contribute to 67–72% (29 GtCO2e) of annual global greenhouse emissions, rendering them major drivers of global warming (Shukla et al. 2022). As of 2024, the global mean temperature was 1.55 °C above pre-industrial conditions (1850–1900) (WMO 2025); however, climatic trend studies indicate certain regions with disproportionately higher warming rates. For instance, in the Middle East and North Africa (MENA), the average temperature increase was 0.45 °C per decade from 1981 to 2019, nearly twice the global trend (Zittis et al. 2022). According to climate models, extreme heatwaves with air temperatures exceeding 50 °C are expected to become increasingly frequent in the MENA region (Christidis et al. 2023), with severe implications for urban populations. Indeed, 600 million people in MENA, over 90% of whom live in cities, are expected to face yearly recurring extreme heatwaves (Zittis et al. 2021).

Global warming intensifies the Urban Heat Island (UHI) effect, whereby urban areas exhibit significantly warmer air temperatures than their rural counterparts. Impervious surfaces such as concrete and asphalt readily absorb shortwave solar radiation during daytime, subsequently releasing longwave thermal energy as ambient temperatures drop in the evening. Consequently, the UHI is most evident during nighttime, under clear and calm weather conditions (Oke 1982). Empirical research

M. Younes (✉)
Independent Researcher, Dubai, United Arab Emirates
e-mail: mohammadyounes08@gmail.com

A. Vavassori · M. A. Brovelli
Department of Civil and Environmental Engineering, Politecnico di Milano, Piazza Leonardo da Vinci 32, 20133 Milano, Italy

M. M. Al Saud (ed.), *Hydroclimate Changes: Implications of the Middle East Region*,
https://doi.org/10.1007/978-3-032-20053-2_6

has found that urban morphology strongly influences the UHI. Indeed, compact building layouts with limited vegetation are associated with the highest temperatures, whereas areas dominated by green spaces tend to remain cooler (Stewart et al. 2014); in particular, tall vegetation significantly influences local climate by buffering high temperatures through evapotranspiration and shading effects (Oke 1989). The UHI effect has a wide range of implications. Urban residents face increased thermal discomfort, alongside higher risks of heat-related morbidity and mortality. Energy demand for cooling also increases, resulting in higher CO_2 emissions and operational costs for buildings (Nuruzzaman 2015).

In view of the above considerations, UHI research has recognised the limitations of a simple urban-rural dichotomy in capturing the complex thermal variability observed in cities (Oke 2004). Indeed, even within the same urban area, thermal differences of up to 5 °C can be observed between parks and mid-rise building zones (Stewart and Oke 2010). To tackle this problem, the Local Climate Zone (LCZ) concept was developed in 2012. LCZs are defined as "regions of uniform surface cover, structure, material, and human activity that span hundreds of meters to several kilometres in horizontal scale" (Stewart and Oke 2012). The LCZ framework was designed for universal applications, avoiding region-specific terminologies and urban definitions. This classification system subdivides urban regions into 17 categories, i.e. 10 "built-up" and 7 "land cover" classes (see Fig. 1). In-situ temperature observations and surface-atmosphere modelling in LCZs concurred that building morphology (height and spacing), pervious surface fraction, tree density and soil wetness significantly contribute to the thermal differences observed in cities (Stewart et al. 2014). The physical and thermal properties of LCZs are described by several Urban Canopy Parameters (UCPs), including, e.g. Sky View Factor (SVF), Building Height, Tree Canopy Height, Aspect Ratio, and Impervious Surface Fraction, all of which can be measured or computed, thereby providing an objective framework for mapping LCZs based on geometrical and physical quantities (Stewart and Oke 2012).

Serving as a new lens through which urban environments can be studied, LCZ research has gained momentum over the past decade. Indeed, the number of yearly LCZ-related publications increased from 17 to 300 between 2013 and 2022. Although originally developed for UHI studies, LCZs are increasingly being utilised for a variety of applications, including urban planning, building energy modelling, and urban thermal comfort (Han et al. 2024).

Accurate LCZ mapping is a growing and evolving field, aiming to provide a standardised methodology for spatially delineating urban landscapes in support of urban climate studies. Since 2012, over 300 articles on the topic have been published, with the majority concentrated in Europe, East Asia and the United States (Han et al. 2024; Wang et al. 2024). In contrast, research related to LCZs in the MENA region remains limited, with individual countries contributing fewer than five publications between 2012 and 2023 (see Fig. 2). Expanding the LCZ mapping in the MENA region is of critical importance, both for advancing the global applicability of urban climate tools and for addressing region-specific challenges. In fact, given the pronounced warming trend across MENA, the application of the LCZ framework can provide

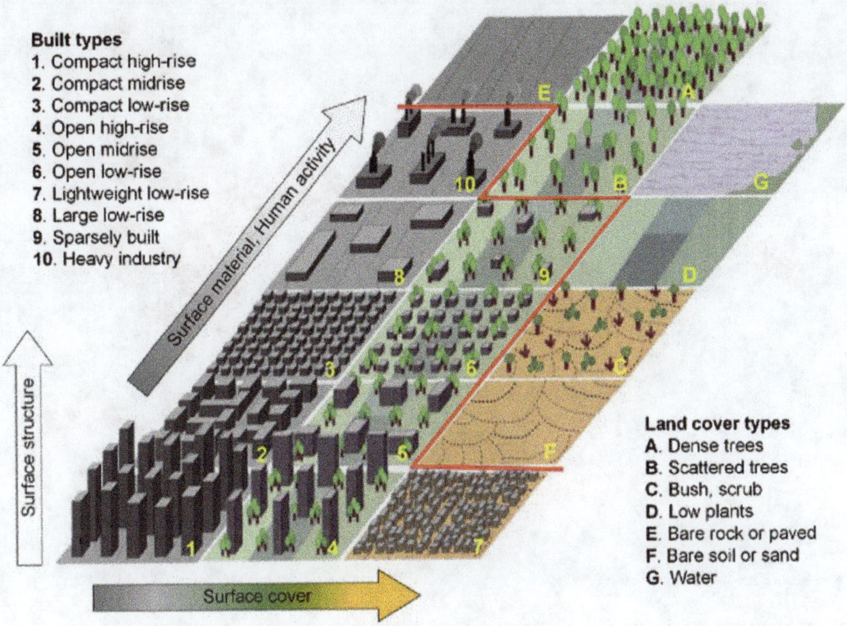

Fig. 1 The local climate zone (LCZ) classification system illustrated in a three-dimensional space. Surface cover (x-axis) describes the dominant land cover type, ranging from impervious, vegetation to bareground. The y-axis represents surface material, indicating the level of human activitiy occuring in the LCZ. Surface structure (z-axis) entails the the average height of built structures in the LCZ. (Huang et al. 2023)

a reference for evidence-based urban planning, thereby enhancing the capacity of cities to adapt to climate change.

State-of-the-art LCZ mapping methods that integrate geospatial data and machine learning (ML) techniques have so far been applied mainly in Europe and North America (Demuzere et al. 2019a, 2020). More recent studies have advanced this field by combining remote sensing (RS) and Geographic Information System (GIS) data to train ML models for LCZ classification. In particular, Vavassori et al. (2024) proposed a hybrid RS-GIS approach that used PRISMA hyperspectral imagery to classify LCZs in Milan, Italy, achieving improved results compared to Sentinel-2 multispectral data. While promising, this method has yet to be tested in regions with diverse urban forms and environmental conditions to evaluate its broader applicability. To address this gap, the present study pursues three objectives:

- to test the effectiveness of the hybrid RS-GIS approach proposed in Vavassori et al. (2024) in three MENA cities, i.e. Cairo (Egypt), Khartoum (Sudan), and Algiers (Algeria);
- to evaluate the contribution of hyperspectral PRISMA data in improving the LCZ classification accuracy compared to Sentinel-2 multispectral imagery;

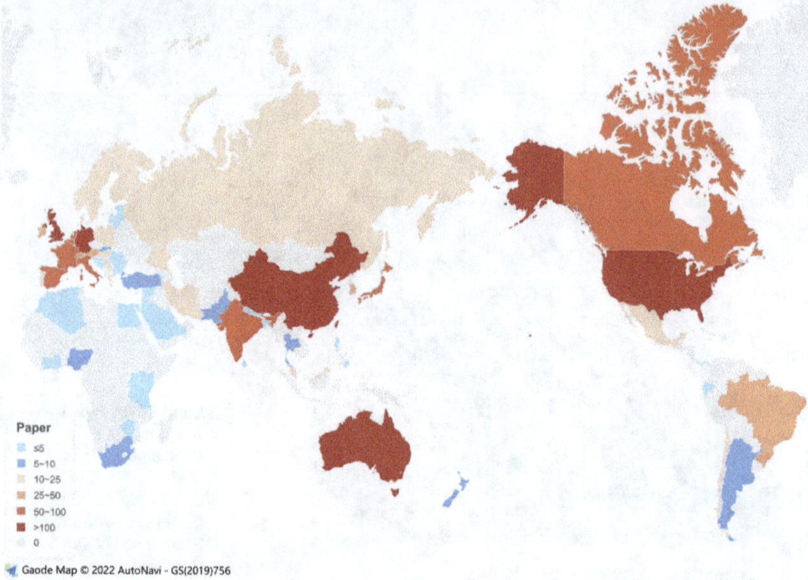

Fig. 2 World map visualizing the number of LCZ publications per country between 2012 and 2024. The countries with over 100 LCZ publications during this time were the China, the United States, the United Kingdom, Germany and Australia. MENA countries tend to have less than 5 LCZ publications, if any. (Han et al. 2024)

- to enhance the original methodology by streamlining the data acquisition and processing, thereby improving the reproducibility.

The paper is structured in the following sections. Section 2 provides an overview of the state-of-the-art on LCZ mapping. Section 3 presents the MENA city case studies. The proposed methodology is described in detail in Section 4, while the results are presented in Section 5. Lastly, a discussion on the future implications of this work is presented in Section 6, followed by concluding remarks in Section 7.

2 State-of-the-Art on LCZ Mapping

Mapping LCZs in urban environments has evolved markedly since 2012, and various methods have been developed over this period. Early efforts primarily relied on field surveys and direct visual assessments of building morphology and surface characteristics. However, with the increasing availability of high-resolution satellite imagery and geospatial datasets, recent research has shifted toward automated classification techniques, often incorporating ML. LCZ mapping methods can be broadly categorised into three main groups: RS-based, GIS-based, and hybrid methods. The following sections provide an overview of each approach.

2.1 RS-Based Methods

Recent advances in geospatial artificial intelligence (GeoAI) have accelerated research on automated LCZ mapping, accounting for 66.2% of related publications between 2012 and 2023 (Huang et al. 2023). Most approaches rely on open-source satellite imagery in combination with expert-defined training areas (TA) for supervised classification across a region of interest. Within this framework, LCZ classification has been implemented using pixel-based, object-oriented, and scene-based techniques. ML algorithms currently dominate existing RS-based methods, although deep learning approaches are increasingly adopted (Wang et al. 2024).

RS-based LCZ mapping was pioneered by Bechtel and Daneke (2012) and formalised into the World Urban Database and Access Portal Tools (WUDAPT) protocol in 2015, which has become a widely applied standard within the urban climate research community (Bechtel et al. 2015). WUDAPT employs pixel-wise classification of open-source satellite data such as Landsat and Sentinel imagery. TAs are delineated manually through expert visual interpretation, which is a critical yet labour-intensive step. Random forest (RF) was established as the standard ML classifier in the WUDAPT protocol, serving as the current benchmark for RS-based methods (Bechtel et al. 2015). Credited to its decision tree ensemble architecture, RF mitigates overfitting while maintaining strong predictive performance in a variety of domains. Nevertheless, supervised ML algorithms require large quantities of data for training (Salman et al. 2024). A major limitation of the WUDAPT methodology lies in its reliance on manually delineated LCZ polygons, typically drawn in GIS software such as Google Earth or QGIS. Furthermore, the HUMINEX project highlighted inconsistencies in crowdsourced LCZ training datasets, which strongly affect classification accuracy (Bechtel et al. 2017). Reported accuracies for global WUDAPT-based LCZ maps typically range between 50 and 60%, with built-up classes (LCZ 1-10) exhibiting lower accuracy than natural classes (LCZ A-G) (Bechtel et al. 2019). To address this imbalance, studies have incorporated synthetic aperture radar (SAR) data from platforms such as Sentinel-1 and PALSAR-2, which have been shown to improve classification accuracy for built-up LCZ types (Wang et al. 2024).

2.2 GIS-Based Methods

GIS-based methods classify spatial units into LCZs using UCP thresholds defined by Stewart and Oke (2012). Typically, these workflows involve six stages: data collection, parameter calculation, spatial unit definition, classification, post-processing, and accuracy assessment (Quan and Bansal 2021). UCPs are commonly derived from open-source datasets and then assigned to spatial units for classification.

Spatial units are generally defined either as parcels or grids (Wang et al. 2024). Parcels represent discrete morphological entities such as lot polygons, urban islets, or blocks, usually spanning 200–500 m (Stewart and Oke 2012). Alternatively, equally sized rectangular grids (so-called fishing nets) provide a simplified means of delineation. The selection of an appropriate spatial scale varies across studies, with methods including correlation analysis between urban morphology and surface temperatures (Cai et al. 2019) and geographic location sensitivity testing (Zheng et al. 2018). Once spatial units are parameterised by UCP values, rule-based algorithms are applied to assign LCZ types. Within the literature, three main rule-based approaches have been identified: the standard rule-based classifier, its improved variants, and fuzzy rule-based classifiers (Quan and Bansal 2021). While less developed than RS-based approaches, some studies have experimented with alternative algorithms such as Naïve Bayes, K-means clustering, and decision-making frameworks (Wang et al. 2024).

As a final step, post-processing is typically used to enhance map homogeneity by reducing noise. A majority filter is commonly applied (Geletič and Lehnert 2016), although alternative methods have been proposed. For example, Estacio et al. (2019) developed a cellular automata-based post-processing technique that iteratively reclassifies neighbourhood cells based on fuzzy membership of LCZ types. It is worth mentioning that accuracy assessment remains a major challenge for GIS-based approaches, as evaluation methods are not standardised. Current practices range from visual inspection and overlays with temperature observations to validation against RS-based LCZ maps (Wang et al. 2024).

2.3 Hybrid Methods

While spectral reflectance data provided by satellite imagery is effective for land cover classification, it does not capture the three-dimensional characteristics of urban morphology, which are essential for distinguishing built LCZs. GIS-derived datasets describing UCPs help to bridge this gap and can be integrated to construct training datasets that enhance the performance of supervised classification, thereby complementing the strengths of RS-based mapping. Moreover, the application of machine learning techniques enables objective evaluation of LCZ classification quality through standard performance metrics and confusion matrices. For these reasons, hybrid methods have been developed to combine the advantages of RS and GIS-based methods.

Building on the WUDAPT protocol, Demuzere et al. (2019a) combined Landsat 8 and Sentinel-1 data with digital elevation, slope, aspect, and forest canopy height, achieving overall accuracies above 80% across all LCZ classes. Similarly, Zhou et al. (2020) integrated Landsat 8 multispectral imagery with UCPs, including building height, building coverage ratio, impervious surface fraction, and pervious surface fraction, reporting classification accuracies of 84-100% for built classes. More recently, Vavassori et al. (2024) proposed a hybrid approach that incorporates PRISMA hyperspectral data alongside UCP layers to improve LCZ classification in Milan, Italy. By applying Principal Component Analysis (PCA) to 239 spectral bands, they reduced computational demands while retaining most spectral information within 10 principal components (PCs). Compared to Sentinel-2 multispectral data, PRISMA improved the accuracy of individual LCZ classes by an average of 16%, leading to a 5% increase in mean overall accuracy (OA).

These findings demonstrate that hybrid RS/GIS approaches represent a promising advancement in automated LCZ mapping, delivering higher accuracies than conventional methods. Given the distinct urban and natural landscapes of the MENA region, further research should assess the applicability and potential benefits of such approaches in this context.

2.4 *Available LCZ Maps*

Since the establishment of LCZs as the level-0 data layer in the WUDAPT protocol, considerable efforts have been made to improve the accessibility and scalability of LCZ mapping (Ching et al. 2018). A notable example is the LCZ Generator, an online platform that allows users to generate LCZ maps solely requiring a TA layer as input from the user. Built on Google Earth Engine (GEE), the tool incorporates 33 globally available input layers and leverages cloud computing to produce LCZ classifications for any region of interest (Demuzere et al. 2021b). Building on this advancement, Demuzere et al. (2022) produced the first global LCZ map using a curated training dataset consisting of 63,847 polygons across 410 regions of interest. Reported F1-scores ranged from 50 to 78% for built-up classes and above 60% for natural classes. Among the built-up categories, Lightweight Low-rise (LCZ 7) was identified as particularly challenging due to its limited representation in training datasets and its mixed spectral signatures. As LCZ 7 often corresponds to informal settlements or slums constructed with lightweight materials, it is more prevalent in developing regions, including countries across the MENA, highlighting the importance of region-specific investigations (Boanada-Fuchs et al. 2024).

Large-scale LCZ mapping has also been undertaken at continental scales. For Europe and the Continental United States (CONUS), Demuzere et al. (2019a) and Demuzere et al. (2020) utilised an extended WUDAPT protocol, integrating Landsat 8, Sentinel-1, and digital elevation and surface models within GEE. In Europe, class-wise F1 scores exceeded 80% for most LCZs, except for classes 4, 5, and 10, which scored below 70% (Demuzere et al. 2019a). For the CONUS, class-wise F1 scores ranged from 55% to 85%. In this case, the lack of a reliable nationwide building height dataset was identified as a key factor limiting the performance of the RF classifier (Demuzere et al. 2020). As an important outcome of the study, a curated training dataset developed by 10 annotation experts was published, supporting future efforts towards accurate LCZ mapping in the region (Demuzere et al. 2021a).

3 MENA City Case Studies

Based on the availability of high-quality UCP data, three major cities in the MENA region were selected for this study: Cairo, Khartoum, and Algiers (Fig. 3). Each city features different climatic conditions and urban architecture, thus testing the performance of our hybrid LCZ mapping method across a diverse range of urban landscapes. The following sub-sections provide a brief overview of each MENA city.

3.1 Cairo, Egypt

Cairo, inhabited by over 22 million people, is Egypt's capital city. It is the largest urban area in MENA and the sixth largest city worldwide (United Nations 2019; World Bank 2024). Since the mid-twentieth century, the city's population has nearly doubled roughly every two decades, leading to rapid urban expansion into peripheral regions now collectively called Greater Cairo (GC) (Sutton and Fahmi 2001). Between 1973 and 2021, the urban footprint of GC grew from 199 km^2 to 2225 km^2,

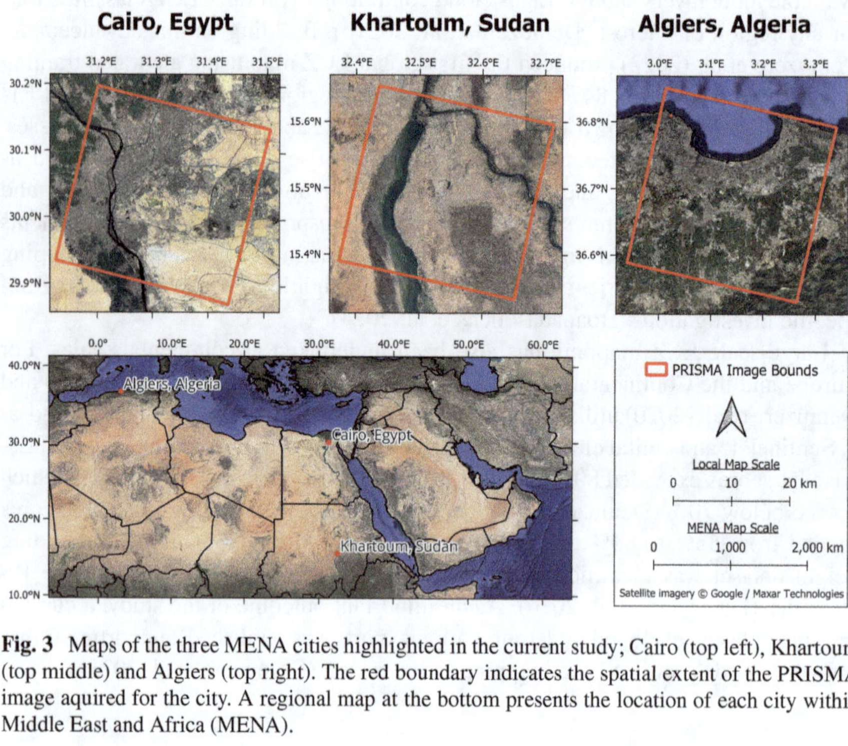

Fig. 3 Maps of the three MENA cities highlighted in the current study; Cairo (top left), Khartoum (top middle) and Algiers (top right). The red boundary indicates the spatial extent of the PRISMA image aquired for the city. A regional map at the bottom presents the location of each city within Middle East and Africa (MENA).

representing an elevenfold increase in built-up areas (Salem et al. 2024). Consequently, green spaces made up only 11.4% of Cairo's total area in 2020, providing a per capita space of 9.6 m^2 (El Araby 2002; UN-Habitat 2024).

Situated at 30°02′ N, 31°14′ E, with an elevation of 23 m, Cairo experiences a hot arid desert climate (BWh) according to the Köppen-Geiger classification, characterised by low annual precipitation and high temperatures (Kottek et al. 2006). The mean annual precipitation is 20.6 mm, with summer temperatures ranging between 20 and 35 °C and winter temperatures between 8 and 20 °C (World Bank 2021). The impacts of anthropogenic climate change are already evident: average temperatures in Egypt increased by approximately 0.1 °C per decade from 1901 to 2013, accelerating to 0.38 °C per decade between 2000 and 2020. Climate projections suggest that by 2081–2100, Egypt's mean temperature could rise 2.5–6 °C above pre-industrial levels, exceeding the anticipated global average increase (IEA 2023).

The urban environment of GC, dominated by thermally intensive construction materials such as clay brick, stone, and concrete (Ali et al. 2013), renders the city particularly susceptible to the UHI effect, with significant consequences for its densely populated areas.

3.2 *Khartoum, Sudan*

The capital of Sudan, Khartoum, hosts 6.5 million inhabitants and represents the country's largest urban settlement (World Bank 2024). Driven by high birth rates and intense rural-to-urban migration, Khartoum's population doubled almost every 20 years between 1956 and 2008, going from 245 thousand to 5.5 million (Hamid and Bahreldin 2013). A large influx of migrants during the mid-1980s gave rise to squatter settlements (LCZ-7), which constituted 38.7% of Khartoum's urban area in 2020 (Team 2020). Urban expansion is evident in land-cover changes; indeed, urbanised areas increased from 45% in 1972 to 85% in 2006, while vegetation cover declined from 29% to 8% over the same period (Yousif et al. 2015). Widespread deforestation has exacerbated local thermal conditions, producing nighttime UHI intensities of up to 6.4 °C during summer (Elagib 2011).

Located at 15°36′ N, 32°33′ E, with an altitude of 380 m, the city experiences a hot arid desert climate (BWh) (Kottek et al. 2006; Hamid and Bahreldin 2013). Summer temperatures peak at 42 °C in May, with minimums around 26 °C, while winter temperatures range from 15 °C to 25 °C. The annual precipitation is limited, averaging roughly 145 mm (World Bank 2021).

The built environment further contributes to the city's thermal profile. Modern structures predominantly use cement blocks and iron sheets, whereas traditional housing is constructed from mud, sticks, and thatch (Ismail Ahmed Ali and Szalay 2019). As a water-stressed region (Kharraz et al. 2012), Khartoum faces additional environmental pressures from contaminated groundwater due to inadequate sewage infrastructure, traffic-related air pollution, and widespread solid-waste accumulation

(Hamid and Bahreldin 2013). In the context of global warming, these conditions present significant challenges for sustainable urban development.

3.3 Algiers, Algeria

Algiers, the capital of Algeria, hosts approximately 2.9 million inhabitants (World Bank 2024). Following independence in 1962, large-scale rural-to-urban migration fostered rapid population growth, with annual increases averaging 3.48% until 1987 (JICA 2005). Since then, the city's population has continued to expand at an annual rate of 2.7%, contributing to an estimated urban growth of 4% per year (Angel et al. 2016). RS analysis using Landsat imagery revealed an urbanisation rate of 7.1 km^2 per year between 1987 and 2016, culminating in a total urban area of 334 km^2. By the end of this period, vegetation cover was estimated at 32% (Bouchachi and Zhong 2017), notably higher than in Cairo and Khartoum.

Algiers is located at 36°45′ N, 3°04′ E, along the Mediterranean coast, at an average elevation of 40–50 m. North of the Atlas Mountains, Algiers' Southern region features hilly slopes and mountains, reaching elevations of 410 m. According to the Köppen-Geiger classification, Algiers has a warm temperate climate with dry, hot summers (Csa) (Kottek et al. 2006). Summer temperatures range from 18 to 33 °C, while average winter temperatures are between 6 °C and 12 °C, with January being the coldest month. The city receives an average annual precipitation of 630.5 mm (World Bank 2021).

The urban fabric of Algiers reflects its historical and modern development. Traditional housing stock, predominantly constructed before the 1950s, employed stone, earth bricks, clay, and timber. More recent construction relies extensively on reinforced concrete and steel, marking a transition to modern architectural practices (Farsi et al. 2002).

4 Methods

In this study, we adopted the hybrid method proposed in Vavassori et al. (2024), which combines RS and GIS data for the supervised classification of LCZs using the RF algorithm. The source code implementing the methodology is fully available in a dedicated GitHub repository (https://github.com/gisgeolab/LCZ-ODC/tree/Processing-Notebooks) as a series of Jupyter Notebooks and Python functions (Capizzi et al. 2024). Building upon this code base, we developed a semi-automated LCZ mapping pipeline. Additional Python scripts were implemented for data acquisition, UCP computation, and TA quality control, streamlining the workflow and reducing the effort required to apply the method to a city or region of interest. The source code with usage instructions can be found on GitHub (https://github.com/medo-younes/lcz-py).

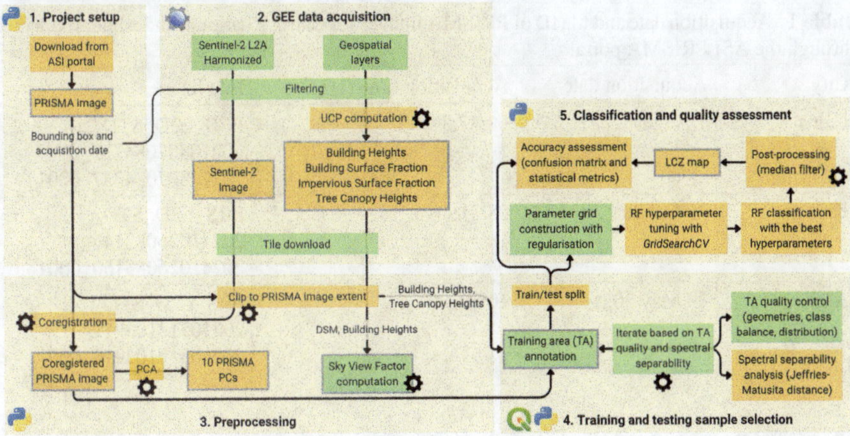

Fig. 4 Workflow of the LCZ mapping workflow. In orange: steps adopted from Vavassori et al. (2024); in green: proposed additional steps

A key component for streamlining the hybrid method was the semi-automated data acquisition pipeline. The PRISMA image bounds are used as input to programmatically access global datasets, including Sentinel-2 imagery and geospatial layers (e.g., building heights and tree canopy height), through the GEE Python API. Most UCPs, except SVF, are computed directly within the GEE cloud environment, leveraging its computational resources. Since the TA quality control is a critical step in the generation of reliable LCZ maps (Bechtel et al. 2017), we incorporated an additional component to verify the compliance of the TAs with the WUDAPT protocol recommendations (Ching et al. 2018), thereby reducing the effort and time needed to carry out the whole process, as well as limiting the subjectivity in the TA generation. Appendix 1 outlines the workflow steps in detail, while Fig. 4 provides a visual representation of the methodological workflow.

4.1 Satellite Image Acquisition

Spectral reflectance provided by satellite imagery constitutes a foundational data layer for automated LCZ mapping, owing to the global coverage and frequent revisit time of satellite sensors. Hyperspectral imagery was sourced from the PRISMA (*PRecursore IperSpettrale della Missione Applicativa*, Hyperspectral Precursor of the Application Mission) mission owned by the Italian Space Agency (ASI). Launched on March 22, 2019, PRISMA operates in a frozen Sun-Synchronous Low Earth Orbit at 615 km, attaining a swath size of 30 km and a nominal revisit time of 29 days at the same look angle, up to < 7 days with variable looking angle. PRISMA offers 30 m spatial resolution hyperspectral imagery as well as a 5 m panchromatic band. An onboard prismatic spectrometer captures VNIR (Visible and

Table 1 Acquisition date and file ID of PRISMA images for Cairo, Khartoum and Algiers collected through the ASI PRISMA portal

City	Acquisition date	Acquisition time (UTC)	File ID
Cairo	7 August 2021	08:47:32	PRS_L2D_STD_ 20210807084732_ 20210807084737_0001
Khartoum	26 October 2024	08:31:46	PRS_L2D_STD_ 20241026083146_ 20241026083150_0001
Algiers	13 May 2021	10:39:11	PRS_L2D_STD_ 20210513103911_ 20210513103916_0001

Near-InfraRed, 400–1010 nm) and SWIR (Short-Wave InfraRed, 920–2500 nm) ranges with an average spectral resolution of 10 nm (ASI 2009). On the other hand, multispectral imagery was sourced from the Copernicus Sentinel-2 mission, a two-satellite constellation in Sun-Synchronous orbit at 786 km. With a 290 km swath and 5-day revisit time, Sentinel-2 captures 13 VNIR and SWIR bands (443–2190 nm). Four bands (RGB and NIR) are at 10 m resolution, while the others are provided at 20 m or 60 m (ESA 2012).

PRISMA imagery was accessed through the ASI data portal, which allows users to filter available acquisitions by area of interest, date range, and cloud cover. In this work, PRISMA images with the least cloud cover (i.e., < 5%) were selected, as listed in Table 1 for each MENA city. Contrarily, Sentinel-2 images were collected programmatically as part of an automated data acquisition pipeline, using the GEE Python API, from the harmonised Sentinel-2 L2A image collection, which contains analysis-ready atmospherically corrected bottom-of-atmosphere (BOA) surface reflectance. Based on the acquisition date and image footprint of the PRISMA image, a spatiotemporal query was constructed in GEE. To ensure high-quality images were collected, a date range of two months before and after the PRISMA acquisition date was set. A bounding box constructed from the PRISMA image footprint was used as a spatial filter. Similarly to PRISMA, a filter was applied to include images with a cloud cover percentage of less than 5%. A final composite image was produced with the resulting Sentinel-2 image collection by calling the quality mosaic function in GEE, returning the highest quality pixels in a single image based on the QA60 band. Following Vavassori et al. (2024), eight bands were downloaded (2, 3, 4, 5, 6, 7, 8A, 11 and 12) at 20-meter resolution.

4.2 UCP Computation

Given the regional relevance of the present study, the geospatial datasets needed for computing the UCPs were selected in such a way that they cover large parts

Table 2 Geospatial datasets retrieved from Google Earth Engine (GEE) for computing the Urban Canopy Parameters (UCPs)

Dataset name	Pixel size	Years	UCPs
Open buildings temporal V1 [Link]	4 m	2016–2023	Building heights, building surface fraction Sky view factor
ESA WorldCover 10 m v200 [Link]	10 m	2021–2022	Impervious surface fraction
ALOS DSM: global 30 m v4.1 [Link]	30 m	2006–2011	Sky view factor
High resolution 1 m global canopy height maps [Link]	1 m	2018–2020	Tree canopy height

of the MENA region; when possible, global datasets were considered to enable the full computation reproducibility at the global level. Furthermore, only datasets that could be programmatically accessed through a unified API were considered in order to foster an automated workflow. Hence, geospatial data were retrieved through the GEE Data Catalogue, where many datasets relevant to urban studies are available and can be easily accessed through a Python API, allowing for spatiotemporal querying and geoprocessing directly within the GEE cloud environment.

Table 2 outlines the data employed for the UCP computation across the three cities. In particular, the Open Buildings Temporal V1 is the only dataset without global coverage, although it extends across large parts of Africa (including North Africa), Asia, and South America, making it a valuable building height dataset for developing countries. The geospatial datasets were queried according to the spatial and temporal extent of the PRISMA image acquisitions. Indeed, for datasets spanning over multiple years, the timestamp closest to the PRISMA acquisition date was selected.

For the computation of UCPs, the geoprocessing steps outlined in Vavassori et al. (2024) were implemented in Python, leveraging the GEE API, and carried out directly in the cloud environment. UCPs were eventually downloaded in the desired resolution (Appendix). It is relevant to highlight that SVF is the only parameter that was not computed through GEE due to the unavailability of dedicated functionality. For this reason, the SVF computation was implemented with the Relief Visualisation Toolbox (RVT) in Python using the *rvt.vis.sky_view_factor* function (Zakšek et al. 2011). Finally, the UCP layers were scaled to the 0–1 range with min-max normalisation, in such a way that they are comparable to the reflectance data provided by the satellite imagery.

4.3 Training Data Annotation and Quality Control

As the effectiveness of automated LCZ mapping largely depends on the quality of the TA used to train the ML classifier (Bechtel et al. 2015), a quality control step was added to the existing workflow. The goal was to limit as much as possible the arbitrariness of the TA selection, which demands an experienced visual interpretation of the satellite imagery.

TAs were developed by the authors to ensure consistency across the three cities. Training data annotation was carried out in QGIS (v3.42), leveraging an array of geospatial layers to achieve accurate and consistent labelling of the training polygons. In particular, the PRISMA image footprint was considered as a reference to guarantee that all training polygons are within its spatial extent. Secondly, the Sentinel-2 RGB image supported the visual interpretation of LCZs. As a further support for LCZ interpretation, high-resolution satellite imagery provided by Google, Airbus and Maxar technologies was added in QGIS. Wherever discrepancies between the two images were found, we relied on Sentinel-2 as it reflected the acquisition date of the PRISMA data.

While these data are effective in differentiating land cover classes, further information is needed to properly distinguish the built-up LCZs. Indeed, previous studies have highlighted the difficulty in accurately labelling built-up classes even when inspecting high-resolution satellite imagery (Betchel et al. 2017). To address this challenge, the building height layer was reclassified using QGIS to visualise three classes, supporting the distinction between low-rise (3–10 m), mid-rise (10–25 m) and high-rise (> 25 m) buildings (Stewart and Oke 2012). In addition, the tree canopy height dataset was used, aiding in discriminating between different degrees of vegetation density. Canopy height data was also beneficial in accurately distinguishing natural classes such as low vegetation (≤ 3 m) and trees (> 3 m). By following this annotation workflow, stricter control was maintained over the UCP ranges of the different LCZ classes in the TA, thereby enhancing the discriminatory capability of the RF classifier.

A series of quality control measures was integrated into the LCZ mapping pipeline to help users refine the TA before proceeding with model development. The LCZ training polygons were evaluated based on three criteria:

- geometric properties: following WUDAPT protocol recommendations, each polygon should have a minimum width and height of 200 m, a surface area greater than 0.04 km^2, and a shape ratio below 3 (Ching et al. 2018);
- class balance: the relative proportion of each LCZ class within the TA should be reasonably balanced; an optimal class distribution is estimated from the total number of classes in the training set, enabling users to identify which classes are over- or under-represented;
- UCP values: for each LCZ class, users can check whether the UCP values of the TA polygons fall within the acceptable ranges defined by Stewart and Oke (2012).

The quality control step is designed to be run iteratively during the annotation process, streamlining the creation of high-quality TA for LCZ mapping. After quality checks, the TA polygons were divided into training and testing subsets, with 70% allocated for training and 30% reserved for testing, before being rasterised for use in the RF algorithm. To capture the spatial variability of LCZ classes, TA polygons were distributed evenly across the study area. This approach reduces spatial dependence between neighbouring pixels and provides a more robust evaluation of the model's generalisation ability (Karasiak et al. 2022).

4.4 Model Training and Assessment

For model training and classification, satellite imagery was combined with the UCP layers following the methodology of Vavassori et al. (2024). Specifically, the first 10 PRISMA PCs (30 m) and the Sentinel-2 bands from B02 to B12 (20 m) were used. RF was selected as the classification algorithm, and its hyperparameters were tuned through cross-validation. To preserve class balance, a stratified repeated 5-fold cross-validation was employed.

Hyperparameter tuning was performed using a grid search implemented with the GridSearchCV class of the scikit-learn Python library, systematically testing combinations of parameters. The parameter grid included the number of estimators (*n_estimators*), the maximum number of features considered at each split (*max_features*), and the splitting criterion (*criterion*). The maximum number of leaf nodes (*max_leaf_nodes*) was also added as a regularisation parameter. As noted by Zlobin and Bazylevych (2025), setting this parameter too low may cause underfitting, while a too high value increases the risk of overfitting. To identify an optimal value, a validation curve of the F1-score was generated by iterating across values from 5 to 200. The optimal threshold was selected at the point where the F1-score stabilised, after which this parameter was added to the grid search.

The best combination of hyperparameters was then applied to predict LCZ classes in the test set, from which confusion matrices, accuracy metrics (OA, precision, recall, and F1-score), and feature importances were derived. Finally, the full image was classified into LCZs, and a 3×3 median filter was applied as a post-processing step to reduce noise in the output.

5 Results

5.1 Training Areas

A total of 711 LCZ TA polygons were created for the MENA region in this study. Figure 5 illustrates the proportion of total TA area per LCZ class for each city, while maps showing the spatial distribution of LCZ TAs are provided in the supplementary material. Figure 6 presents the statistical distribution of UCP values of TA data in each city, with reference to UCP ranges prescribed by Stewart and Oke (2012).

Each city contained more than 200 training polygons, with total TA areas ranging from 514.22 to 1381.03 km^2. The distribution and size of LCZs reflected the distinctive urban and natural environments of each city. In Khartoum, a defining feature is

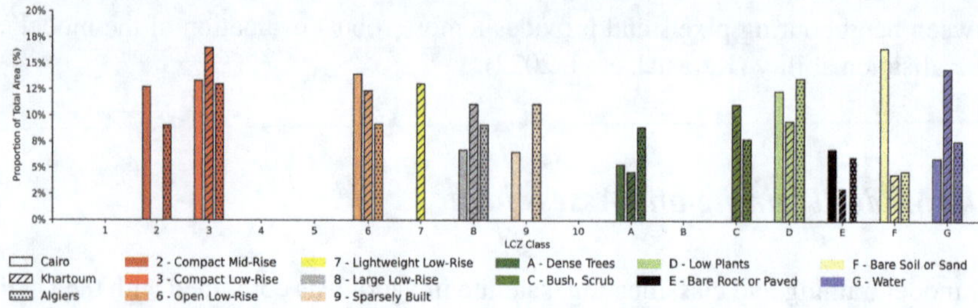

Fig. 5 Proportion of total area for each LCZ class within the training dataset of Cairo (plain), Khartoum (slashes) and Algiers (dots)

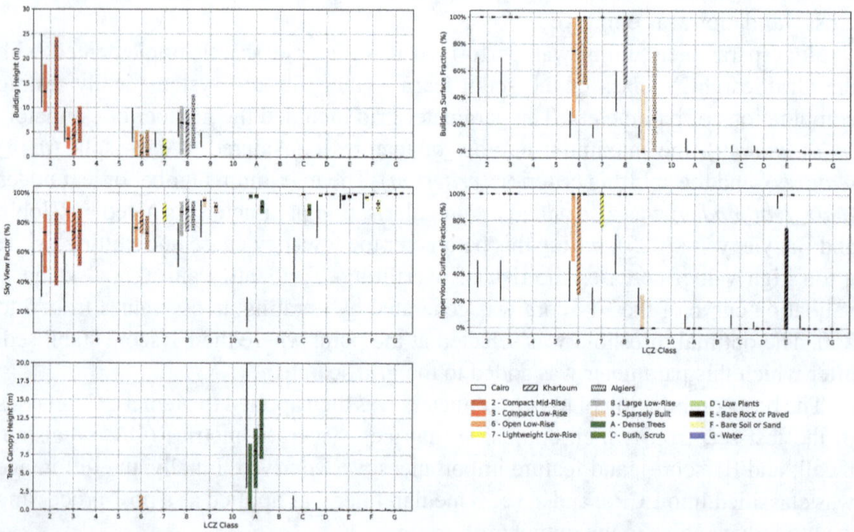

Fig. 6 Box plots indicating median, lower quartile and upper quartile of urban canopy parameters (UCPs) per LCZ class within TA datasets for Cairo (plain), Khartoum (slashed) and Algiers (dotted). The highest available resolution for each UCP layer was utilised: building heights (5 m), building surface fraction (20 m), sky view factor (5 m), impervious surface fraction (20 m), and tree canopy height (2 m). Vertical black lines indicate prescribed UCP ranges per LCZ according to Stewart and Oke (2012)

the prevalence of informal lightweight settlements (LCZ 7) on the urban periphery. Conversely, Khartoum lacks extensive zones of mid-rise buildings (> 200 m), which explains the omission of LCZ 2. Among natural classes, Bush/Scrub (LCZ 13) was present in both Khartoum and Algiers. Due to the spectral and morphological similarity of LCZ 13 and LCZ 14, this dataset provides an opportunity to assess the potential of hyperspectral imagery for distinguishing between these classes. Overall, a good correspondence between TA data and assigned UCP value ranges was obtained, although building surface and impervious surface fractions of the TA data showed discrepancies with the literature-based value ranges.

5.2 *Spectral Separability*

Figure 7 shows the Jeffries-Matusita (JM) distances between LCZs in each city, computed from pixel-wise spectral signatures of the TA data using Sentinel-2 and PRISMA bands. JM values close to 2 indicate high spectral separability, whereas values near 0 indicate poor separability. Multispectral and hyperspectral signature plots for all LCZ TAs are provided in the supplementary material.

Across the three cities, JM distances derived from Sentinel-2 ranged between 0.85 and 2.00, with lower separability generally observed among built classes. By contrast, hyperspectral data from PRISMA produced consistently higher separability, with distances ranging from 1.48 to 2.00. Notable improvements were found for LCZs 2 (Compact Mid-Rise), 3 (Compact Low-Rise), 6 (Open Low-Rise), and 7 (Lightweight Low-Rise). For instance, in Cairo, the JM distance between LCZ 2 and LCZ 3 increased by 0.81 when using hyperspectral data. In Khartoum, the distance between LCZ 2 and LCZ 7 rose from 1.11 with Sentinel-2 to 1.91 with PRISMA (an increase of 0.80). Similarly, in Algiers, the lowest separability observed with Sentinel-2 was between LCZ 3 and LCZ 6 (1.18), which improved to 1.97 with hyperspectral data.

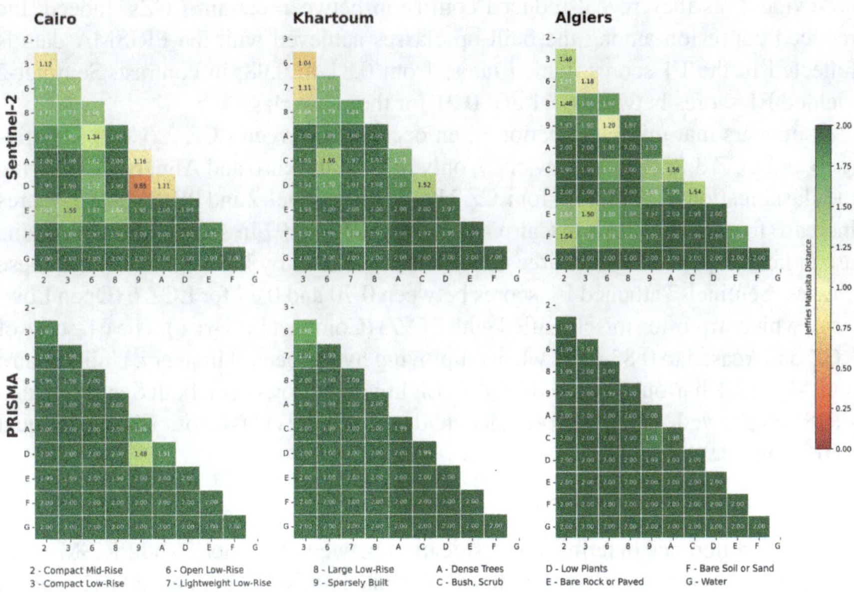

Fig. 7 Pairwise JM distances between pixel-wise spectral signatures of LCZ training polygons. The top row shows distances computed from Sentinel-2 multispectral bands, while the bottom row shows distances derived from 239 PRISMA hyperspectral bands. Columns correspond to the three MENA cities: Cairo (left), Khartoum (centre), and Algiers (right)

5.3 Model Assessment

Two RF classifiers were trained for each city, one using Sentinel-2 bands and the other using the first 10 PRISMA PCs; ancillary UCP data were included in both models. The optimal hyperparameters obtained through grid search cross-validation are reported in Appendix 5. Classification accuracies obtained for the testing samples are shown in Table 3, while Fig. 8 presents the confusion matrices, which show relative frequencies (i.e., the percentage of testing pixels). Feature importances were also assessed: permutation importances, expressed in terms of mean decrease in accuracy, are provided in Appendix 3, and the mean decrease in impurity values from the decision tree ensembles are given in Appendix 4. In the following, we discuss the classification results in detail.

Built-up classes

The classification results obtained with the PRISMA imagery, in combination with the ancillary UCP layers, attained OA of 0.94–0.96, corresponding to an increase of 0.10–0.12 compared to their Sentinel-2 counterparts. Regarding the F1-score, values were 0.93–0.96 for PRISMA and 0.70–0.84 for Sentinel-2. Examining the confusion matrices across all three cities, the advantages of hyperspectral data for LCZ mapping are evident, as they reveal reduced confusion between certain LCZs. Indeed, the reduced confusion among the built-up classes achieved with the PRISMA data is reflected in the F1-scores, which range from 0.81 to 0.98; in contrast, Sentinel-2 yielded F1-scores between 0.41 and 0.91 for the same classes.

It appears that misclassification often occurred between LCZ 2 (Compact Mid-rise) and LCZ 3 (Compact Low-rise), only present in Cairo and Algiers. Comparing the classification performance for LCZ 2 between Sentinel-2 and PRISMA, F1-scores increase from 0.73 to 0.98 in Cairo, and from 0.75 to 0.97 in Algiers. In general, the urban landscape among the three cities was dominated by low-rise LCZs. For these classes, Sentinel-2 attained F1-scores between 0.70 and 0.74 for LCZ 6 (Open Low-rise), which was often misclassified with LCZ 3 (Compact Low-rise). The F1-score of LCZ 6 increased to 0.85–0.97 when employing hyperspectral imagery. Unlike Cairo and Algiers, Khartoum featured LCZ 7 (Lightweight Low-rise); both Sentinel-2 and PRISMA showed a good performance in identifying this class, with F1-scores equal to 0.91 and 0.98, respectively.

Built-up and vegetation classes

A notable difference in terms of classification between Sentinel-2 and PRISMA can be observed in the quantity of built-up LCZ pixels misclassified as natural classes, particularly the LCZ 15 (Bare Rock or Paved) in the case of Cairo and Khartoum. In Algiers, built-up classes were sometimes misclassified as LCZ 16 (Bare Soil or Sand). PRISMA included many fewer of these cases, if at all, for some classes. LCZ 9 (Sparsely Built) appears to be a challenging class to properly identify, as it comprises approximately 20% of built surfaces, while it is dominated by vegetation. In Cairo, 2.5% of Sentinel-2 pixels labelled as LCZ 9 were in fact classified as LCZ

Table 3 Test performance of the random forest classifiers for Cairo, Khartoum, and Algiers, reported as precision, recall, F1-score, and overall accuracy

	Cairo—07 August 2021						Khartoum—26 October 2024						Algiers—13 May 2021					
	Sentinel-2			PRISMA			Sentinel-2			PRISMA			Sentinel-2			PRISMA		
	Precision	Recall	F1	Precision	Recall	F1	Precision	Recall	F1	Precision	Recall	F1	Precision	Recall	F1	Precision	Recall	F1
2—Compact mid-rise	0.75	0.75	0.75	0.98	0.98	0.98	–	–	–	–	–	–	0.71	0.76	0.73	0.97	0.98	0.97
3—Compact low-rise	0.54	0.69	0.61	0.95	0.94	0.94	0.78	0.74	0.76	0.93	0.85	0.89	0.82	0.88	0.85	0.97	0.99	0.98
5—Open low-rise	0.70	0.69	0.7	0.98	0.95	0.97	0.70	0.79	0.74	0.81	0.90	0.85	0.67	0.73	0.7	0.86	0.96	0.91
7—Lightweight low-rise	–	–	–	–	–	–	0.90	0.91	0.91	0.98	0.98	0.98	–	–	–	–	–	–
8—Large low-rise	0.49	0.40	0.44	0.76	0.87	0.81	0.82	0.79	0.81	0.95	0.97	0.96	0.82	0.6	0.69	0.98	0.88	0.93
9—Sparsely built	0.47	0.40	0.43	0.8	0.83	0.81	–	–	–	–	–	–	0.83	0.77	0.8	1.00	0.93	0.96
11—Dense trees	0.63	0.57	0.6	0.9	0.64	0.75	0.76	0.66	0.71	0.89	0.84	0.87	0.95	0.97	0.96	0.97	0.99	0.98
13—Bush, scrub	–	–	–	–	–	–	0.79	0.83	0.81	0.97	0.94	0.95	0.74	0.73	0.73	0.91	0.90	0.90
14—Low plants	0.70	0.69	0.69	0.91	0.95	0.93	0.67	0.60	0.64	0.89	0.98	0.93	0.95	0.98	0.96	0.99	1.00	0.99
15—Bare rock or paved	0.61	0.61	0.61	0.89	0.89	0.89	0.92	0.95	0.94	1.00	1.00	1.00	0.99	0.95	0.97	0.97	0.98	0.98
16—Bare soil or sand	0.96	0.96	0.96	0.99	0.97	0.98	0.96	0.98	0.97	1.00	1.00	1.00	0.76	0.91	0.83	0.83	0.96	0.89
17—Water	1.00	0.98	0.99	0.97	1.00	0.99	1.00	1.00	1.00	1.00	1.00	1.00	0.99	1.00	1.00	1.00	0.93	0.96
Overall accuracy	0.70	0.70	0.70	0.93	0.93	0.93	0.84	0.84	0.84	0.94	0.94	0.94	0.84	0.84	0.84	0.96	0.96	0.96

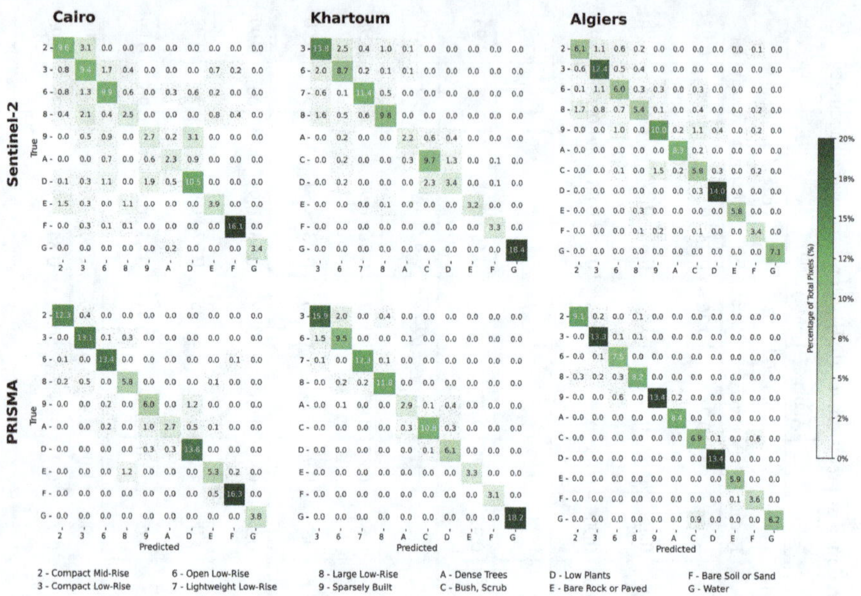

Fig. 8 Confusion matrices of LCZ classification using random forest models. The top row shows results from Sentinel-2 data, and the bottom row shows results from the first 10 PRISMA principal components. UCP data were included in all models. Columns correspond to the three MENA cities

14 (Low Plants), attaining an F1-score of 0.43. In comparison, only 0.3% of pixels of the same class were misclassified using PRISMA, resulting in an F1-score of 0.81. LCZ 9 in Algiers was most likely to be misclassified for LCZ 13 (Bush, Scrub) in the Sentinel-2 dataset, with an F1-score of 0.81. Finally, no misclassification was reported between LCZ 9 and LCZ 13 in Algiers using PRISMA, resulting in a higher performance with an F1-score of 0.96.

Vegetation classes

For the vegetation classes, such as LCZ 11 (Dense Trees), 13 (Bush, Scrub), and 14 (Low Plants), PRISMA generally performed better, with F1-scores ranging from 0.75 to 0.99. A reduced performance is evident for Sentinel-2, as proved by F1-scores in the 0.60–0.96 range. On the other hand, substantial increases in accuracy were evident for LCZ 14 using PRISMA compared to Sentinel-2, particularly in Cairo (+24%) and Khartoum (+29%). Only a 3% increase in F1-score was observed in Algiers for this class. Similar values were attained for LCZ 11. LCZ 13 was only present in Khartoum and Algiers, where PRISMA achieved 14% and 17% increases in F1-scores, respectively.

Bare and water classes

A good performance was observed for LCZ 15 (Bare Rock or Paved), 16 (Bare Soil or Sand), and 17 (Water) across all cities and classifiers. F1-scores ranged from

0.61 to 1.00 for Sentinel-2 and 0.89 to 1.00 for PRISMA. The highest performance improvement in F1-score, equal to 28%, was recorded for LCZ 15 in Cairo. The increase in accuracy using hyperspectral data for LCZs from 15 to 17 was less pronounced compared to the built-up and vegetation classes, between 0 and 6%. One notable change in performance was for LCZ 17 in Algiers, which decreased from 1.00 to 0.96 when comparing Sentinel-2 and PRISMA.

Feature Importances

When comparing permutation feature importances between the PRISMA and Sentinel-2 RF classifiers, clear differences were observed across cities. In Cairo, permuting PRISMA's PC02 led to a 23% mean decrease in classification accuracy, making it the most important feature for predicting LCZs. It was followed by building height (12.8%) and PC01 (4.6%). In Khartoum, SVF showed the highest importance (8.2%), closely followed by PRISMA's PC02 (7.4%) and PC03 (4.6%). Similarly, in Algiers, SVF (12.2%) and PC02 (11.7%) were the two most influential features, with tree canopy height also ranking highly (9.9%). When averaging the mean accuracy decrease across all cities, the three most important features overall were PC02, SVF, and building height.

5.4 *LCZ Maps*

LCZ classification across the full geographic extent of the PRISMA and Sentinel-2 images was implemented with the best classifiers for each city. A median filter with a 3 × 3 kernel size was applied to the output classification maps, reducing noise associated with pixel-wise classification methods. Figure 9 shows the classified LCZ maps in Cairo, Khartoum and Algiers for Sentinel-2 and PRISMA PCs. Spatial differences were computed by down-sampling the PRISMA LCZ map to match the Sentinel-2 LCZ map resolution (20 m) and comparing the LCZ classes; the results are shown in the supplementary materials.

On a global level, a strong correspondence was observed between the Sentinel-2 and PRISMA LCZ maps, although the RF classifier trained on PRISMA data produced visually smoother and more continuous LCZ regions. Pixel-level differences were distributed across the entire study area of all cities, although discrepancies were more pronounced among urban sites. Large regions of natural cover were generally in agreement when comparing the PRISMA and Sentinel-2 RF classifiers. Notable discrepancies were observed in specific areas. For instance, in Eastern Cairo, where the PRISMA RF model consistently classified LCZ-6 across the urban region, in line with annotator observations, whereas the Sentinel-2 map showed a highly mixed classification of LCZ-2 and LCZ-6. The mid-eastern region of Cairo is predominantly residential areas; however, the PRISMA LCZ map contains large sections of LCZ-8, indicating some confusion occurred between industrial and residential zones. In eastern Khartoum, the Sentinel-2 RF model misclassified LCZ-8

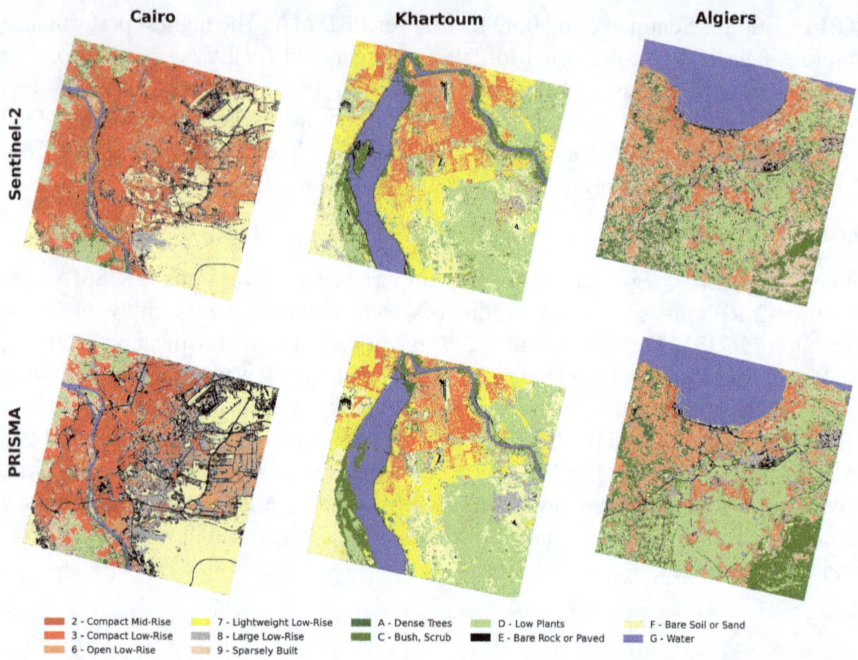

Fig. 9 LCZ maps produced by the best-performing random forest classifiers for Cairo (top), Khartoum (middle), and Algiers (bottom). Maps on the left were classified using Sentinel-2 data, while maps on the right were classified using the first 10 PRISMA principal components. A 3 $\times$ 3 median filter was applied to all outputs. Sentinel-2 classifications were masked to match the spatial extent of the PRISMA imagery

alongside other built classes, while PRISMA produced a more coherent classification. Substantial differences were found in the southwest region of the Algiers study area, where LCZ-9 was predicted over large regions that were considered natural scrubland (LCZ-C) by the annotator. The southeast region of Algiers featured a large degree of confusion between farmland (LCZ-D) and dense trees (LCZ-A) for the Sentinel-2 RF classifier. To a lesser extent, farmland in this region was confused for scrubland (LCZ-C) with the PRISMA classifier.

6 Discussion

Large-scale, automated LCZ mapping is an essential step toward making robust urban climate studies more accessible, particularly in developing countries such as those in the MENA region. The emergence of geospatial cloud-computing platforms such as GEE has significantly expanded the potential for automation; however, limitations remain regarding the availability of hyperspectral data with global coverage. Hyperspectral data, especially when combined with high-resolution UCPs, offer strong potential for enhancing LCZ mapping. The integration of these datasets and technologies in future efforts will greatly contribute to advancing the field.

Building on the work of Vavassori et al. (2024), we automated key stages in data collection and preprocessing using GEE, substantially reducing the time required for users. Additionally, we developed a series of quality-control tools to facilitate the production of reliable, high-quality TAs, a crucial step since the selection of TAS strongly influences both model training and accuracy assessment.

Overall, we observed substantial improvements in LCZ classification performance when employing the first 10 PRISMA PCs along with UCP data as opposed to Sentinel-2 multispectral imagery. Our results are consistent with Vavassori et al. (2024), who reported a 5% increase in OA and a 16% mean increase in class-wise accuracies. Across the three MENA cities studied, OA increased by an average of 11%, and class-wise F1-scores improved by 14% when RF models were trained with PRISMA PCs and UCP layers. The value of hyperspectral data was further underscored by the JM distance, which showed higher spectral separability for built-up TAs when PRISMA bands were used. This was reflected in a 22% average improvement in F1-scores for built LCZs, while natural classes showed a mean increase of 11% in F1-scores when using PRISMA combined with UCP layers. These findings further support the conclusions of Vavassori et al. (2024).

LCZ maps obtained with PRISMA LCZ classifiers contained less noise, resulting in cleaner, continuous LCZ areas throughout the study area. Although PRISMA's increased spectral resolution may aid in reducing this noise, another factor is the large pixel size (30 m), which reduces pixel-wise variability in UCP data. It is worth noting that pixel-wise discrepancies between output PRISMA and Sentinel-2 LCZ maps largely occurred in urbanised regions, reflecting the challenge of classifying built LCZs in dense urban environments (Bechtel et al. 2019).

This study also investigated two underrepresented LCZs (LCZ-7 and LCZ-C) that are sparsely represented in the globally curated TA dataset of Demuzere et al. (2022). Khartoum contained both classes, while Algiers only included LCZ-C. In particular, informal settlements and slums (LCZ-7) constitute a large proportion of urban areas in developing countries, including those in MENA (Diab et al. 2020). Recent LCZ mapping efforts highlighted the challenge of mapping LCZ-7, consisting of lightweight materials that result in a mixed spectral signature (Demuzere et al. 2022). Increased focus for this LCZ is therefore essential to facilitating regional LCZ mapping. While Sentinel-2 RF classifiers achieved strong performance for LCZ-7 (F1: 0.91), other built classes were less accurate (F1 < 0.8). By contrast, the F1-score of LCZ-7 increased by 7% when using PRISMA PCs compared to Sentinel-2, while other built classes attained scores above 0.84. These findings highlight the enhanced discriminatory power that hyperspectral imagery offers for built LCZ classification. Furthermore, the strong performance in classifying LCZ-7 emphasises the value of hyperspectral data for MENA-wide LCZ mapping. LCZ-C closely resembles LCZ-D, making it another challenging class. Both classes are similar in terms of UCPs (Fig. 6); therefore, spectral data are critical for distinguishing these classes. We observed notable increases in JM distances between LCZ-C and LCZ-D by utilising PRIMSA imagery (Fig. 7). LCZ-C exhibited a notable improvement in classification performance when using PRISMA data in both Khartoum and Algiers, with the increase amounting to 14% and 17%, respectively. Sentinel-2 RF models showed a higher degree of confusion between LCZ-C and LCZs 9 and D, resulting in a lower classification accuracy for these classes. Contrarily, PRISMA's higher spectral separability produced F1-scores of 0.90 or higher for LCZs C, D, and 9 in both cities, demonstrating that hyperspectral data supports more accurate classification of diverse natural LCZs.

Permutation feature importance across the three MENA cities indicated that the contribution of PRISMA features was not consistent across urban regions. Nevertheless, PRISMA's first three PCs, particularly PC02, were consistently influential for RF classification performance across all three cities, corroborating the findings of Vavassori et al. (2024). It is worth noting that while PRISMA PCs had the strongest impact on model performance in Cairo and Khartoum, UCPs were more important in Algiers. With extensive vegetation cover compared to the other cities, tree canopy height was the third most important feature in Algiers. SVF remained a critical UCP layer in LCZ classification across all three MENA cities. Generally, these results are in quite good agreement with those attained in Milan. The variability in feature importance between cities underscores the diversity of urban morphology in MENA and the RF algorithm's adaptability in feature selection. In Cairo, for example, PC02 was twice as important as building height, whereas in Algiers, PC02 and SVF were equally significant. These results highlight the importance of considering ecoregional differences (Demuzere et al. 2019b): cities such as Algiers (temperate) and Cairo (arid) may require tailored TA subsets or even unsupervised clustering approaches to group urban areas with similar morphologies and spectral characteristics, thereby improving classification performance.

Some limitations must also be acknowledged. First, we did not assess seasonal variability in LCZ classification, which is an important consideration for classes such as cropland (LCZ-D). Future work should explicitly evaluate the role of seasonality in hyperspectral-based LCZ mapping. Second, the lack of high-resolution, open-source building height data remains a constraint for the MENA region. Although Google's Open Buildings Temporal V1 dataset is valuable, it does not cover several major countries in the Levantine and Gulf region. The recently released Global Building Atlas (Zhu et al. 2025) may help address this limitation in regional workflows. Third, PRISMA hyperspectral data currently requires manual downloads through a user interface, which is a common challenge for hyperspectral satellites, thus limiting automation. However, forthcoming missions such as the Copernicus Hyperspectral Imaging Mission for the Environment (CHIME), scheduled for launch in 2028, will provide API access to hyperspectral data (ESA 2021). The growing number of publicly funded hyperspectral missions will likely enable planetary-scale, accurate LCZ mapping in the near future.

Another important consideration is transferability. Given that curated TA datasets are often limited to a subset of cities, exploring the transferability of hyperspectral data is critical for upscaling LCZ maps in MENA. Demuzere et al. (2019b) demonstrated that LCZs can be accurately classified using training data from nearby cities, provided they fall within the same ecoregion. This finding reflects our observation that feature importance varied substantially between the three cities studied, which have different urban morphologies and natural landscapes. Future research should test the hybrid approach proposed here, combining hyperspectral imagery with UCPs, for transferring TA data across cities.

Finally, recent advancements in deep learning offer new opportunities. Vision transformer-based foundation models can produce local token embeddings that capture spatial-spectral relationships in hyperspectral data, enabling transfer learning in classification tasks (Wang et al. 2025). As a potential alternative to PCA, token embeddings could provide a more powerful feature representation for regional LCZ mapping, making this a compelling direction for future research.

7 Conclusions

This study evaluated a hybrid LCZ mapping method that integrates PRISMA hyperspectral imagery with ancillary UCP layers for three MENA cities, i.e. Cairo, Khartoum and Algiers. Compared with RF classifiers trained on Sentinel-2 multispectral data, PRISMA PCs yielded substantial improvements in terms of classification accuracy. OA increased by 10% for Khartoum, 12% for Algiers and 23% for Cairo, attaining values of 93% or higher in all three cities. These findings clearly demonstrate the value of hyperspectral data for enhancing LCZ mapping performance in MENA, where urban areas are often characterised by heterogeneous morphologies and complex natural landscapes.

The analysis of permutation feature importances revealed that, while the first three PRISMA PCs consistently contributed to classification performance, the relative influence of UCPs varied across the cities. SVF was found to be a crucial UCP layer in supporting LCZ classification across all three case studies, along with building height and tree canopy height. The variability observed suggests that a geographically weighted model training, which assigns greater weight to TAs near the target city, may further improve the classification outcomes. Alternatively, spatially subsetting the training datasets according to ecoregions, as recommended by Demuzere et al. (2019a, b), offers another promising strategy for building a strong, region-specific classifier. Future development may investigate the integration of local token embeddings and Vision Transformer models as an alternative to PCA, which could also support generalisation across larger regions such as MENA. In fact, these methods could enable spectral-spatial clustering of TAs, thereby helping produce regionalised LCZ classifiers that balance accuracy and scalability.

To encourage reproducibility and future improvements of the proposed pipeline, the source code was made available on GitHub (https://github.com/medo-younes/lcz-py). The workflow requires only an active GEE account and a valid PRISMA license.

Appendices

Appendix 1. Detailed steps of the LCZ mapping pipeline.

#	Step	#	Action
1	PRISMA image acquisition	1.1	PRISMA data portal registration
		1.2	Query images by area of interest, date range, and cloud cover percentage (< 5%)
		1.3	Download the level L2D product
2	Sentinel-2 image acquisition	2.1	Filter harmonised Sentinel-2 L2A image collection with PRISMA metadata: bounding box, date range (± 2 months), and cloud cover percentage (< 5%)
		2.2	Select Sentinel-2 bands (B02, B03, B04, B05, B06, B07, B8A, B11, and B12)
		2.3	Build a quality mosaic composite with the QA60 band
		2.4	Download composite image at 20 m and 30 m resolution
3	GIS data acquisition	3.1	Filter the GIS layers with the PRISMA bounding box and acquisition date
		3.2	Clip GIS layers to the PRISMA bounding box
4	UCP computation	4.1	**Building Heights**: average resampling of the 4 m building heights to 20/30 m
		4.2	**Building Surface Fraction**: computation of the building binary mask (building height > 0); average resampling of the building binary mask to 20/30 m
		4.3	**Impervious Surface Fraction**: computation of the impervious binary mask using the built-up class value (50); average resampling of the impervious binary mask to 20/30 m
		4.4	**Tree Canopy Height**: average resampling to 20/30 m
		4.5	**Sky View Factor** (from DSM, 30 m): average resampling to 20 m resolution (GEE); computation of SVF using DSM at 20/30 m resolution using *rvt.vis.sky_view_factor*
		4.6	**Sky View Factor** (from Building Heights, 5 m): computation of SVF using building heights at 5 m using *rvt.vis.sky_view_factor*; average resampling of 5 m SVF layer to 20/30 m
		4.7	**Sky View Factor** (combined): stack the two SVF rasters (obtained from Building Heights and DSM); computation of the minimum of the raster stack
5	Sentinel-2 pre processing	5.1	Merge and stack Sentinel-2 bands
		5.2	Clip to PRISMA image bounding box
6	PRISMA image co registration and PCA	6.1	Coregister PRISMA image with Sentinel-2 image using the GeFolki algorithm

(continued)

(continued)

#	Step	#	Action
		6.2	Export full hyperspectral image as GeoTIFF (239 bands)
		6.3	Export RGBN image as GeoTIFF
		6.4	Export 5 m panchromatic band as GeoTIFF
		6.5	Principal component analysis (PCA) on coregistered PRISMA image (n_components = 10)
7	UCP layer pre processing	7.1	Merge UCP layer tiles (20 and 30 m)
		7.2	Reproject match 20 m UCP layers to the Sentinel-2 image
		7.3	Reproject match 30 m UCP layers to the PRISMA image
8	Training area annotation	8.1	Create new QGIS project
		8.2	Load data: Sentinel-2 image, PRISMA panchromatic band, building heights and tree canopy height. Add google satellite XYZ tile
		8.3	Reclassify building heights layer to 3 classes (in Style settings)
		8.4	Reclassify tree canopy height layer to 3 classes (in Style settings)
		8.5	Create a new GeoJSON layer, create a numeric column named "class"
		8.6	Create LCZ polygons with class labels
9	Training area quality control	9.1	Run LCZ quality control python notebook
		9.2	Run spectral separability python notebook
		9.3	Review QC results and revise training areas accordingly
10	Prepare train/test sets	10.1	Split training polygons 70/30, stratified by class and polygon area (small/large)
		10.2	Rasterise train/test sets at 20 and 30 m
11	RF hyperparameter tuning	11.1	Validation curve on *max_leaf_nodes*, select optimal parameter
		11.2	Generate parameter grid based on optimal *max_leaf_nodes*
		11.3	Grid search with stratified k-fold cross-validation (k = 5)
	RF classifier assessment	11.4	Predict LCZs in the test set with the best RF classifier
		11.5	Compute performance metrics, confusion matrix, feature importances and permutation importances
12	LCZ classification	12.1	Classify LCZs in the full PRISMA + UCPs dataset
		12.2	Classify LCZs in the full Sentinel-2 + UCPs dataset

Appendix 2. UCP layers for Cairo, Khartoum and Algiers. Values are rescaled to the 0–1 range.

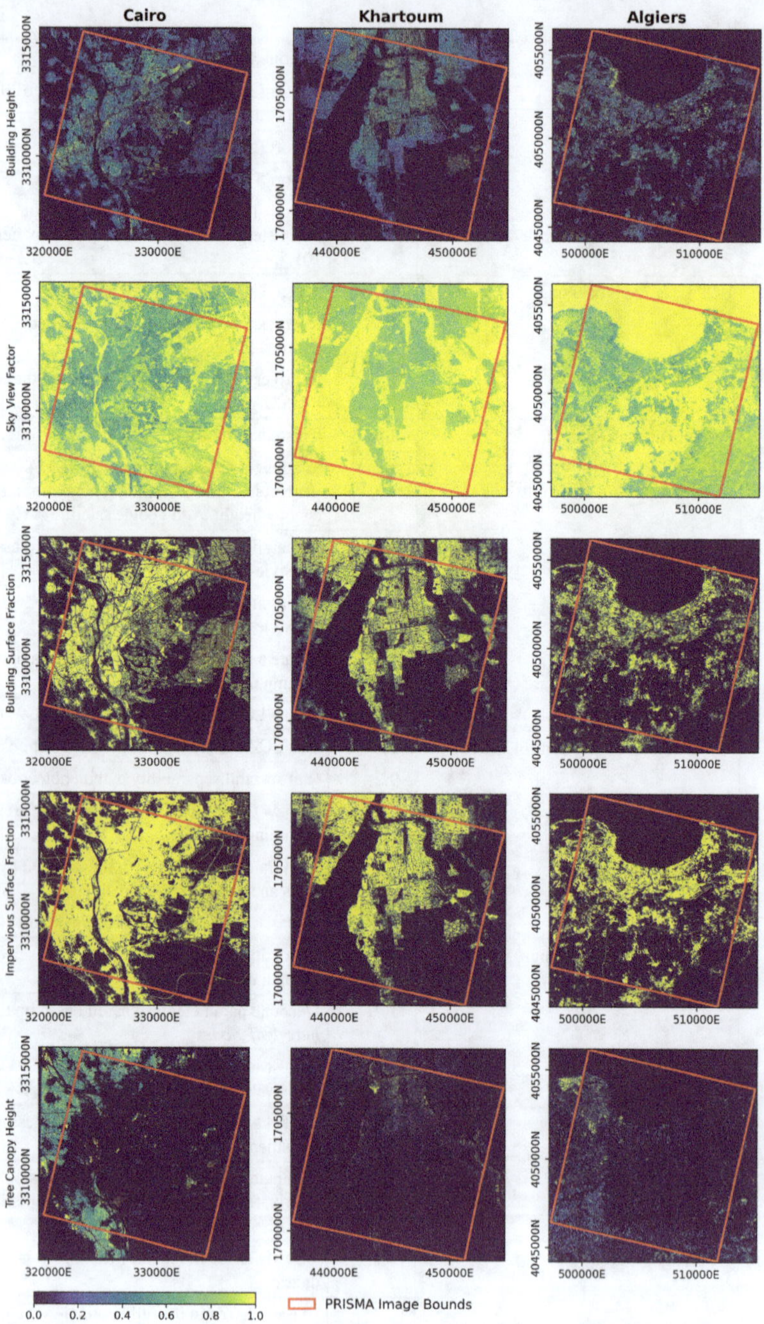

Appendix 3. Permutation feature importances from the Random Forest classifier for Cairo, Khartoum, and Algiers. Features include the 10 principal components (PCs) of PRISMA hyperspectral data, Sentinel-2 spectral bands, and Urban Canopy

Parameter layers. Importances are expressed as the mean decrease in accuracy across the decision tree ensemble. Black bars indicate the average importance of each feature across all three cities.

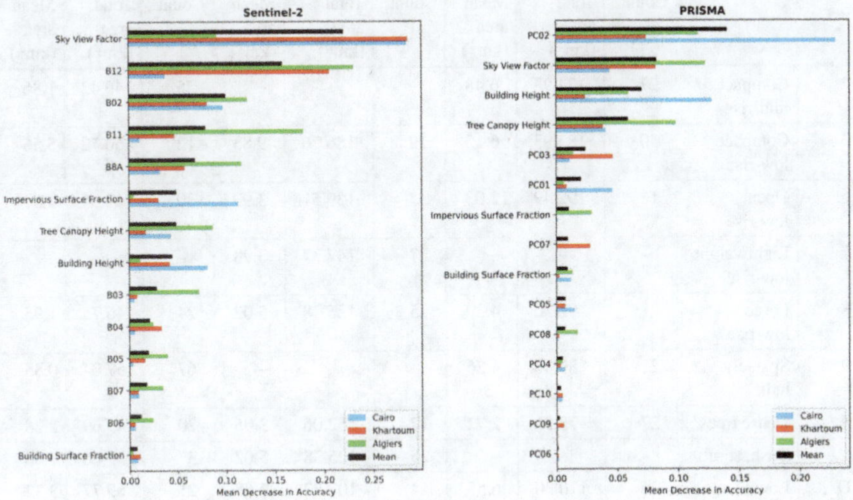

Appendix 4. Feature importances, expressed as mean decrease in impurity from the Random Forest classifier. Features include the 10 principal components (PCs) of PRISMA hyperspectral data, Sentinel-2 spectral bands, and Urban Canopy Parameter layers. Importances are measured as the mean decrease in impurity of the feature across the decision tree ensemble. Black bars indicate the average importance of each feature across all three cities.

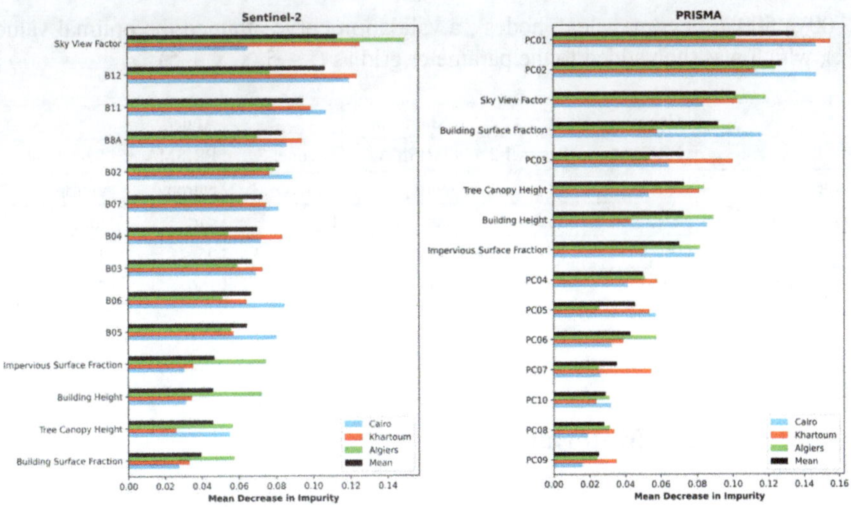

Appendix 5. Best Random Forest parameters resulting from grid search: 'criterion' ('entropy', 'gini'), 'max_features' ('auto', 'log2', 'sqrt'), 'n_estimators' ('150',

Table 1 Count, total area and mean area of training polygons across all LCZ classes in Cairo, Khartoum and Algiers

LCZ	LCZ name	Cairo			Khartoum			Algiers		
		Count	Total area (km^2)	Mean area (km^2)	Count	Total area (km^2)	Mean area (km^2)	Count	Total area (km^2)	Mean area (km^2)
2	Compact mid-rise	27	174.98	6.48	–	–	–	25	46.42	1.86
3	Compact low-rise	30	183.93	6.13	19	186.76	9.83	12	66.72	5.56
6	Open low-rise	16	192.49	12.03	35	139.51	3.99	30	47.05	1.57
7	Lightweight low-rise	–	–	–	37	147.37	3.98	–	–	–
8	Large low-rise	21	92.20	4.39	25	125.38	5.02	24	46.75	1.95
9	Sparsely built	21	89.43	4.26	–	–	–	67	57.04	0.85
A	Dense trees	27	73.31	2.72	17	52.06	3.06	20	45.69	2.28
C	Bush, scrub	–	–	–	25	125.38	5.02	37	39.86	1.08
D	Low plants	26	170.30	6.55	14	107.53	7.68	21	69.77	3.32
E	Bare rock or paved	24	93.94	3.91	14	35.51	2.54	18	31.64	1.76
F	Bare soil or sand	16	227.93	14.25	14	49.82	3.56	21	24.20	1.15
G	Water	8	82.52	10.31	9	164.40	18.27	11	39.08	3.55
	Total	216	1381.03	7.10	209	1133.72	6.30	286	514.22	2.27

'200', '500'). For 'max_leaf_nodes', a validation curve attained the optimal value (x), which was then added to the parameter grid as (x – 5, x, x + 5).

	Cairo		Khartoum		Algiers	
	PRISMA	Sentinel-2	PRISMA	Sentinel-2	PRISMA	Sentinel-2
criterion	entropy	gini	entropy	gini	entropy	entropy
max_features	log2	log2	sqrt	log2	sqrt	log2
max_leaf_nodes	90	190	85	110	95	125
n_estimators	150	500	500	200	150	200

Supplementary Materials

Local Climate Zone Mapping in MENA: A Streamlined Remote Sensing and GIS Method with Hyperspectral Imagery

See (Figs. 10, 11 and 12).

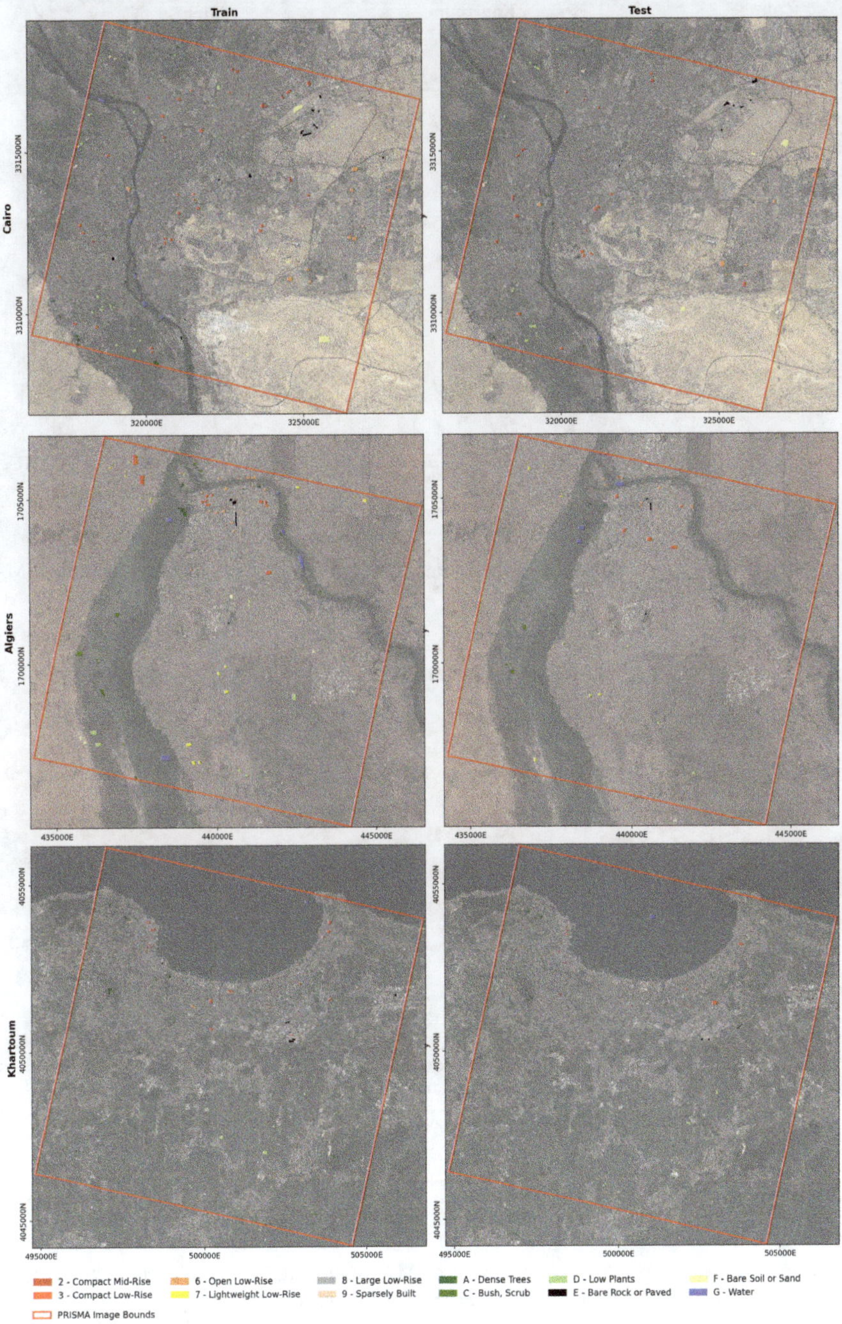

Fig. 10 Distribution of training and testing polygons in Cairo, Khartoum and Algiers. PRISMA image bounds are outlined in red

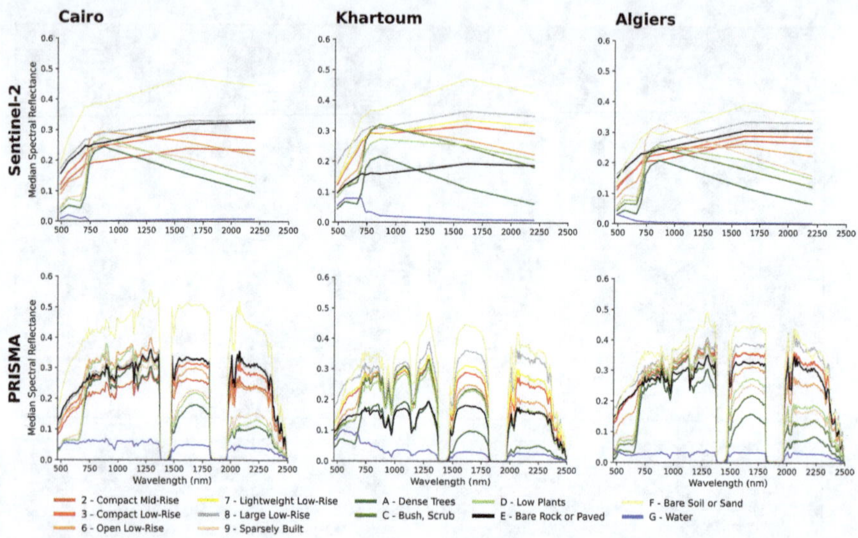

Fig. 11 Median spectral signatures of LCZ training polygons for Sentinel-2 (left) and PRISMA (right)

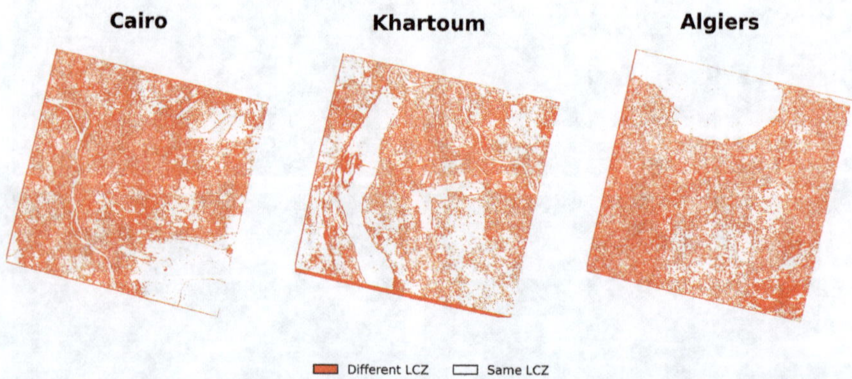

Fig. 12 Spatial differences in LCZ classification between PRISMA and Sentinel-2 RF classifiers. Red pixels indicate that different LCZ classes were predicted between the models, while white pixels indicate the models predicted the same LCZ class

References

Agenzia Spaziale Italiana (ASI) (2009) The PRISMA mission. Italian space agency. https://www.asi.it/wp-content/uploads/2020/03/The-PRISMA-mission.pdf

Ali AAM, Hagishima A, Abdel-Kader M, Hammad H (2013) Vernacular and modern building: estimating the CO_2 emissions from the building materials in Egypt. Building simulation Cairo 2013—towards sustainable & green life, Cairo, June 23rd—24th topic name: climate change & architecture. https://d1wqtxts1xzle7.cloudfront.net/32337003/IBPSA_Egypt-libre.pdf?1391593403=&response-content-disposition=inline%3B+filename%3DVernacular_and_Modern_Building_Estimatin.pdf&Expires=1759659107&Signature=YWhgQBQa8Oegvaek1HsrWgie6aHA72P8k2nsBaPqs-HHimBvMzcZZV7LnVxP9UXGtnTi8hDR7TydMw5KbmNkOzEvy1-ijSkeDxKU68oFlKorA-m3k4AYqn78JK2M4uEotZubEXtY4OmUIRuUzMIelY9-hRfu7blxly81IdK55g1YgovGIiu60xbNaAllMAjiOGToCuuTFCqxplVFm8PnLJCQREWSFEhwaRg2P-cdw4GPPFYbjftjIW013zs9hhKU3GECXFWRWOmkbOU-QiyBeUWwFD0twBLtl3QumVcHhOB30QFB439-Wjd3OtPa8vIjuvNsTBYaJiBC5v6H6Oz1qQ__&Key-Pair-Id=APKAJLOHF5GGSLRBV4ZA

Angel S, Blei AM, Parent J, Lamson-Hall P, Sánchez NG, Civco DL, Lei RQ, Thom K (2016) Atlas of urban expansion. Atlas of urban expansion—Algiers. http://atlasofurbanexpansion.org/cities/view/Algiers

Bechtel B, Daneke C (2012) Classification of local climate zones based on multiple earth observation data. IEEE J Sel Top Appl Earth Obs Remote Sens 5(4):1191–1202. https://doi.org/10.1109/JSTARS.2012.2189873

Bechtel B, Alexander PJ, Böhner J, Ching J, Conrad O, Feddema J, Mills G, See L, Stewart I (2015) Mapping local climate zones for a worldwide database of the form and function of cities. ISPRS Int J Geo Inf 4(1):199–219. https://doi.org/10.3390/ijgi4010199

Bechtel B, Alexander PJ, Beck C, Böhner J, Brousse O, Ching J, Demuzere M, Fonte C, Gál T, Hidalgo J, Hoffmann P, Middel A, Mills G, Ren C, See L, Sismanidis P, Verdonck M-L, Xu G, Xu Y (2019) Generating WUDAPT level 0 data—current status of production and evaluation. Urban Climate 27:24–45. https://doi.org/10.1016/j.uclim.2018.10.001

Bechtel B, Demuzere M, Sismanidis P, Fenner D, Brousse O, Beck C, Van Coillie F, Conrad O, Keramitsoglou I, Middel A, Mills G, Niyogi D, Otto M, See L, Verdonck M-L (2017) Quality of crowdsourced data on urban morphology—the human influence experiment (HUMINEX). Urban Sci 1(2). https://doi.org/10.3390/urbansci1020015

Boanada-Fuchs A, Kuffer M, Samper J (2024) A global estimate of the size and location of informal settlements. Urban Sci 8(1):18. https://doi.org/10.3390/urbansci8010018

Bouchachi B, Zhong Y (2017) Monitoring urban land cover/land use change in algiers city using landsat images (1987–2016). Int Arch Photogramm, Remote Sens Spat Inf Sci XLII-2/W7:1083–1090. https://doi.org/10.5194/isprs-archives-XLII-2-W7-1083-2017

Cai Z, Tang Y, Chen K, Han G (2019) Assessing the heat vulnerability of different local climate zones in the old areas of a chinese megacity. Sustainability 11(7):2032. https://doi.org/10.3390/su11072032

Capizzi E, Vavassori A, Cedeno R (2024) Processing, co-registration, and classification in LCZ of PRISMA and Sentinel-2 imagery (Version v1.0). Zenodo. https://doi.org/10.5281/ZENODO.13884806

Ching J, Mills G, Bechtel B, See L, Feddema J, Wang X, Ren C, Brousse O, Martilli A, Neophytou M, Mouzourides P, Stewart I, Hanna A, Ng E, Foley M, Alexander P, Aliaga D, Niyogi D, Shreevastava A, Theeuwes N (2018) WUDAPT: an urban weather. Clim Environ Model Infrastruct Anthropocene. https://doi.org/10.1175/BAMS-D-16-0236.1

Christidis N, Mitchell D, Stott PA (2023) Rapidly increasing likelihood of exceeding 50 °C in parts of the Mediterranean and the Middle East due to human influence. Npj Clim Atmos Sci 6(1):45. https://doi.org/10.1038/s41612-023-00377-4

Demuzere M, Bechtel B, Middel A, Mills G (2019a) Mapping Europe into local climate zones. PLoS ONE 14(4):1–27. https://doi.org/10.1371/journal.pone.0214474

Demuzere M, Bechtel B, Mills G (2019b) Global transferability of local climate zone models. Urban Climate 27:46–63. https://doi.org/10.1016/j.uclim.2018.11.001

Demuzere M, Hankey S, Mills G, Zhang W, Lu T, Bechtel B (2020) Combining expert and crowd-sourced training data to map urban form and functions for the continental US. Scientific Data 7(1):264. https://doi.org/10.1038/s41597-020-00605-z

Demuzere M, Kittner J, Martilli A, Mills G, Moede C, Stewart ID, van Vliet J, Bechtel B (2022) A global map of local climate zones to support earth system modelling and urban-scale environmental science. Earth Syst Sci Data 14(8):3835–3873. https://doi.org/10.5194/essd-14-3835-2022

Demuzere M, Hankey S, Mills G, Zhang W, Lu T, Bechtel B (2021a) CONUS-wide LCZ map and training areas (p. 432705665 Bytes). figshare. https://doi.org/10.6084/M9.FIGSHARE.11416950

Demuzere M, Kittner J, Bechtel B (2021b) LCZ generator: a web application to create local climate zone maps. Front Environ Sci 9–2021. https://doi.org/10.3389/fenvs.2021.637455

Diab Y, El Shaarawy B, Yousry S (2020) Informal settlements in the Arab region. United Nations Human Settlements Programme. https://unhabitat.org/sites/default/files/2020/12/regional_is_report_final_dec_2020.pdf

El Araby M (2002) Urban growth and environmental degradation. Cities 19(6):389–400. https://doi.org/10.1016/S0264-2751(02)00069-0

Elagib NA (2011) Evolution of urban heat island in Khartoum. Int J Climatol 31(9):1377–1388. https://doi.org/10.1002/joc.2159

European Space Agency (ESA) (2021) Copernicus hyperspectral imaging mission for the environment: mission requirements document (ESA-EOPSM-CHIM-MRD-3216,Version 3.0). https://sentiwiki.copernicus.eu/__attachments/1678167/ESA-EOPSM-CHIM-MRD-3216%20-%20CopEX%20Hyperspectral%20Imaging%20Mission%20for%20the%20Environment%202021%20-%203.0.pdf

European Space Agency (ESA) (2012) Sentinel-2: ESA's optical high-resolution mission for GMES operational services (ESA SP-1322/2). ESA Communications. https://sentinels.copernicus.eu/documents/247904/349490/S2_SP-1322_2.pdf

Estacio I, Babaan J, Pecson NJ, Blanco AC, Escoto JE, Alcantara CK (2019) GIS-based mapping of local climate zones using fuzzy logic and cellular automata. Int Arch Photogramm, Remote Sens Spat Inf Sci XLII-4-W19:199–206. ISPRS TC IV PhilGEOS x GeoAdvances 2019 "Geomatics and Data Science: Towards Adaptive Management in a Changing World" (Volume XLII-4/W19)—14–15 November 2019, Manila, Philippines. https://doi.org/10.5194/isprs-archives-XLII-4-W19-199-2019

Farsi M, Lazzali F, Ait-Méziane Y (2002) World housing encyclopedia report. Earthquake Engineering Research Institute. https://www.eeri.org/lfe/pdf/algeria_stone_masonry.pdf

Geletič J, Lehnert M (2016) GIS-based delineation of local climate zones: the case of medium-sized central European cities. Morav Geogr Rep 24(3):2–12. https://doi.org/10.1515/mgr-2016-0012

Hamid GM, Bahreldin IZ (2013) Environmental sustainability in greater Khartoum between natural assets and human interventions. Int J Sustain Build Technol Urban Dev 4(2):100–110. https://doi.org/10.1080/2093761X.2013.801804

Han J, Mo N, Cai J, Ouyang L, Liu Z (2024) Advancing the local climate zones framework: a critical review of methodological progress, persisting challenges, and future research prospects. Hum Soc Sci Commun 11(1):538. https://doi.org/10.1057/s41599-024-03072-8

Huang F, Jiang S, Zhan W, Bechtel B, Liu Z, Demuzere M, Huang Y, Xu Y, Ma L, Xia W, Quan J, Jiang L, Lai J, Wang C, Kong F, Du H, Miao S, Chen Y, Chen J (2023) Mapping local climate zones for cities: a large review. Remote Sens Environ 292:113573. https://doi.org/10.1016/j.rse.2023.113573

IEA (2023) National climate resilience assessment for Egypt. IEA. https://www.iea.org/reports/national-climate-resilience-assessment-for-egypt

Ismail Ahmed Ali S, Szalay Z (2019) Towards developing a building typology for Sudan. IOP Conf Ser: Earth Environ Sci 323(1):012012. https://doi.org/10.1088/1755-1315/323/1/012012

Japan International Cooperation Agency (JICA) (2005) Final report on the study for the master plan on urban development in the Wilaya of Algiers, Democratic and Popular Republic of Algeria, vol 2, Chapter 2. Japan International Cooperation Agency (JICA). https://openjicareport.jica.go.jp/pdf/12084554_01.pdf

Karasiak N, Dejoux J-F, Monteil C, Sheeren D (2022) Spatial dependence between training and test sets: another pitfall of classification accuracy assessment in remote sensing. Mach Learn 111(7):2715–2740. https://doi.org/10.1007/s10994-021-05972-1

Kharraz JE, El-Sadek A, Ghaffour N, Mino E (2012) Water scarcity and drought in WANA countries. Procedia Eng 33:14–29. https://doi.org/10.1016/j.proeng.2012.01.1172

Kottek M, Grieser J, Beck C, Rudolf B, Rubel F (2006) World map of the Köppen-Geiger climate classification updated. Meteorol Z 15(3):259–263. https://doi.org/10.1127/0941-2948/2006/0130

Nuruzzaman Md (2015) Urban heat island: causes, effects and mitigation measures - a review. Intern J Environ Monitor Anal 3(2):67. https://doi.org/10.11648/j.ijema.20150302.15

Oke TR (1982) The energetic basis of the urban heat island. Q J R Meteorol Soc 108(455):1–24. https://doi.org/10.1002/qj.49710845502

Oke TR (1989) The micrometeorology of the urban forest. Philos Trans R Soc London B Biol Sci 324(1223):335–349. https://doi.org/10.1098/rstb.1989.0051

Oke TR (2004) Initial guidance to obtain representative meteorological observations at urban sites (Instruments and Observing Methods Report No. 81 No. WMO/TD-No. 1250). World Meteorological Organization

Quan SJ, Bansal P (2021) A systematic review of GIS-based local climate zone mapping studies. Build Environ 196:107791. https://doi.org/10.1016/j.buildenv.2021.107791

Salem M, Tsurusaki N, Xu X, Xu G (2024) Revealing the transformation of spatial structure of greater Cairo: insights from satellite imagery and geospatial metrics. J Urban Manag 13(4):565–579. https://doi.org/10.1016/j.jum.2024.10.008

Shukla PR, Skea J, Reisinger AR, IPCC (eds) (2022) Climate change 2022: mitigation of climate change. IPCC

Stewart ID, Oke TR (2010) Thermal differentiation of local climate zones using temperature observations from urban and rural field sites. Preprints, 9th Symposium, on urban environment, keystone, CO, 2–6

Stewart ID, Oke TR, Krayenhoff ES (2014) Evaluation of the "local climate zone" scheme using temperature observations and model simulations. EBSCOhost. https://doi.org/10.1002/joc.3746

Stewart ID, Oke TR (2012) Local climate zones for urban temperature studies. Bull Am Meteor Soc 93(12):1879–1900

Sutton K, Fahmi W (2001) Cairo's urban growth and strategic master plans in the light of Egypt's 1996 population census results. Cities 18(3):135–149. https://doi.org/10.1016/S0264-2751(01)00006-3

Team TCLNR (2020) Investigating the urban land nexus and inclusive urbanisation in Dares Salaam, Mwanza, and Khartoum. https://opendocs.ids.ac.uk/articles/report/Investigating_the_Urban_Land_Nexus_and_Inclusive_Urbanisation_in_Dar_es_Salaam_Mwanza_and_Khartoum/26433388

UN-Habitat (2024) World cities report 2024. UN-Habitat. https://unhabitat.org/sites/default/files/2024/11/wcr2024_-_full_report.pdf

United Nations (2019) World urbanization prospects: the 2018 revision. United Nations

Vavassori A, Oxoli D, Venuti G, Brovelli MA, Siciliani De Cumis M, Sacco P, Tapete D (2024) A combined remote sensing and GIS-based method for local climate zone mapping using PRISMA and Sentinel-2 imagery. Int J Appl Earth Obs Geoinformation 131:103944. https://doi.org/10.1016/j.jag.2024.103944

Wang D, Hu M, Jin Y, Miao Y, Yang J, Xu Y, Qin X, Ma J, Sun L, Li C, Fu C, Chen H, Han C, Yokoya N, Zhang J, Xu M, Liu L, Zhang L, Wu C, … Zhang L (2025) HyperSIGMA: hyperspectral intelligence comprehension foundation model. IEEE Trans Patt Anal Mach Intel 47(8):6427–6444. https://doi.org/10.1109/TPAMI.2025.3557581

Wang R, Wang M, Ren C, Chen G, Mills G, Ching J (2024) Mapping local climate zones and its applications at the global scale: a systematic review of the last decade of progress and trend. Urban Climate 57:102129. https://doi.org/10.1016/j.uclim.2024.102129

World Bank (2021) Climate change knowledge portal. https://climateknowledgeportal.worldbank.org/country/egypt-arab-republic/climate-data-historical

World Bank (2024) Population in largest city. World Bank Open Data. https://data.worldbank.org/indicator/EN.URB.LCTY?end=2024&name_desc=false&start=2024&view=bar

World Meteorological Organization (WMO) (2025) State of the global climate 2024 (Nos. 978-92-63-11368–5). World Meteorological Organization. https://wmo.int/sites/default/files/2025-03/WMO-1368-2024_en.pdf

Yousif TA, Mohamed AA, Ibrahim EM (2015) Land use/land cover change detection due to urbanization case study: southern part of Khartoum. J Nat Res Environ Stud 3:1–8

Zakšek K, Oštir K, Kokalj Ž (2011) Sky-view factor as a relief visualization technique. Remote Sens 3(2):398–415. https://doi.org/10.3390/rs3020398

Zheng Y, Ren C, Xu Y, Wang R, Ho J, Lau K, Ng E (2018) GIS-based mapping of Local Climate Zone in the high-density city of Hong Kong. Urban Climate 24:419–448. https://doi.org/10.1016/j.uclim.2017.05.008

Zhou X, Okaze T, Ren C, Cai M, Ishida Y, Mochida A (2020) Mapping local climate zones for a Japanese large city by an extended workflow of WUDAPT Level 0 method. Urban Climate 33:100660. https://doi.org/10.1016/j.uclim.2020.100660

Zhu XX, Chen S, Zhang F, Shi Y, Wang Y (2025) GlobalBuildingAtlas: an open global and complete dataset of building polygons, heights and LoD1 3D models. arXiv. https://doi.org/10.48550/ARXIV.2506.04106

Zittis G, Hadjinicolaou P, Almazroui M, Bucchignani E, Driouech F, El Rhaz K, Kurnaz L, Nikulin G, Ntoumos A, Ozturk T, Proestos Y, Stenchikov G, Zaaboul R, Lelieveld J (2021) Business-as-usual will lead to super and ultra-extreme heatwaves in the Middle East and North Africa. NPJ Clim Atmos Sci 4(1):20. https://doi.org/10.1038/s41612-021-00178-7

Zlobin M, Bazylevych V (2025) A data-driven approach for balancing overfitting and underfitting in decision tree models. https://dspace.kntu.kr.ua/handle/123456789/16348

Zittis G, Almazroui M, Alpert P, Ciais P, Cramer W, Dahdal Y, Fnais M, Francis D, Hadjinicolaou P, Howari F, Jrrar A, Kaskaoutis DG, Kulmala M, Lazoglou G, Mihalopoulos N, Lin X, Rudich Y, Sciare J, Stenchikov G, … Lelieveld J (2022) Climate change and weather extremes in the eastern mediterranean and middle east. Rev Geophys 60(3):e2021RG000762. https://doi.org/10.1029/2021RG000762

Climate Change and Water Stress in the Middle East: Transforming Conflict into Cooperation

Mohammed Torki Bani Salameh

1 Introduction

The Middle East, one of the world's most water-scarce regions, faces heightened challenges due to political instability, military conflicts, and the impacts of climate change. Climate change intensifies these challenges by altering precipitation patterns, increasing temperatures, and causing more frequent extreme weather events, thereby affecting water availability and quality (Wessels, 2009). Over 60% of the population lives under high surface water stress, significantly impacting public health and economic development (Varady et al., 2023). This article explores the potential for transforming water from a conflict source to a cooperation catalyst in the Middle East, focusing on the Nile, Euphrates, and Tigris basins.

The arid climate of the Middle East has historically constrained the availability of freshwater resources, but the current trajectory of climate change is exacerbating these constraints at an alarming rate. Political instability and ongoing military conflicts further compound these issues, making effective water resource management increasingly difficult. This complexity is particularly evident in the transboundary river basins of the Nile, Trans-Jordan, Euphrates, and Tigris, where multiple nations share water resources that are becoming progressively scarcer (Salameh et al., 2021) (Fig. 1).

Climate change is predicted to have severe impacts on the hydrological cycle in these basins, altering precipitation patterns and increasing the frequency and severity of droughts. These changes are expected to decrease river flows significantly, thereby intensifying water scarcity. Furthermore, rising temperatures and increased evaporation rates will further diminish water availability. The resultant water stress is not only a challenge for environmental sustainability but also poses significant threats to public health and economic development in the region (Pemunta et al., 2021).

M. T. B. Salameh (✉)
Political and International Studies Department, Yarmouk University, Irbid, Jordan

M. M. Al Saud (ed.), *Hydroclimate Changes: Implications of the Middle East Region*,
https://doi.org/10.1007/978-3-032-20053-2_7

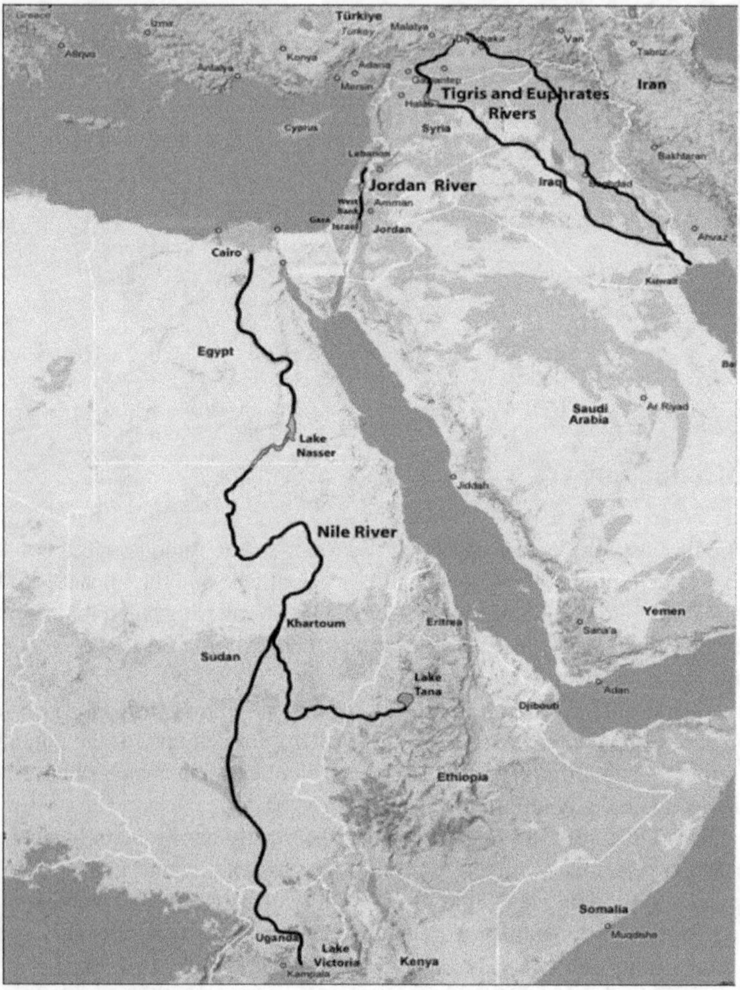

Fig. 1 Map of the Middle East highlighting major transboundary river basins (Nile, Tigris, Euphrates, Jordan)

Addressing these challenges requires a multifaceted approach that integrates climate change projections into water management and diplomacy strategies. It is crucial to shift the perspective from viewing water as a source of conflict to recognizing its potential as a catalyst for cooperation. This transformation can be achieved through adaptive measures, innovative frameworks, and robust water governance institutions. By fostering collaboration among riparian states and involving diverse stakeholders in decision-making processes, the region can move towards sustainable water management and regional stability (Vij et al., 2020).

2 The Current State of Water Resources in the Middle East

The Middle East and North Africa (MENA) region has long been recognized as the most water-stressed region globally. The MENA region's arid climate, limited freshwater supplies, and growing demand for water have led to elevated levels of water stress in virtually all countries. According to data from 2019, sixteen of the twenty-five most water-stressed countries in the world are in the MENA region, with Bahrain ranked as the world′s most water-stressed country (Susskind, 2017). The situation is becoming increasingly dire due to the exacerbating impacts of climate change, which threaten to further destabilize the region's already precarious water balance.

The differences in water vulnerability across the region are highly correlated to the level of access each country has to various water resources, both from freshwater and non-conventional water supplies. Access to renewable supplies from surface water systems, such as rivers, can place nations at an advantage compared to those with limited ability to draw on freshwater sources, including groundwater extracted from subsurface aquifers. Conversely, countries with extremely limited opportunities to leverage freshwater supplies can mitigate this risk by expanding their use of non-conventional water resources, such as desalination and water recycling, contingent on their financial and developmental capacity to invest in the necessary water infrastructure (Michel et al., 2012). Sixteen of the twenty-five most water-stressed countries are located in this region, underscoring its disproportionate vulnerability to climate change impacts on water security and reflecting the interplay of arid climate, scarce freshwater resources, and uneven access to non-conventional water supplies such as desalination and recycling (Fig. 2).

3 Impact of Climate Change on Water Stress

Climate change continues to exacerbate vulnerabilities across the Middle East by increasing the frequency and severity of droughts, thereby widening the gap between water supply and demand. Projections suggest that by 2025, flows in the Tigris and Euphrates rivers will decline substantially, resulting in severe water shortages, particularly in Iraq. These reductions are expected to intensify existing imbalances, with adverse implications for water quality, the resilience of critical water infrastructure, and the prospects for transboundary cooperation—factors that together compound the region's socioeconomic challenges. The anticipated contraction in river flows underscores the acute vulnerability of the Tigris–Euphrates basin to climate change impacts, with Iraq especially exposed due to its heavy reliance on these rivers. Such projections highlight the urgent need for adaptive water management strategies and cooperative governance frameworks to mitigate worsening regional risks (Fig. 3).

The Middle East's susceptibility to drought is a major concern. Droughts not only reduce the availability of water but also degrade the quality of the remaining water.

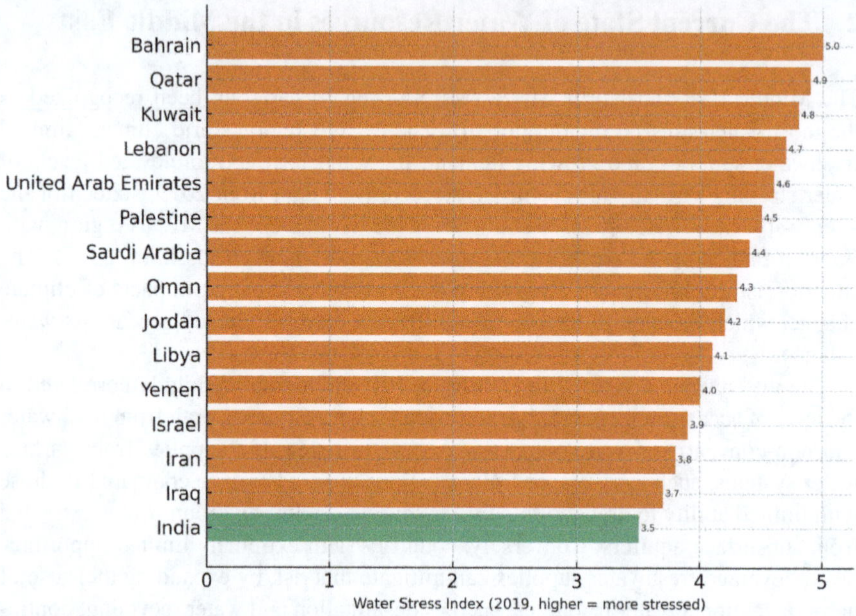

Fig. 2 World's most water-stressed countries (MENA countries highlighted)

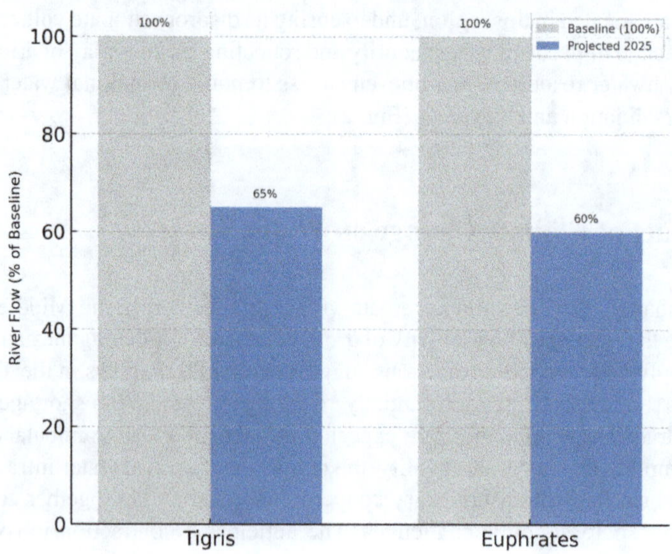

Fig. 3 Projected decline in Tigris- Euphrates flow by 2025

Reduced river flows lead to higher concentrations of pollutants and salinity levels, making the water less suitable for drinking, agriculture, and industry. In countries like Iraq, where agriculture relies heavily on the Tigris and Euphrates rivers, the anticipated reductions in water flow will significantly impact food production, potentially leading to increased food insecurity and higher food prices.

The primary climate impact driving water insecurity is extreme heat and warming. Recent years have seen a significant rise in daily temperature highs, especially during summer periods, breaking historical records for daily temperature highs in many countries and cities in the region. This extreme heat has coincided with the hottest months ever recorded globally, amplifying the effects of climate change on water resources. Higher temperatures increase evaporation rates from reservoirs, rivers, and soils, reducing the overall availability of water. This is particularly problematic in an already arid region where water is a precious and limited resource (Center for Strategic and International Studies, 2024).

In addition to extreme heat, altered precipitation patterns are a significant consequence of climate change that affects water availability in the Middle East. Some areas may experience reduced rainfall, further exacerbating drought conditions, while others might face intense and unpredictable rainfall events leading to flash floods. These floods can cause significant damage to water infrastructure, contaminating water supplies and disrupting the delivery of clean water to communities. The variability and unpredictability of precipitation make it challenging to manage water resources effectively and plan for future water needs.

Sea level rise, another impact of climate change, poses a threat to the coastal aquifers in the region. As sea levels rise, saltwater intrusion into freshwater aquifers can occur, contaminating the freshwater supplies and making them unusable for drinking and irrigation. This is a critical issue for coastal cities and agricultural areas that depend on groundwater.

The compounded effects of climate change on water resources also threaten the region's socio-economic stability. Water scarcity can lead to increased competition for limited resources, heightening tensions between different user groups and between countries sharing transboundary water sources. This competition can exacerbate existing conflicts and contribute to new ones. For instance, disputes over water rights in the Tigris-Euphrates basin have the potential to escalate if the available water becomes even scarcer due to climate change.

Transboundary water cooperation is essential to address these challenges, but it is made more difficult by the political complexities and historical grievances in the region. Effective water management and cooperation require strong governance structures, transparent data sharing, and collaborative frameworks that consider the needs and rights of all riparian countries. However, achieving such cooperation is challenging in a region marked by political instability and mistrust.

Adapting to the impacts of climate change on water resources in the Middle East will require significant investments in resilient infrastructure and innovative water management practices. This includes enhancing water storage capacities, improving irrigation efficiency, and developing technologies for water recycling and desalination. Additionally, integrating climate change projections into national and regional water policies will be crucial for planning and preparing for future water scenarios (Kehl, 2017).

4 Transboundary Water Disputes

Transboundary water disputes are a major challenge in the MENA region, where several countries share critical water resources. The Nile, Tigris, and Euphrates rivers, for example, flow through multiple countries, each with its own water needs and management practices. The construction of the Grand Ethiopian Renaissance Dam (GERD) on the Blue Nile has been a contentious issue, leading to tensions between Ethiopia, Egypt, and Sudan. Egypt, which relies heavily on the Nile for its water needs, fears that the dam will significantly reduce its water supply, while Ethiopia views the dam as essential for its economic development and energy needs (Salameh, 2021). These competing interests underscore the geopolitical complexity of shared river basins and highlight the urgent need for cooperative governance mechanisms (Fig. 4).

To address the severe water scarcity, many MENA countries have turned to non-conventional water resources. Desalination, as mentioned earlier, is a key technology in the region, particularly in the GCC countries. However, desalination is energy-intensive and expensive, which can be a significant barrier for less wealthy nations (Tesfaye, 2013).

Another important non-conventional water resource is wastewater reuse. Several countries in the region have implemented wastewater treatment and recycling programs to provide an additional source of water for agriculture and industrial use. For example, in Israel, treated wastewater is used extensively for irrigation, helping to conserve freshwater resources (Jungudo, 2021).

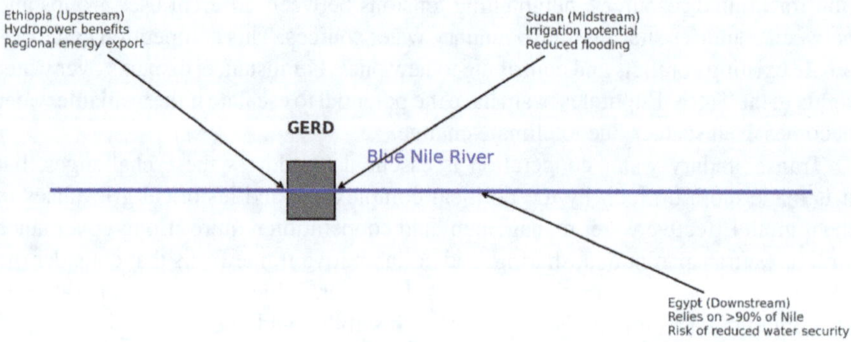

Fig. 4 Schematic illustration of the Grand Ethiopian Renaissance Dam and its primary stakeholders

5 Climate Change Projections and Impacts

Climate change is significantly exacerbating the vulnerabilities in the Middle East by increasing the frequency and severity of droughts. This intensifies the already existing gap between water supply and demand, leading to severe water shortages. Projections indicate that by 2025, the flows in the Tigris and Euphrates rivers will decrease significantly, particularly affecting Iraq, which heavily relies on these rivers for its water supply. The amplifying effects of climate change not only reduce water availability but also degrade water quality. Reduced river flows lead to higher concentrations of pollutants and salinity, making the water less suitable for drinking, agriculture, and industrial use.

Furthermore, climate change impacts critical water infrastructure and transboundary water cooperation, compounding the region's existing socio-economic challenges. Rising temperatures and increased evaporation rates further diminish water availability, while altered precipitation patterns result in both reduced rainfall and intense, unpredictable rainfall events. These changes can cause significant damage to water infrastructure, contaminate water supplies, and disrupt the delivery of clean water to communities.

The compounded effects of climate change on water resources threaten the region's socio-economic stability. Water scarcity can lead to increased competition for limited resources, heightening tensions between different user groups and countries sharing transboundary water sources. Effective water management and cooperation require strong governance structures, transparent data sharing, and collaborative frameworks to ensure regional stability and sustainable development. Investing in resilient infrastructure, enhancing water-use efficiency, and integrating climate change projections into water management strategies are crucial for addressing these challenges and mitigating the impacts of climate change on water resources in the Middle East (Mbaku, 2020).

6 Transboundary Water Governance and Diplomacy

Effective management of transboundary water resources is essential to ensure regional stability and sustainable development in the Middle East. The region's internal water mismanagement and ongoing conflicts necessitate robust water governance institutions and diverse diplomatic engagement platforms. Water diplomacy in the Middle East involves dynamic processes aimed at developing sustainable and peaceful solutions to water management issues while promoting cooperation among riparian stakeholders.

Water diplomacy entails negotiating agreements on water allocation, sharing hydrological data, and implementing joint water management projects. It requires the involvement of all relevant stakeholders, including governments, local communities, and international organizations, to address the complexities of transboundary water challenges. Effective water diplomacy fosters mutual understanding, trust, and cooperation among countries sharing water resources, thereby reducing the potential for conflict.

Additionally, innovative frameworks and adaptive measures, such as the Water Diplomacy Framework, emphasize flexibility and stakeholder involvement in negotiations. By viewing water as a shared resource and focusing on mutual benefits, rather than zero-sum outcomes, the Middle East can transform water from a potential source of conflict into a catalyst for regional cooperation and sustainable development (Jolk & Adel, 2005).

7 The Grand Ethiopian Renaissance Dam (GERD) and its Diplomatic Implications

The construction of the Grand Ethiopian Renaissance Dam (GERD) on the Blue Nile is one of the most contentious issues in the Nile Basin, significantly impacting diplomatic relations between Egypt, Ethiopia, and Sudan. GERD, poised to become Africa's largest hydroelectric project, promises to provide Ethiopia with much-needed electricity and foster substantial economic growth. The dam is expected to generate over 6000 megawatts of power, transforming Ethiopia into a major energy exporter and significantly boosting its development prospects.

However, the project has been a focal point of diplomatic tensions. Egypt, which relies on the Nile for over 90% of its freshwater needs, fears that the dam will drastically reduce the flow of Nile waters downstream, posing a severe threat to its water security and agricultural productivity. The reduction in water availability could impact millions of farmers, lead to food insecurity, and affect Egypt's overall economy, which is heavily dependent on agriculture (Salameh et al., 2021).

Sudan also shares concerns about the potential reduction in water availability and its impacts on its water infrastructure and agriculture. While Sudan stands to benefit from the regulation of water flow and reduced flooding due to GERD, it remains apprehensive about the overall changes in water distribution and the lack of comprehensive agreements on the dam's operation (Salameh, 2021).

The GERD crisis highlights the need for cooperative and inclusive water management strategies in the Nile Basin. Effective diplomacy, transparent negotiations, and mutual agreements are essential to ensure that the benefits of the dam are maximized while minimizing adverse impacts on downstream countries. Regional cooperation and scientifically informed strategies can help navigate these complex water challenges, promoting sustainable development and stability in the Nile Basin.

8 Diplomatic Dynamics of the Tigris and Euphrates Basin

The situation in the Tigris and Euphrates basin exemplifies the complex interplay of water security and diplomacy amid rising regional tensions and environmental challenges. The power differential between Turkey and its neighbours Syria and Iraq, alongside its NATO membership, has allowed Turkey to manage its water issues through faits accomplis. Turkey's South-eastern Anatolia Project (GAP), which includes 22 dams and 19 power plants, significantly affects the region. The newly constructed Ilisu Dam grants the Turkish government substantial control over water flows in the Tigris. The impact of Turkish dams on Iraq's water resources has been substantial, affecting agriculture, drinking water, and overall economic stability (Turkey Studies Unit, 2024).

Iran's role in stifling Iraqi water flow adds another layer of complexity. The construction of dams on the Sirwan (Diyala) and Little Zab Rivers has significantly reduced water flows into Iraq, impacting over 8000 acres of farmland in the Sulaymaniyah Governorate alone. This has potential implications for drinking water supplies in towns like Qalat Daza and Raniyah (Al-Harahsheh & Salameh, 2023).

9 Adaptive Measures and Innovative Frameworks

Adaptive measures such as developing drought-resistant crops, improving water-use efficiency, and investing in water infrastructure are vital. Furthermore, the water-energy nexus requires integrated management, and renewable energy sources, such as solar and wind, can reduce the water footprint of energy production (McCaffrey, 1997).

Innovative frameworks such as the Water Diplomacy Framework (WDF) emphasize the flexibility and adaptability needed in negotiations over water resources (Fig. 5). These frameworks advocate for involving all stakeholders in decision-making processes and focusing on mutual gains rather than zero-sum outcomes. By transforming water from a finite resource into a flexible one through technologies like desalination and wastewater reuse, it is possible to create additional value and reduce conflicts over water allocation (Al Gharaybeh & Bani Salameh, 2024).

The model integrates scientific knowledge and stakeholder engagement as inputs, channels them through negotiation and cooperation processes, and aims to produce outcomes such as conflict prevention and enhanced regional collaboration.

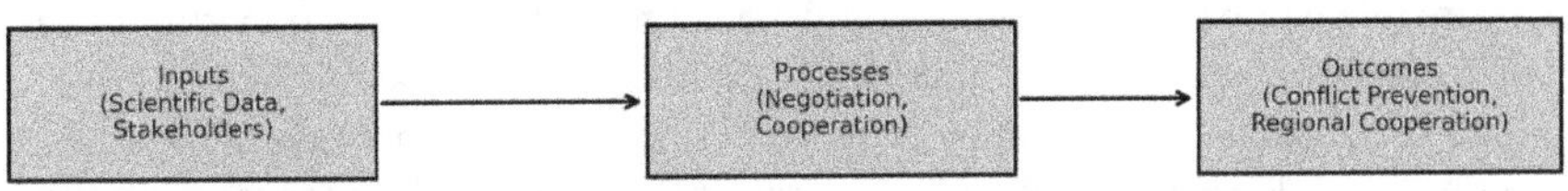

Fig. 5 Conceptual representation of the Water Diplomacy Framework (WDF)

10 De-Securitizing Water and Multi-Track Diplomacy

De-securitizing water by treating it as a shared resource rather than a strategic asset is crucial for fostering cooperation and collaborative management. This approach shifts the focus from competition over water resources to mutual benefits and shared responsibilities. By viewing water as a common good, countries can work together to manage and protect their shared water supplies more effectively.

Multi-track water diplomacy, which involves engaging diverse stakeholders and employing flexible negotiation frameworks, is essential for enhancing transboundary water governance and ensuring regional stability. This method includes governments, local communities, NGOs, and international organizations in the dialogue, allowing for a more comprehensive and inclusive approach to water management. By incorporating scientific data and evidence-based strategies into the negotiation processes, stakeholders can make informed decisions that address the complexities of transboundary water issues.

Scientifically informed, multi-track water diplomacy helps break down barriers to cooperation and fosters the political will necessary to enhance transboundary water governance. It encourages transparency, data sharing, and collaborative problem-solving, which are critical for building trust among riparian countries. Implementing these strategies can transform water from a potential source of conflict into a catalyst for cooperation and sustainable development.

In the face of climate change, which exacerbates water scarcity and variability, such cooperative frameworks are more important than ever. Effective transboundary water governance can mitigate the impacts of climate change on water resources, ensure equitable distribution, and promote regional stability. By treating water as a shared resource and leveraging multi-track diplomacy, the Middle East can navigate its water challenges and achieve sustainable development and lasting peace (Bani Salameh, 2024).

11 Water Security Challenges and Opportunities

Water security and water diplomacy in the Middle East present both significant challenges and unique opportunities. The region's inherent water scarcity, exacerbated by political instability and mismanagement, necessitates a multifaceted approach to ensure sustainable water management and regional stability. This article has examined how transboundary water governance, particularly in the Nile, Euphrates, Tigris, and Jordan River basins, can transition from sources of conflict to opportunities for cooperation through well-structured water diplomacy (Bani Salameh, 2024).

The key to transforming water-related tensions into collaborative efforts lies in the concept of de-securitization of water resources. This involves recognizing water as a shared resource critical to all parties and fostering a cooperative mindset among riparian states. By engaging in scientifically informed, multi-track diplomacy, stakeholders can build the political will necessary to overcome historical grievances and power asymmetries. The experiences from various international river basins underscore that effective water diplomacy can lead to tangible cooperation and sustainable management practices (Jungudo, 2021).

Despite these challenges, there are clear opportunities for progress. Regional initiatives like the Nile Basin Initiative (NBI) and international involvement from entities such as the World Bank provide platforms for dialogue and cooperation. Moreover, innovative frameworks that emphasize flexibility, mutual gains, and stakeholder involvement can help navigate the complexities of water diplomacy. The implementation of benefit-sharing mechanisms, as seen in other regions, could serve as a model for the Middle East, promoting sustainable development and equitable resource distribution (Bani Salameh, 2024).

By prioritizing cooperative approaches, leveraging scientific knowledge, and involving a broad range of stakeholders, the region can transform water from a source of conflict into a catalyst for peace and sustainable development. Addressing the water crisis requires a comprehensive and integrated approach that balances national interests with regional stability and environmental sustainability (Bani Salameh, 2024).

12 Conclusion

The Middle East faces a multifaceted water crisis exacerbated by climate change. Desalination has alleviated some of the region's water deficit, but it presents environmental and economic challenges. The region also contends with transboundary water disputes, groundwater depletion, and vulnerable water infrastructure susceptible to extreme weather events. Addressing water scarcity and improving water management is critical for the stability, sustainability, and well-being of its populations.

Climate-induced water stress in the Middle East necessitates comprehensive and adaptive strategies. While desalination provides a crucial alternative water source, it is energy-intensive and environmentally taxing, particularly in less affluent nations. Enhancing other non-conventional water resources, such as wastewater reuse and rainwater harvesting, is essential (Bani Salameh, 2024).

Transboundary water disputes over the Nile, Tigris, and Euphrates rivers highlight the need for robust water governance and cooperative frameworks. Collaborative management of shared water resources is vital to prevent conflicts and ensure equitable distribution. Initiatives like the Nile Basin Initiative and international mediation efforts are positive steps but require sustained political will and mutual trust among riparian states.

Groundwater depletion is another critical issue, with many countries over-extracting aquifers faster than they can be replenished. This unsustainable practice poses long-term risks to water security. Implementing groundwater management plans that balance extraction with recharge is vital for maintaining these crucial reserves.

The vulnerability of water infrastructure to extreme weather events necessitates significant investment in resilient infrastructure. Upgrading existing systems and designing new ones to withstand increased floods, droughts, and heatwaves are essential to minimize disruptions to water supply and protect public health and economic stability.

Addressing the water crisis in the Middle East necessitates a multifaceted approach that integrates climate change projections into comprehensive water management strategies. Adopting innovative frameworks, such as the Water Diplomacy Framework, is essential. This framework emphasizes flexibility, stakeholder involvement, and mutual gains in water negotiations, fostering cooperation and reducing conflicts. By involving all relevant parties, from local communities to national governments, in the decision-making process, these strategies can lead to more sustainable and equitable water management practices.

Treating water as a shared resource rather than a strategic asset is crucial for regional stability. This perspective encourages collaborative management and joint investment in water infrastructure, ensuring that all countries benefit from shared water resources. Such cooperation is vital to address the challenges posed by climate change and ensure long-term water security in the Middle East.

References

Al Gharaybeh MH, Bani Salameh MT (2024) An analytical study of Iran's ambitions in the Arabian gulf region (1979-2022). Dirasat: Hum Soc Sci 51(2):360–371

Al-Harahsheh STAQ, Salameh MTB (2023) Challenges of the water crisis in the Arab gulf countries (2011–2021). In: Jeon HY (ed) Sustainable Development of Water and Environment. ICSDWE 2023. Environmental Science and Engineering. Springer, Cham

Bani Salameh MT (2024) The economic, political, and security dimensions of the water crisis in the Middle East. Dirasat Human Soc Sci 51(3):443–459

Center for Strategic and International Studies (CSIS) (2024) Water and the future of the Middle East. https://www.csis.org/programs/middle-east-program/major-projects/middle-east-transformation-initiative/water-and-future

Jolk, B., Adel, D.: Water wars: upcoming conflicts in the Middle East. Supreme council for Culture, p. 75. (2005)

Jungudo MM (2021) The role of Nile Basin initiative (NBI) in the Nile water conundrum: 1999-2015. Gombe J Hist Stud 1(1):200

Kehl JR (2017) Water security in transboundary systems: cooperation in intractable conflicts and the Nile system. Water Secur Middle East: Essays Sci Soc Coop:39

Mbaku JM (2020) The controversy over the grand Ethiopian renaissance dam, Brookings

McCaffrey, S. C. (1997) Convention on the law of the non-navigational uses of international watercourses. https://legal.un.org/avl/ha/clnuiw/clnuiw.html

Michel D et al (2012) Water challenges and cooperative response in the Middle East and North Africa. In: US–Islamic World Forum Papers, p 4

Pemunta NV et al (2021) The grand Ethiopian renaissance dam, Egyptian National Security, and human and food security in the Nile River basin. Cogent Soc Sci 7(1). https://doi.org/10.1080/23311886.2021.1875598

Salameh et al (2021) The water crisis and the conflict in the Middle East. Sustain, Water Resour Manag 7(69)

Salameh M (2021) Dam wars: are Ethiopia, Turkey, and Iran leading to water Armageddon? Middle East Policy 28:147–157

Susskind LE (2017) The political and cultural Dimensions of water diplomacy in the Middle East. In: Cahan J (ed) Water Security in the Middle East: Essays in Scientific and Social Cooperation. Anthem Press, pp 185–206

Tesfaye A (2013) The politics of the imposed and negotiation of the emerging Nile Basin regime. Int J Ethiop Stud 7(1 & 2):57–76

Turkey studies unit (2024) Turkish-Iraqi agreement: security in exchange for trade water. EPC https://epc.ae/en/details/featured/turkish-iraqi-security-agreement-security-in-exchange-for-trade-water

Varady et al (2023) Transboundary water governance scholarship: a critical review. Environments 10(2):27. https://doi.org/10.3390/environments10020027

Vij et al (2020) Power in water diplomacy. Water Int 45(4):249–253. https://doi.org/10.1080/02508060.2020.1778833

Wessels J (2009) Water crisis in the Middle East: an opportunity for new forms of water governance and peace. Whitehead J Diplom Int Relat:131

Will the Climate of Saudi Arabia Be More Humid? Remote Sensing-Based Analysis for Climate, Vegetation and Drought Indices

Mashael M. Al Saud

1 Introduction

The Kingdom of Saudi Arabia, the largest territory of the Arabian Peninsula, is surrounded by the Red sea, Arabian Gulf, and the Arabian Sea from west, east and south; respectively (Fig. 1). Desert land is the main territory with mountain range parallel to the coast from the west, while the interior areas are largely occupied by a plateau and only in the south-east is a flat area. The entire Arabian Peninsula is characterized by desert climate with hot temperatures during the day and low temperature at night. According to the Köppen-Geiger climate classification, the entire Arabian Peninsula is considered as an arid climate zone (or desert climate zone) including some mountainous patches with semiarid climate, and the aridity is attributed to the geographic location in the subtropical high-pressure zone where the blowing downward movement of warm, dry air hinders the formation of clouds. The climate of Saudi Arabia induces sharp water scarcity and dominant drought conditions with annual rainfall rates below 120 mm (at most), while the average air temperature is about 35 °C (GAMEP 2012); with potential evapotranspiration exceeding 2000 mm/year (Baban 2022). The prevailing climate affects the limited water resources, which are mainly fossil groundwater (non-renewable) as the main water body in Saudi Arabia.

Drought spells in Saudi Arabia in stressing water availability which does not exceed 73 m^3/capita/year; while water demand is estimated at 706 m^3/capita/year (UNESCO 2021). This has been reflected on the aspects of water consumption and notably in the agriculture sector; where uncontrolled groundwater pumping exhausted the existing aquifers of fossil water which are often exceeding 1000 m and more in depth. The agricultural lands are limited to some regions along flood plains in the large-scale wadis which carry storm water over limited time period and the

M. M. Al Saud (✉)
King Saud University, Riyadh, Kingdom of Saudi Arabia
e-mail: geomm5.mashael@gmail.com

M. M. Al Saud (ed.), *Hydroclimate Changes: Implications of the Middle East Region*,
https://doi.org/10.1007/978-3-032-20053-2_8

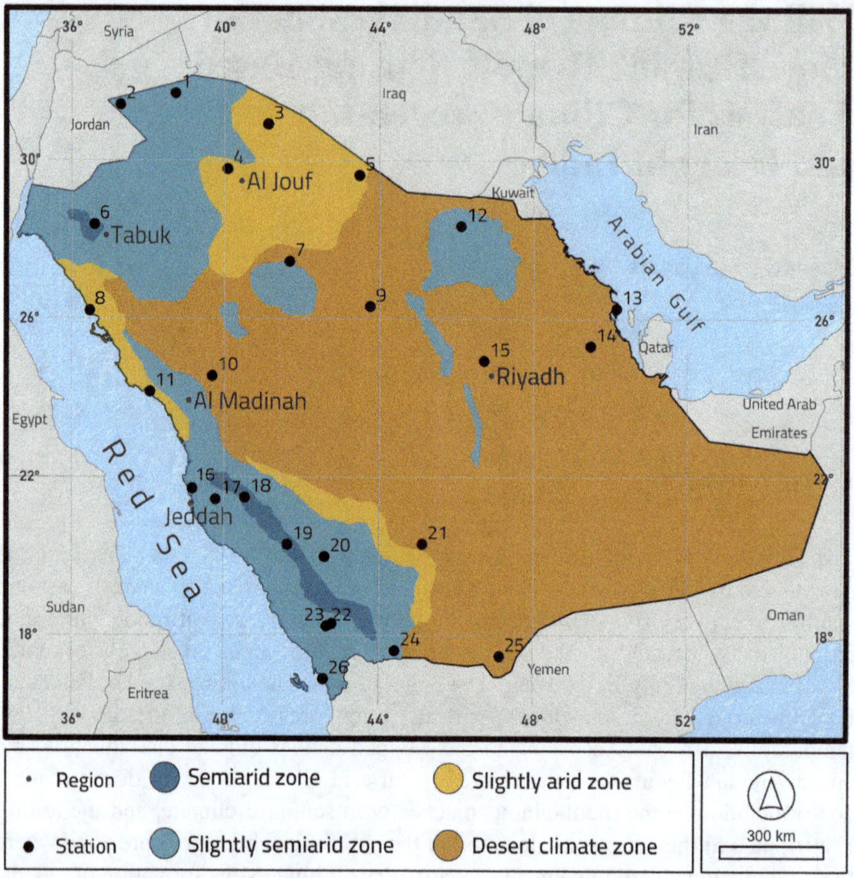

Fig. 1 Climatic zones of Saudi Arabia

runoff often does not exceed few days. From the natural point of view, groundwater depletion (*i.e.*, lowering in water table) is estimated between -6.9×10^{-2} and -8.6×10^{-2} cm/month as detected by TWS (Total Water Storage) which is observed in GRACE (Gravity Recover and Climate Experiment), a twin satellite for monitoring a long-term hydrological conditions, in northern Saudi Arabia, particularly in agricultural areas (Wehbe 2022).

As a negative influential factor on the development, water deficit catalyzed Saudi Arabia to adopt the United Nations Sustainable Development Goals (SDGs) to enable equal access to clean and safe water. Yet, there is still uncertainty in defining the climatic trends in Saudi Arabia; especially unprecedented climatic conditions that have been lately observed. They are represented mainly by torrential rain, and increased number of warm days. There is also a sort of climatic alteration with various geographic zones that this is noticeable in northern regions of the Middle East with warmer climate in conjunction with decreased precipitation, while the southern regions (*i.e.*, Arabian Peninsula) have a higher precipitation rate. In this respect, rainfall patterns for the entire Arabian Peninsula lately changed towards torrential rain revealing an increasing trend in climatic extremes, and evidencing an aspect of climatic variability in the region as it was presumed by (IPCC 2022). The dense precipitation occurs over a short time period (*i.e.* torrential rain) and reflects on the increased number of flash floods in many wadis of Saudi Arabia resulting in severe damages to the infrastructure and the environment (Al Saud 2015; ESCWA 2023).

There are several studies characterizing the climate of Saudi Arabia; however, these studies pointed out inconsistent conclusions and findings. For example, ESCWA (Economic and Social Commission for Western Asia) (2023) stated that the average annual air temperature was increased by 0.5 °C between 1975 and 2000, and average annual precipitation was increased by 30–40 mm between 1995–2014. McCabe et al. (2023) reported that precipitation will increase by 5–10% by 2050, coupled with a temperature rise between 2 and 3 °C. While Hasanean and Almazroui (2015) stated that the southern part of the Arabian Peninsula and regions along the coast of the Red Sea, rainfall has been increased, whilst it was decreased over most areas of Saudi Arabia during the 2000–2009 decade, and this was estimated between 6.2 and 47.8 mm per decade compared to 1980–1989.

El-Samra et al. (2017) showed evidence of a significantly drier climate over the northeast Arabian Peninsula from 2016 until the midcentury, but Almazroui (2020) projected an increase in the precipitation rate during the midcentury and by the end of the century to the present climate over most parts of the peninsula. The analyzed period between 1978–2009 showed an increase in precipitation by 13.9 mm/decade as well as air temperature increase by 0.71 °C per decade. In this respect, World Bank (WB) (2021) registered an overall increase in air temperature between 1900 and 2020 by 1.8 °C, and decrease in precipitation for the same period by 2.02 mm. The aforementioned studies, with obvious contradictory, motivated performing the current study covering the longest time period available for the entire Saudi Arabia, and using various drought indices.

Based on the previous discussion and the diverse estimates whether for the analyzed or the projected precipitation and air temperature rates, there is obvious contradictory and uncertainty in defining the major climatic variables for Saudi Arabia. This creates a misunderstanding about the climate in the region and highlights the need for an in-depth climatic analysis and drought indices. A detailed analysis for major climatic indices are required for Saudi Arabia, which the main objective of this study, and this needs to utilize data from ground-based stations as well as on data from remote sensing products and using statistical analysis to investigate of drought indices which reflects precipitation and air temperature for the past and projected time periods.

The novelty of this study includes the analysis of climate, drought and vegetation indices for the entire Saudi Arabia and over a longest time period available, which was not covered before. In addition, there is an integration of different data sources and the support of satellite-based data and adopted climatic platforms, which were previously applied for the region but they used sole methods and analyzed limited time periods.

2 Data and Methods

2.1 *Study Area*

Saudi Arabia is the largest country in the Middle East and the 14th largest country in the World. It composes the largest territory of the Arabian Peninsula (Al-Jazeera Al-Arabia) which is bounded by the Red Sea from the west, the Arabian Sea from the south and the Arabian Gulf from the east, and thus totaling surface area of about 2,217,000 km^2. The Kingdom has a perimeter of about 6749 km where 4530 km is a terrestrial border and 2219 km shoreline. Thus, 1709 km extends along the Red Sea (including the Gulf of Aaqba) and 510 along the Arabian Gulf (Al Saud 2020).

Deserts occupy wide lands of Saudi Arabia where the Arabian Shield, including Najed Region, curvedly extends in the western side of the Peninsula forming Al-Hejaz and Asir Mountain Ranges. The coastal plain along the Red Sea is named as Al-Tohama. While, An-Nafood Desert represents the northern part between Saudi Arabia, Iraq and Jordan. Besides, Al-Ruba'a Al-Khali Desert is located in the southeast of the Kingdom.

2.2 *Climate of Saudi Arabia*

The climate of Saudi Arabia is characterized by warm weather, with high temperature during the day and cold temperature at night and it averages 32 °C between the coastal plain and mountain regions (Youssef et al. 2016); while the precipitation is low enough and it is almost less than 100 mm/year. In addition, climatic extremes are well pronounced the Arabian Peninsula with frequent sand storms, torrential rain and heat waves. The geomorphological aspects and latitudinal extent control the climate of the entire Arabian Peninsula. Deserts are dominant; and therefore, desert climate with high aridity exists except in some mountainous regions; especially at the southern west part which has a semiarid climate. Yet, there is significant air temperature and humidity variance due to the influence of a subtropical high-pressure system.

According to the IPCC 6th Assessment Report (IPCC 2022), over the last 40 years, the average air temperature in the Arabian Peninsula has been increasing at a faster rate than the global trend. During the same period extreme rainfall has been increasing in the elevated areas. The report shows that precipitation has exhibited spatio-temporal variability in the region where, approximately every 10 years, the general trend moves from increasing to decreasing rainfall, and the precipitation has demonstrated a correlation with air temperature such that warmer years are drier than relatively cooler years (ESCWA 2023). Moreover, precipitation spatially varies across the country and it is relatively high in Asir Highland which is a mountainous region exceeding 2300 m in some localities. While, the rainfall in the Rub' al-Khali is nearly nil; and low rainfall has also been detected in the northern part, An Nafud and coastal Red Sea regions.

In this study, a map for climatic zones in Saudi Arabia has been produced (Fig. 1). Thus, a systematic integration, in the geographic information system (GIS), for a number of climatic and geographic components has been carried out to determine the climatic zones of Saudi Arabia. The resultant map shows four main climatic zones (Fig. 1), and the integrated components included: (i) datasets from several previous climatic records and remote sensing platforms (Table 1), (ii) geomorphologic units, (iii) dominant arid lands and deserts, (iv) north-westerly winds existing due to the distribution of high and Low-pressure ridge over various regions from Saudi Arabia. The resulted climatic zones can be classified as follows:

- Semiarid zones: These are the most humid regions in Saudi Arabia due to their mountainous geomorphology (average altitude = 1100 m). The reported air temperature and precipitation in these regions are averaging at < 20 °C and > 300 mm; respectively. The produced map (Fig. 1) shows that the semi-arid zones in Saudi Arabia is 32,078 km^2 (1.5% of the total area), and they exist in Asir Highland and on the mountains west to Tabuk.
- Slightly semiarid zones: These zones are characterized by spatio-temporal climatic variability between various years, but they are in general changing from low humidity to aridity conditions. These zones are located in the northwest, southwest and a patch in the northeast of the country totaling an area of about 563,112 km^2 (26% of the total area). The average air temperature ranges between 20 and 23 °C; and precipitation average between 90 and 140 mm.
- Slightly arid zones: These zones are almost transitional zones towards the arid climate where fringes of desert lands occur. They are obviously located near Hafer Al-Baten and among the topographic belts in the northern coast of the Red Sea, as well as to the east of Asir Highland totaling an area of about 146,327 km^2 (7% of the total area). In the slightly arid zones, the average air temperature ranges between 23 and 25 °C; while precipitation averages between 80 and 120 mm.
- Desert zones: These arid zones comprise the largest area of Saudi Arabia with 1,405,173 km^2 (65.5% of the total area), and they include the main desert lands of the Arabian Peninsula (Fig. 1). These zones have an average air temperature exceeding 28 °C, but it may exceed 50 °C in summer; while precipitation is often nil and it averages between 20 and 50 mm.

Table 1 Precipitation datasets for Saudi Arabia, retrieved from various ground-based and remote sensing sources

#	Station	Ground-based data			Remote sensing data		Average annual precipitation (mm)
		Almazroui et al. (2013)	Hasaean and Almazroui (2015)	NCM (2022)	TRMM (2022)	CMORPH (2023)	
		1978–2009	1978–2009	1992–2022	1998–2015	1998–2023	
1	Turaif	84.2	85.5	79.0	80.0	83.0	82.34
2	Guriat	46.7	46.7	52.27	53.0	59.0	51.53
3	Arar	58.6	58.6	56.41	69.0	71.0	62.72
4	Al-Jouf	54.6	56.4	59.75	60.0	69.0	59.95
5	Rafha	85.1	86.9	77.81	106.0	113.0	93.76
6	Tabuk	28.7	29.3	31.33	33.0	37.0	31.86
7	Hail	113.7	116.4	84.27	114.00	112.0	108.07
8	Wejh	25.0	25.3	35.64	44.0	46.0	35.18
9	Qassim	143.9	145.6	130.70	150.0	144.0	142.84
10	Madina	63.6	64.7	59.0	66.0	65.0	63.66
11	Yenboa	30.3	31.0	32.38	56.0	38.0	37.53
12	Al-Qaysumah	125.5	126.5	104.13	103.0	131.0	118.02
13	Dhahran	90.2	92.5	95.2	99.0	96.0	94.58
14	Al-Hessa	84.8	84.8	91.63	106.00	96.0	92.64
15	Riyadh	110.6	110.6	104.36	113.0	109.0	109.51
16	Jeddah	52.1	52.1	109.0	87.0	113.0	82.64
17	Makka	110.6	110.6	68.05	115	108.0	102.45
18	Taif	172.2	174.6	168.72	179.0	187.0	176.30
19	Al-Baha	142.6	142.6	118.12	120.0	149.0	134.46
20	Bisha	88.7	88.5	83.08	111	94.0	93.05
21	W. Aldawasser	25.5	26.1	28.62	50.0	48.0	35.64
22	Abha	229.1	230.3	202.79	210.0	224.0	219.23
23	Khamis Mushait	190.6	110.6	181.14	201.0	207.0	178.06
24	Najran	60.4	60.1	74.20	65.0	71.0	66.14
25	Sharurah	70.3	70.3	64.22	54.0	63.0	64.36
26	Jizan	128.6	131.8	153.26	135	150.0	139.73
Average		92.7	90.64	87.8	99.0	103.0	94.63

2.3 Data Sources

Precipitation and Temperature Data

Spatio-temporal analysis and elaboration of climatic trends often requires precise datasets, and this can be obtained from ground-based stations (*e.g.*, in-situ rain gauges) which give reliable and accurate results as the amount of rainfall is only known at a point scale, but not for the surrounding. In order to have precipitation measuring networks, these ground-based stations should be uniformly distributed considering various geomorphological units and geographic grids. However, it is usually a challenge to find sequential datasets for precipitation and air temperature with long time series, and this is always the case where ground-based stations are lacking, insufficient or non-functional.

As complementary to ground-based stations, satellite-based precipitation products offer quasi-global coverage of rainfall and oceans satellite products can provide rainfall estimates at high spatial (i.e., 0.25° latitude × 0.25° longitude grid size) and temporal (even hourly) resolution (Youssef et al. 2016). Therefore, remote sensing products and global simulated models are used in case of data lacking, and the interpolated global models for climatic variables are frequently used to study climatic regimes and validate numeric products obtained from remote sensing and model simulations (Xie et al. 2007; Harris et al. 2014; Youssef et al. 2016; Chen and Wang 2022). In addition, there are several climate-based platforms where climatic datasets and products can be retrieved. However, the accuracy assessment of the resulting data is always performed to calibrate and validate the datasets retrieved from remote sensing and climate models platforms.

In this study, a detailed survey was carried out from various data sources to compose a comprehensive and a variety of precipitation and air temperature datasets from ground-based stations and from remotely sensing data (example in Table 1). Most of the collected data was calculated on monthly basis; however, that was for different time periods depending on data availability from the available sources. For data reliability; a comparative analysis has been performed between the results from these sources (*i.e.*, ground-based stations and remote sensing sources) and it was resulted that the variance is almost low (Table 2).

Table 2 Summary on the collected precipitation and air temperature datasets for Saudi Arabia

Climatic parameter	Annual average	Ma. (mm)	Min. (°C)	σ	σ^2 (remote sensing/ground data)
Precipitation	94.63 mm	103.0	87.0	4.12	16.95
Air temperature	25.58 °C	28.23	24.63	1.47	2.18

σ Standard deviation
σ^2 Variance

In this study, datasets for precipitation and air temperature were utilized from various sources as shown in Table 3. In addition, the Estimated Precipitation (EP) and Precipitation Anomaly Index (PAI) were retrieved, where EP reflects the cumulative rainfall over a dekad (a 10-day period) and PAI evidences the deviation of temporal precipitation patterns over long-term average. These two variables were retrieved form remote sensing products (Table 3).

Table 3 Sources of collected precipitation and air temperature datasets for Saudi Arabia

Data source	Data description	Period
ECMWF (European Centre for Medium-Range Weather Forecasts), 2024 Available at: https://www.ecmwf.int/	Estimated precipitation (EP)	1997–2024
	Precipitation anomaly index (PAI)	
WCG (World Climate Guide) (2024). Available at: https://www.climatestotravel.com/temperatures/saudi-arabia	Monthly temperature datasets	1995–2024
EN (Environmental Horizon) (2023). Available at: https://envirodata.horizons.govt.nz/	Average monthly air temperature	1992–2022
CMORPH (Center Morphing Technique) (2023). Available at: https://www.ncei.noaa.gov/data/cmorph-high-resolution-global-precipitation-estimates/access/	Daily rainfall data with 8 km × 8 km spatial resolution	1998–2023
TRMM (Tropical Rainfall Mapping Mission) (2022). Available at: http://trmm.gsfc.nasa.gov	Daily and monthly rainfall data with 0.25° × 0.25° gridded spatial resolution microwave imager	1998–2015
CPC-NOAA (Climate Prediction Center-National Oceanic and Atmospheric Administration) (2022). Available at: http://www.cpc.ncep.noaa.gov/	Monthly temperature datasets	1978–2010
NCM (National Center for Meteorology) (2022)	Average daily and monthly rainfall and air temperature	1992–2022
Hasanean and Almazroui (2015)	Monthly rainfall data supported by stochastic models of ARIMA and Providing Regional Climates for Impact Studies of Hadley Centre (PRECIS)	1978–2009
Almazroui (2011, 2013, 2020)	Annual air temperature and rainfall from the National Centers for Environmental Prediction (NCEP) gridded with ground stations	1978–2009

Drought Indices

Other than drought indicators which describe the changing variables such as precipitation, temperature, streamflow, ground and reservoir water levels, soil moisture, and snowpack; drought indices are calculated through numerical applications using climactic/or hydro-meteorological datasets to provide quantitative measurements for drought severity including the determination of temporal and geographic dimensions, and this can be elaborated through a historical reference (or long-term average) in order to evaluate future drought trending. In this respect, remote sensing products are often of useful for deducing the required variables for calculating drought indices.

Given that the calculated climatic datasets (precipitation and air temperature) have been collected, and for the purpose of this study in identifying vegetation and drought-related indices, three vegetation and two drought indices were also analyzed from satellite-based products where various sources were identified (Table 4). These datasets include.

Normalized Difference Vegetation Index (NDVI)

Normalized Difference Vegetation Index (NDVI) measures the green cover (e.g., trees, shrubs, grass, etc.) and it evidences the density and health of vegetation. Ranges of the NDVI values are shown in Table 5, with high positive values corresponding to dense and healthy vegetation and vice versa. NDVI is calculated according with the formula (USGS 2010):

$$\mathrm{NDVI} = NIR - RED/NIR + RED$$

Table 4 Sources of collected vegetation and drought indices for Saudi Arabia

Data source	Data description	Period
FAO (Food and Agriculture Organization), METOP-AVHRR, Global Information and Early Warning System (GIEWS), 2024. Available at: https://www.fao.org/giews/earthobservation/asis/index_2.jsp?lang=en	NDVI Vegetation Condition Index (VCI) Vegetation Health Index (VHI)	1984–2024 1984–2024 1984–2024
LCSC (Laboratory of Climate Services and Climatology), 2024a. Available at: https://spei.csic.es/spei_database/#map_name=spei01#map_position=1463	Standardized Precipitation Evapotranspiration Index (SPEI)	1955–2024
LCSC (Laboratory of Climate Services and Climatology), 2024b. Available at: https://sedi.csic.es/?var=SEDI&ti=515&sp=Raster&tp=undefined	Standardized Evapotranspiration Deficit Index (SEDI)	1980–2022

Table 5 Categorization of the used vegetation indices in this study

Category	NDVI	Category	VCI (%)	Category	VHI
Water	– 0.28–0.015	Extreme drought	0–10	Extremely dry	0–10
Built-up	0.015–0.14	Severe drought	10–20	Severely dry	10–20
Barren land	0.14–0.18	Moderate drought	20–30	Moderately dry	20–30
Shrub and grassland	0.18–0.27	Light drought	30–40	Mid dry	30–40
Spare vegetation	0.27–0.36	Sporadic light drought	40–70	Normal	40–50
Dense vegetation	0.36–0.74	No drought	80–100	Good	50–60
				Excellent	60–100

where NIR is the light reflected in the near-infrared spectrum; and RED is the light reflected in the red range of the spectrum.

Vegetation Condition Index (VCI)

Vegetation Condition Index (VCI) indicates vegetation health in comparison to the historical trends. The lower and higher VCI values indicate bad and good vegetation conditions; respectively. VCI is expressed by percentage and it varies from 0 for extremely unfavorable to 100 for optimal condition (Table 5). According to Kogan (1995), VCI is expressed by the following formula:

$$\text{VCI} = (NDVI - NDVI\ \text{min})/(NDVI\ \text{max} - NDVI\ \text{min}) * 100$$

where NDVI max and NDVI min are multiple year maximum NDVI and minimum NDVI for a pixel with reference to a specific climatology.

Vegetation Health Index (VHI)

Vegetation Health Index (VHI) is the proxy of drought severity evidenced by vegetation health and the influence of air temperature on plant conditions. The VHI combines the Temperature Condition Index (TCI) and the VCI (Kogan 1995). The VHI indices range from 0 to 100 characterizing changes in vegetation conditions from extremely poor to excellent at 100 (Table 5). The TCI, with a constant of (α) 0.5, is calculated using a similar equation to the VCI, but relates the current air temperature to the long-term maximum and minimum with higher air temperatures tend to cause a deterioration in vegetation conditions, as in the formula below (Kogan 1995):

$$\mathrm{VHI} = \alpha^{*} VCI + (1 - \alpha) * TCI$$

Standardized Precipitation Evapotranspiration Index (SPEI)

Standardized Precipitation Evapotranspiration Index (SPEI) is an extension of the Standardized Precipitation Index (SPI), and it is designed to account precipitation and potential evapotranspiration (PET). The SPEI emphasis on the main impact of increased air temperatures on water demand The SPEI is calculated from precipitation and temperature in the form of climatic water balance (Vicente-Serrano et al. 2015; Tirivarombo and Eliasson 2018). The SPEI was computed according to the following formula (Vicente-Serrano et al. 2010):

$$\mathrm{SPEI} = P - PE/\sigma$$

where P is the observed annual average precipitation rate, PE the monthly average of potential evapotranspiration, and σ is the standard deviation of the distribution of P-PET. The SPEI values were categorized into different categories considering various studies (Li et al. 2017; Vicente-Serrano et al. 2018; Mehr 2021); as follows: extremely wet (> 2), very wet (1.50 to 1.99), moderately wet (1 to 1.49), and near normal (0.99 to – 0.99). Conversely, moderately dry (– 1 to – 1.49), severely dry (– 1.50 to – 1.99), and extremely dry (> – 2).

Standardized Evapotranspiration Deficit Index (SEDI)

Standardized Evapotranspiration Deficit Index (SEDI) is attributed to the standardized anomalies of the difference between actual evapotranspiration (AET) and potential evapotranspiration (PET). The calculation of SEDI is based on complementary hypothesis and used an approach to calculate a drought index comparable spatially and temporally SPEI (Vicente-Serrano et al. 2018; Zhang et al. 2019).

The SEDI is classified into eight categories as: <– 2, – 1.5 to – 2, – 1 to – 1.5, – 0.5 to – 1, – 0.5 to 0.5, 0, 0.5 to 1, 1 to 1.5 and > 1.5 for extreme drought, severe drought, moderate drought, mild drought, near normal, moderate moist, severe moist and exceptional moist; respectively. SEDI is calculated as the standardized difference between AET and PET (Zeng et al. 2023) as shown in the following formula:

$$\mathrm{SEDI} = \frac{Ed - Ed_{average}}{Ed_{standard}}$$

where Ed = $AET - PET$ (mm/year).

3 Results

In order to determine the climatic trends (towards drought or humidity) for Saudi Arabia over the last few decades, seven datasets were retrieved from various sources including climatic data (*i.e.*, precipitation and air temperature) and the vegetation and drought indices (*i.e.*, NDVI, VCI, VHI, SPEI and SEDI).

3.1 Precipitation

As mentioned in the previous sections, the precipitation datasets have been collected from different ground-based stations (GbSs) and from climate-based (remote sensing) platform (NCM 2022; TRMM 2022, CMORPH 2023; ECMWF 2024). Precipitation datasets were collectively manipulated from five different sources (Table 1) in order to determine the average annually precipitation rates for the entire Saudi Arabia from these sources; especially there are 26 ground stations covering different (& representative) regions. The retrieved datasets from these stations were supported by data from remote sensing products which have wide spatial coverage. Even though, there is some data sources available from 1978, but the accounted datasets for the five sources were from 1992 to 2023 (*i.e.*, 32-years) because they have an overlap in dates. Thus, the average annual precipitation rate was estimated at 95 mm/year.

Resulted precipitation datasets from different sources is shown in (Fig. 2). The overall trend shows no remarkable descending or ascending trend in the precipitation rate for the time series between 1992 and 2023. The calculated annual average precipitation was 97.4 mm, 96.8 mm and 102.4 mm for selected time interval 1992–2000, 2001–2010, 2011–2023; respectively. While, the overall precipitation increase is about 2.5 mm which is equivalent to 2.6%. However, there is a sort of cyclic trending which can be estimated at about 5–7 years' cycle (Fig. 2).

In order to diagnose the overall trending (*i.e.*, positive or negative slope behavior) from each data source, trend analysis was applied using linear regression for the five sources, and the results were slightly different between these sources. The linear regression for the plotted datasets can be summarized in Table 6.

Considering the regression line for the five data sources in Table 6; therefore, the average will be 0.0545x, which shows a positive trend and then indicating an increase in the precipitation rate.

$$(-0.0854x) + (0.0942x) + (-0.0183x) + (-0.0163x) + (0.0803x) = 0.0545x$$

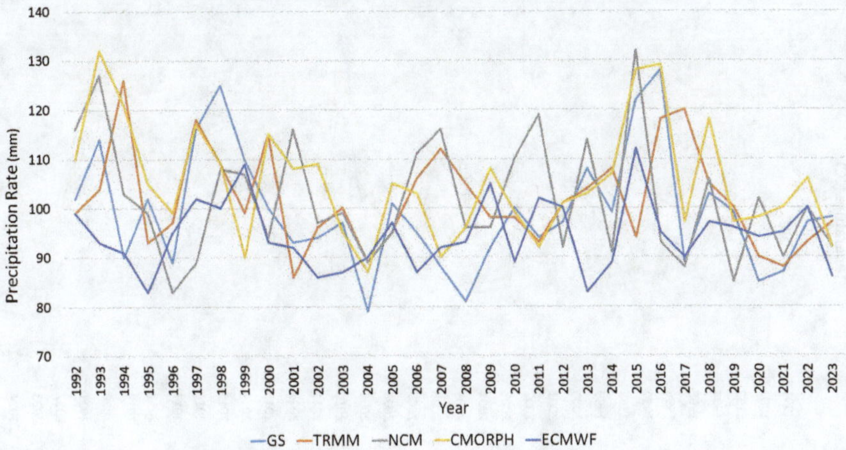

Fig. 2 Average annual precipitation rates of Saudi Arabia as extracted from different sources for the period between 1992 and 2023

Table 6 Summarized results of the regression line calculated for the precipitation retrieved from different sources

Data source	Calculated regression linear	Precipitation trend	Description
Ground-based stations (GbSs)	– 0.0854x + 101.1	Negative	Slight decrease in precipitation rate
TRMM	0.0942x + 101.6	Positive	
NCM	– 0.0183x + 102.9	Negative	Very low decrease in precipitation rate
CMORPH	– 0.0163x + 104.6	Negative	
ECMWF	0.0803x + 94.1	Positive	Slight increase in precipitation rate

3.2 Temperature

Air temperature datasets were also collected from ground stations and from climate-based platforms including: CPC-NOAA (2022), EN (2023) and WCG (2024). The datasets cover 34 years (*i.e.*, 1991–2024). The selection of this time period is also based on the most overlapped dates from the available sources. The average monthly rate of air temperature was available in many of these sources, and they were converted into annual rate to perform the trending illustration (Fig. 3). The calculated average temperature was 24.8 °C, 25.4 °C and 26.1 °C for selected time intervals 1991–2000, 2001–2010, 2011–2024; respectably. From all available data sources, it is obvious that there is an increasing trend in the air temperature over the last 34 years which is estimated at about 0.5 °C, while the average annual air temperature was estimated at 25.7 °C.

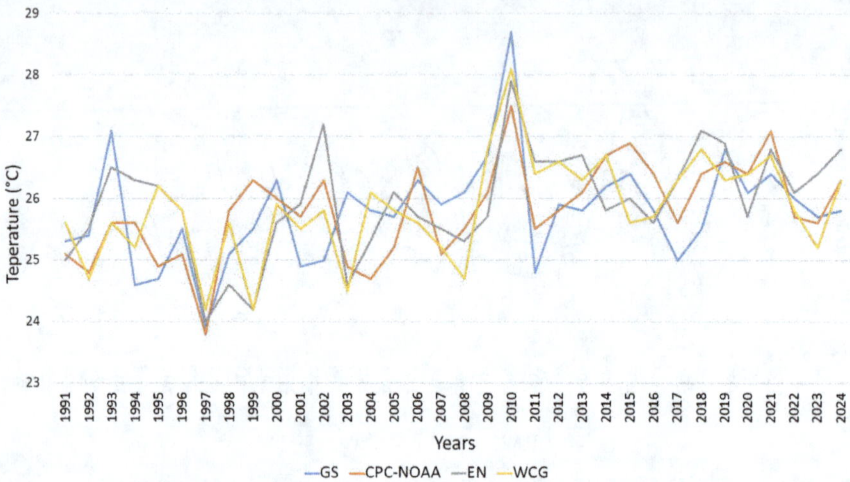

Fig. 3 Average annual air temperature of Saudi Arabia extracted from different sources for the period between 1992 and 2023

Table 7 Summarized results of the regression line calculated for the air temperature retrieved from different sources

Data source	Calculated linear regression	Temperature trend	Description
Ground-based stations (GbSs)	0.0273x + 25.31	Positive	Very low increase in air temperature rate
CPC-NOAA	0.0429x + 25.06		Slight increase in temperature rate
EN	0.0395x + 25.259		
WCG	0.0377x + 25.169		

Linear regression analysis was also performed to determine the overall trending in air temperature for each data source separately. However, all sources show remarkable increase in air temperature even though there are diverse values between these sources. The results of the linear regression averages to about 0.03685x, indicating an overall increase of about 1.94% and this can be summarized in Table 7.

3.3 Normalized Difference Vegetation Index (NDVI) Anomaly

NDVI for Saudi Arabia has been retrieved from the FAO (GIEWS) (Xie et al. 2007) platform for the period between 1984 and 2024. Data was downloaded for each year separately, and then it was compared with the long-term average (LTA) for Saudi

Arabia. The average NDVI is 0.19 which characterizes a light drought condition. NDVI relative difference (*i.e.*, anomaly) with respect to the LTA was also calculated to figure out whether the annual values are aligned above or below the average (Fig. 4). This was also supported by the extraction of NDVI anomaly map to visualize geographic extent NDVI (examples are shown in Fig. 5).

NDVI peaks are clearly observed in the NDVI trends (Fig. 4), such as the peak between 1998 and 2003, yet the overall trend is ascending and it is aligned above the LTA (*i.e.*, 0.19), evidencing more humidity condition. The remarkable increase

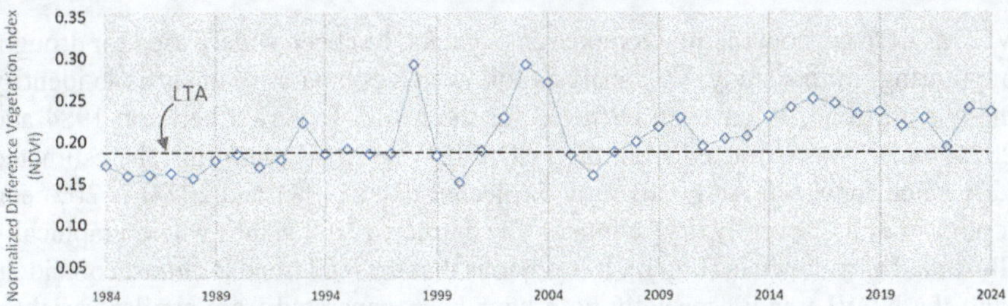

Fig. 4 Diagram showing the changing NDVI with respect to LTA between 1984 and 2024

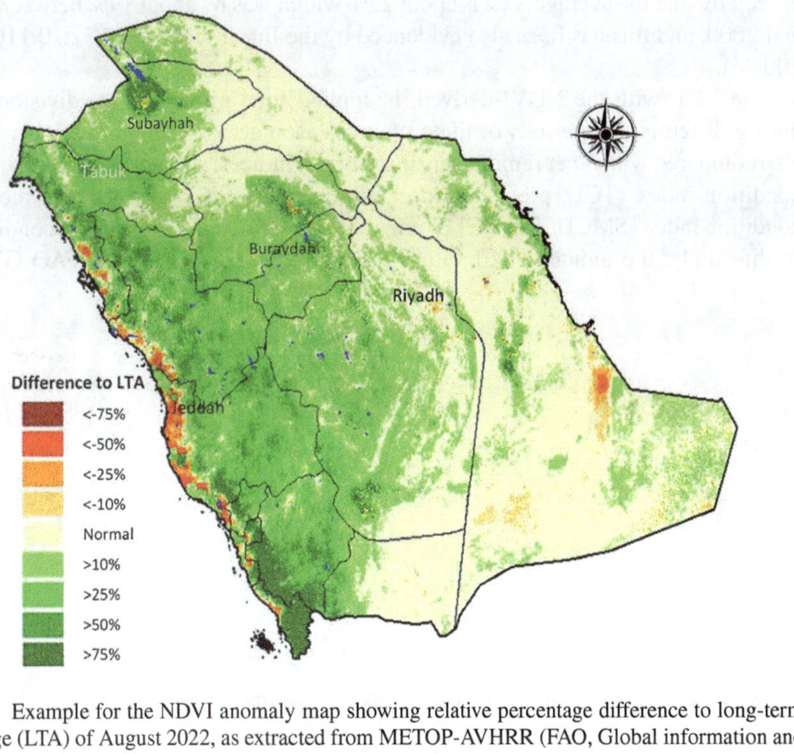

Fig. 5 Example for the NDVI anomaly map showing relative percentage difference to long-term average (LTA) of August 2022, as extracted from METOP-AVHRR (FAO, Global information and Early Warning System-GIEWS 2024)

in the NDVI in Saudi Arabia was obvious since 2009 (Fig. 6) which average about 0.22 with estimated increase of about 13.6% with the previous time before 2010, and this can be attributed to the increased agricultural activities that has been reflected in the NDVI values; especially that land areas with known agricultural activities (*e.g.*, Subayhah, Buraydah, *etc.*) are clearly observed on the recent NDVI maps (Fig. 5).

3.4 Vegetation Condition Index (VCI)

VCI, as derived from thermal remote sensing data, has been widely used for drought monitoring. In this study, VCI analysis followed a comparative analysis to identify the overall trend. It has been retrieved for the available period between 1984 and 2024, which was extracted from the FAO GIEWS (2024) platform. The estimated LTA value for Saudi Arabia over the projected dates (1984 and 2024) is 20% and it points out to "severely dry" climate. The extracted VCI values where graphically illustrated as a shown in (Fig. 6). It is obvious that the VCI trend is almost coincident with the NDVI trend; especially both have been generated from similar variables and this is anticipated since VCI compares the current NDVI to the range of values observed in previous years. (Fig. 6) shows an increase over the projected time series; especially that the average VCI is about 23% which was by about 19% between 1984 and 2000, and this has been also evidenced by the linear regression (*i.e.*, 0.1106x – 200.36).

The VCI, with the NDVI-derived, is applied for various climate divisions and during different sub-periods of the growing season across different regions. It was also compared with other remote-sensing drought indices, such as the air temperature condition index (TCI), precipitation condition index (PCI) and the soil moisture condition index (SMCI). The VCI is remarkably interlinked with climatic conditions. In this study, the monthly VCI values were also retrieved from the FAO GIEWS

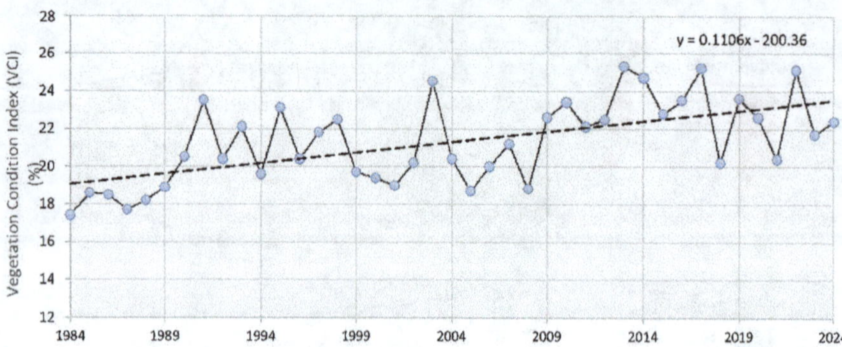

Fig. 6 Annual changes in VCI over Saudi Arabia between 1984 and 2024

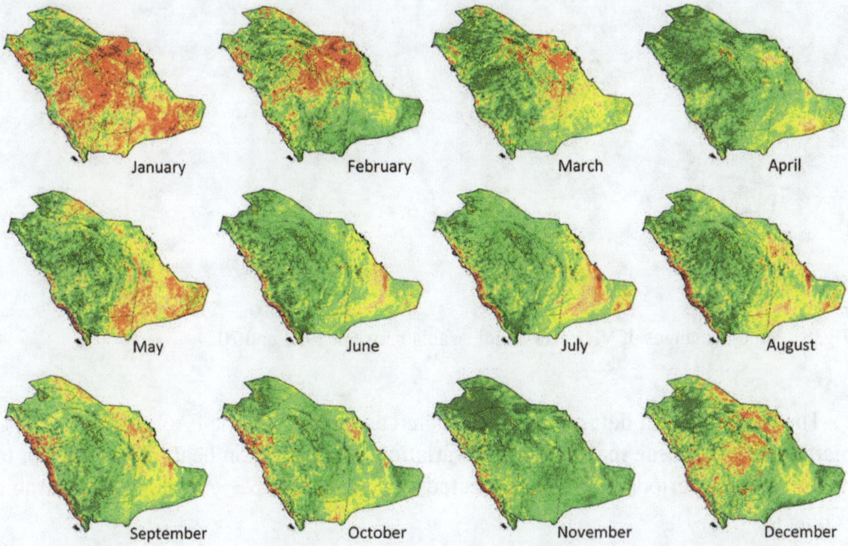

Fig. 7 Representative monthly variation in VCI over Saudi Arabia for the year 2023

(2024) platform where it shows variations of in vegetation conditions with respect to different climate and time periods over a selected year 2023 for Saudi Arabia (Fig. 7).

3.5 Vegetation Health Index (VHI)

The VHI has been instrumental in assessing the impact of drought on agricultural, forestry, ecological and environmental sectors, aiding in the development of mitigation and adaptation strategies to reduce losses caused by drought events (Youssef et al. 2016). It is an elementary indicator used to compute the seasonal drought indicators in the Agricultural Stress Index System (ASIS) including Agricultural Stress Index (ASI), Drought Intensity and Weighted Mean Vegetation Health Index (Mean VHI).

Similar to VCI, the VHI for Saudi Arabia has been retrieved over the period between 1984 and 2024 and this was also extracted from the FAO GIEWS (2024) platform. Annual data retrieve was performed, and the calculated long-term average (LTA) was calculated at 44.07%, which characterizes Saudi Arabia with "normal" drought in terms of vegetation health index. The datasets were graphically illustrated to show the overall trend (Fig. 8). The resulting diagram shows a remarkable increasing in the trend of VHI values for Saudi Arabia between 1984 and 2024, and this increase reports the average of about 52.6% which equals to about 16%, and this was also evidenced by the linear regression (*i.e.*, 0.5136x − 982.72).

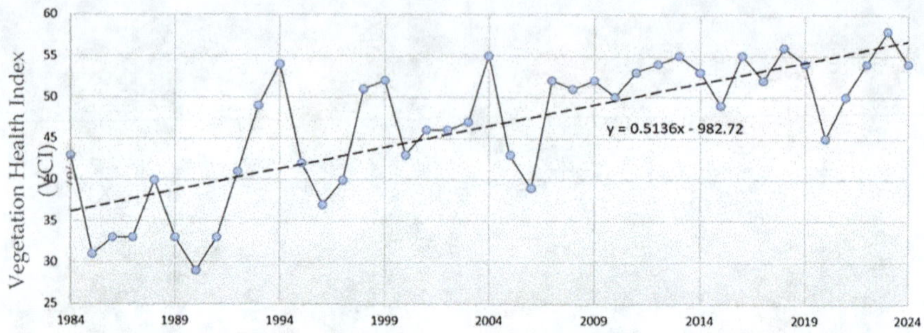

Fig. 8 Annual changes of VHI over Saudi Arabia between 1984 and 2024

The monthly VHI datasets were extracted as maps from the FAO GIEWS (2024) platform where these maps show the variations in vegetation health with respect to different time periods over the projected typical year 2023 year for Saudi Arabia (Fig. 9).

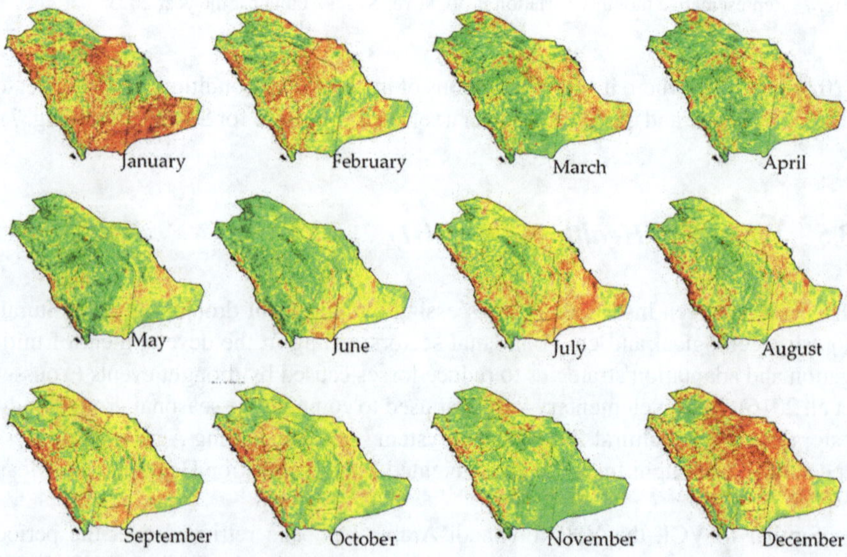

Fig. 9 Representative monthly VHI variation over Saudi Arabia for the selected year 2023

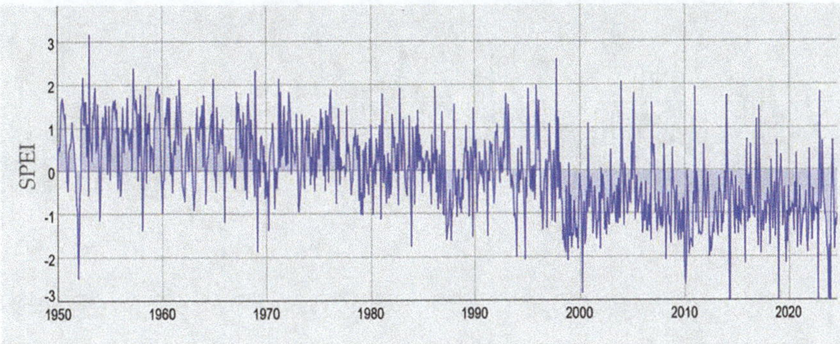

Fig. 10 Temporal evolution of SPEI for Saudi Arabia between 1950 and 2024 (LCSC 2024)

3.6 Standardized Precipitation Evapotranspiration Index (SPEI)

In this study, the SPEI datasets were retrieved from the Laboratory of Climate Services and Climatology (LCSC 2024). Recently, SPEI has become the most adopted drought indices; especially it integrates the input/output (i.e., precipitation and evapotranspiration) of the hydrologic system. The investigated period for SPEI in this study was between 1950 and 2024, and this represents a significant time series to figure out the climatic conditions of Saudi Arabia over this long-term time series.

Results show obvious oscillations in the SPEI (Fig. 10), with an average rate of about 0.39 for the period between 1950 and 2000 (Fig. 10), and this characterizes drought of Saudi Arabia as "near normal" according to Vicente-Serrano et al. (2010). However, from the year 2000 up to date the average has been changed to about − 0.51 which indicated moving towards a "mild drought" Fig. 10). The interval in the oscillating (*i.e.*, maximum or minimum SPEI) has been widened from about 1.9 before 2000 (1950–2000) to about 2.7 after 2000; indicating abrupt climatic change over different time periods, and this can be attributed to the existence of extreme climatic events in the region. While, the overall SPEI trend over the period between 1950 and 2024 is remarkably descending by an estimated percentage of about 43.5%.

3.7 Standardized Evapotranspiration Deficit Index (SEDI)

Compared with other indices, SEDI can reasonably detect drought and climatic variability, and it can more sensitively capture the biological changes of the ecosystems in response to the dynamics of drought intensity. SEDI is also more practical than the precipitation and air temperature indices and specifically in determining signals of biological effects in climate. Hence, it has potential use in assessments of climate change and its impact on ecosystems (Zhang et al. 2019). In this study, SEDI datasets

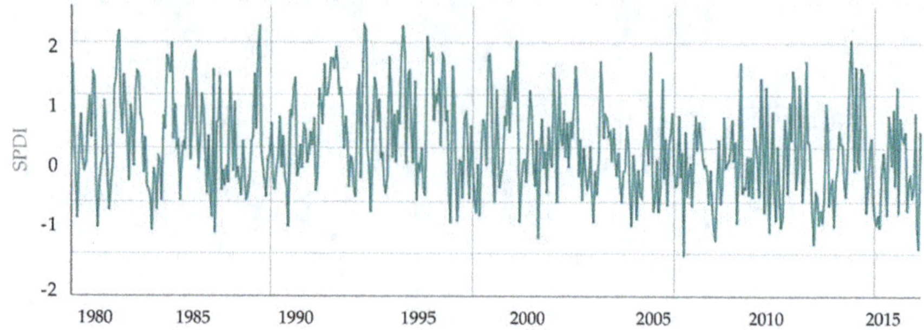

Fig. 11 Temporal evolution of SEDI for Saudi Arabia between 1982 and 2022 (LCSC 2024)

were retrieved from the Laboratory of Climate Services and Climatology (LCSC 2024) for the available period between 1982 and 2022.

Results are almost coincident with those of SPEI notably there are many oscillations over the projected time period (Fig. 11), but the oscillating intervals remained the same over the whole projected time period, and this interval was estimated at 2.3. The average SEDI is about − 0.15 which characterizes Saudi Arabia as "near normal" drought. The overall SEDI trend shows a change from 0.22 before 2000 to − 0.13 indicating very slight descending change over the period between 1982 and 2022 which was estimated at about 20.5%.

4 Discussion

Climate trends are always anticipated to change to more dryness due to the impact of greenhouse gasses (GHGs) and then the global warming which is represented by decreased precipitation and increased air temperature. This is often considered for humid or sub-humid regions or for regions with mild aridity; however, this is not quite accurate if climatic trends are applied for arid regions like the Arabian Peninsula, which is already a dry region.

In Saudi Arabia, water availability remains a challenge with the existing arid climate, and this imposes implementing development projects focusing on the use of non-conventional water resources (*e.g.*, desalinated water, harvested water, etc.) to assure sustainable water supply. Yet, there are numerous studies done to characterize the climate of Saudi Arabia which is known as dry; especially desert lands cover more than 65% of the Arabian Peninsula. Reversibly, there are frequent spells of torrential rainfall resulted in damaging floods in Saudi Arabia, and the volume of precipitated water has remarkably increased; in addition, the area of the green cover has been rapidly increased.

There are many examples evidencing the increased vegetation cover besides least water availability in Saudi Arabia. For instant, the average NDVI surface area has been increased since 1984 by about 8% (about 172.000 km^2) of the Saudi Arabia, while the vegetation cover has been increased from 1.4 to 8.5% between 2018 and 2024 in Saudi Nature Reserve, as well as the vegetation cover has been increased by 360% during five months' rainfall of 2023 in Madinah (NCVC 2024). From the other side, water supply in Saudi Arabia was estimated at 706 m^3/capita/year and this ranks it as one of the first five Arab countries with sufficient water supply, but water availability remains few enough and does not exceed 70 m^3/capita/year (UNESCO 2021).

The aforementioned examples do not reflect any abrupt drought conditions in Saudi Arabia, or at least they conflict with the concept of dominant desert climate represented by low precipitation and high actual evapotranspiration. This in turn raised a question: Will the climate of Saudi Arabia (& the entire Arabian Peninsula) is moving towards more humidity? This study was initiated in order to answer this question, especially that drought indices show either slight drought impact, negligible increase in drought, or decreased drought. This is a clear output from this study where diverse resultant exposed from performed climatic, vegetation and drought indices. Table 8 summarizes the resulted finding in this study where diverse outcomes have been reached.

Data illustrated in Table 8 shows more weighting towards least drought and much deviation towards humid climate than usual in Saudi Arabia, and this conflicts with the common understanding, and this matches with the IPCC sixth Assessment Report on Climate Change (IPCC 2022) which anticipated changing climate conditions in the entire Arabian Peninsula towards more precipitation rate associated with decrease in drought. Therefore, Table 8 shows a slight increase precipitation increase bedside an increase of 0.5 °C in air temperature, and this evidences no remarkable change in the climatic conditions in Saudi Arabia which are already dry. However, all investigated vegetation indices (NDVI, VCI, VHI) showed changes towards more green cover and indicating more wetness. Nevertheless, the performed drought indices (SPEI, SEDI)

Table 8 Summary of the analyzed variables and indices for Saudi Arabia for the period between 1992 to 2023

Variables/ indices	Precipitation	Air temperature	NDVI	VCI	VHI	SPEI	SEDI
Measured value	2.5 mm increase	0.5 °C increase	0.22	23%	52.6%	– 0.51	– 0.13%
Description	Very slight precipitation increase	Temperature increase	Light drought	Moderately dry	Good	Mild drought	Near normal
Result	2.6% increase in precipitation	1.94% increase in temperature	13.6% increase	13% increase	16% increase	43.5% increase above normal (more dry)	20.5% increase above normal (more dry)

showed increase above normal which points out to slight drought that was descried as "mild drought" for SPEI and "near normal drought" for the SEDI (Table 8).

5 Conclusion

Even though, there are many studies done (Almazroui 2011, 2013), yet he climate of Saudi Arabia is not well characterized notably that many studies ended up with contradictory results (Al Saud 2024). This study has been raised to shed light on this issue and to provide a comprehensive conclusion on the climate of Saudi Arabia, as well as to highlight the recent concepts of climate change done by international and regional organizations. Thus, the trends in climatic of Saudi Arabia are not exactly as it is believed, and it is not moving towards more aridity. In this respect, ESCWA (2023) stated that precipitation will increase in the Arabian Peninsula by about 25% in 2060; while McCabe et al. (2023) estimated an increase between 7 and 33% over the next few decades. In addition, the IPCC 6th Assessment Report on Climate Change (IPCC 2022) anticipating an increase in precipitation rate the next few decades. Evidences of this increase has been raised and it is marked through the graphically established trends established in this study.

This study, with performed climatic, vegetation and drought indices, concludes that precipitation over the last few decades has not been remarkably regressed; however, a slight increase in air temperature was reported. This, in general, does not point out to tangible changes in the climate of Saudi Arabia. Nevertheless, the vegetation cover, showed an obvious geographic increase; especially the investigated indices obviously showed this increase which is estimated between 13% and 16% according to Table 8. This remarkable increase in the vegetation cover does not match with the identified climatic conditions; and therefore, the increased vegetation can be attributed to developed water and agricultural projects whether for the use of non-conventional or the increasing agricultural activities that have been executed by many initiatives and programmes such as Saudi Green Initiative (SGI) and the National Center for Vegetation Development and Combating Desertification (NCVC).

From the other side, the calculated drought indices show a gradational trend towards more aridity; especially for the SPEI which was analyzed over a long-term time series (1950 to date), but less trend with SEDI between 1980 and 2022. This deviation towards aridity might be attributed to the increased air temperature rather than any noticeable change in precipitation. The changing SPEI and SEDI towards aridity might be interpreted by the slight increase in temperate besides the increased precipitation. In this respect, the retrieved SPEI evidenced that the change has remarkably begun since1995s, but the SEDI shows more cyclic trends, and this in turn refers to an expected general cyclic trends for both drought indices.

It can be generally concluded that over the last three decades, a very gentle increase in the precipitation rate and a slight increase in the air temperature rate have been reported. Besides, a remarkable widening in the vegetation cover in the entire Saudi Arabia as deduced by the performed vegetation indices. Nevertheless, the applied drought indices (*i.e.*, SPEI and SEDI) showed a continuous ascending towards aridity since the last seven decades, but this may not be significant since the region is already dry and the temperature might play a role in the increased dryness which is well pronounced as a result of global warming.

Based on the above discussion and the reached results, the understanding of deviated climate towards more dry in Saudi Arabia is not accurate, and there must be confidence that the overall climate of the entire Arabian Peninsula is slightly moving to-wards humid than the normal. This must be considered while proposing management plans and strategies; especially for water and agricultural management. Moreover, it should be always considered that the precipitated volume of water on Saudi Arabia is significant enough to be managed notably along wadis and this can be integrally invested by groundwater artificial recharge to feed the depleted aquifers instead of resulting in numerous floods and torrents as the current case in many regions of Saudi Arabia.

Acknowledgments The author extends her acknowledgement to the supportive data sources where from many climatic variables and drought indices have been retrieved; namely, NASA, FAO, GAMEP, ESCWA, ECMWF, WCG, TRMM, CPC-NOAA, CMORPH, LCSC, EN and NCM.

References

Al Saud M (2015) Flood control management for the city and surroundings of Jeddah, Saudi Arabia. Springer. New York, London, p 177. https://doi.org/10.1007/978-94-017-9661-3. 978-94-017-9660-6

Al Saud M (2020) Sustainable land management for NEOM region. Springer International Publishing, p 219. 978-30-305-7633-2

Al Saud M (2024) Mapping drainage basins of the Kingdom of Saudi Arabia. 20 map sheets (1:500.000). Technical Study (Under production)

Almazroui M (2011) Calibration of TRMM rainfall climatology over Saudi Arabia during 1998–2009. Atmos Res 99(3):400–414. https://doi.org/10.1016/j.atmosres.2010.11.006

Almazroui M (2013) Simulation of present and future climate of Saudi Arabia using a regional climate model (PRECIS). Int J Climatol 33:2247–2259. https://doi.org/10.1002/joc.3721

Almazroui M (2020) Changes in temperature trends and extremes over Saudi Arabia for the period 1978–2019. Adv Meteorol 2020(1):8828421. https://doi.org/10.1155/2020/8828421

Baban S (2022) The suitability of satellite remote sensing and GIS technologies for mapping, monitoring and managing water resources in the Middle East. In: Shaban A (ed) Satellite monitoring of water resources in the Middle East. Springer water. Springer, Cham. https://doi.org/10.1007/978-3-031-15549-9_3

Chen H, Wang S (2022) Accelerated transition between dry and wet periods in a warming climate. Geophys Res Lett 49(19):e2022GL099766

CMORPH (Center Morphing Technique) (2023) Precipitation data. https://www.ncei.noaa.gov/data/cmorph-high-resolution-global-precipitation-estimates/access/

CPC-NOAA (Climate Prediction Center (CPC) at the National Oceanic and Atmospheric Administration (NOAA)) (2022). http://www.cpc.ncep.noaa.gov/

ECMWF (European Centre for Medium-Range Weather Forecasts) (2024) Precipitation data. https://www.ecmwf.int/

El-Samra R, Bou-Zeid E, Bangalath H, Stenchikov G, El-Fadel M (2017) Future intensification of hydro-meteorological extremes: downscaling using the weather research and forecasting model. Climate Dyn 49:3765–3785. https://doi.org/10.1007/s00382-017-3542-z

EN (Environmental Horizon) (2023) Temperature data. https://envirodata.horizons.govt.nz/

ESCWA (2023) Regional climate modelling outputs for Saudi Arabia: key findings. In: Technical report. Regional initiative for the assessment of climate change impacts on water resources and socio-economic vulnerability in the Arab Region (RICCAR), 25 p

FAO (Food and Agriculture Organization) (2024) Global information and early warning system (GIEWS). https://www.fao.org/giews/earthobservation/asis/index_2.jsp?lang=en

GAMEP (The General Authority of Meteorology and Environmental Protection) (2012) Information and statistics. https://www.mewa.gov.sa/en/InformationCenter/Pages/default.aspx

Harris I, Jones P, Osborn T, Lister D (2014) Updated high-resolution grids of monthly climatic observations—the CRU TS3.10 datasets: updated high-resolution grids of monthly climatic observations. Int J Climatol 34:623–642. https://doi.org/10.1002/joc.3711

Hasanean H, Almazroui M (2015) Rainfall: features and variations over Saudi Arabia, a review. Climate 3(3):578–626. https://doi.org/10.3390/cli3030578

IPCC (2022) The working group II contribution to the IPCC 6th assessment report on climate change 2022: impacts, adaptation and vulnerability

Kogan F (1995) Application of vegetation index and brightness temperature for drought detection. Adv Space Res 15:91–100

LCSC (Laboratory of Climate Services and Climatology) (2024) Standardized precipitation evapotranspiration index. https://spei.csic.es/spei_database/#map_name=spei01#map_position=1463

Li X, Ju H, Sarah G, Yan C-R, Batchelor W, Liu Q (2017) Spatiotemporal variation of drought characterstics in the Huang-Huai-Hai plain, China under the climate change scenario. J Integr Agric 16(10):2308–2322

McCabe M, Al Shalan M, Al Saud N, Hejazi M, Wada W, Peixoto R, Duarte C, Beck H, Maestre F, Guirado E (2023) Climate futures report: Saudi Arabia in a 3-degrees warmer world. KAUST King Abdullah University of Science and Technology, AEON Collective, p 133

Mehr A (2021) Drought classification using gradient boosting decision tree. Acta Geophys 69(3):1–10

NCM (National Center for Meteorology) (2022). https://ncm.gov.sa/Ar/Climate/Pages/default.aspx

NCVC (National Center for Vegetation Cover Development and Combating Desertification) (2024). https://spa.gov.sa/en/N2028797

Tirivarombo S, Eliasson D (2018) Drought monitoring and analysis: standardised precipitation evapotranspiration index (SPEI) and standardised precipitation index (SPI). Phys Chem Earth Parts A/B/C 106:1–10. https://doi.org/10.1016/j.pce.2018.07.001

TRMM (Tropical Rainfall Mapping Mission) (2022) Rainfall data. http://trmm.gsfc.nasa.gov

UNESCO (2021) Assist strengthening the knowledge base of the IHP National Committees in the member states in the Arab region. Unpublished Technical Report by UNESCO-IHP-Cairo Office, 44 p

USGS (US Geological Survey) (2010) Landsat normalized difference vegetation index. https://www.usgs.gov/landsat-missions/landsat-normalized-difference-vegetation-index

Vicente-Serrano S, Miralles D, Domínguez-Castro F, Azorin-Molin C, El Kenawy A, McVicar T, Tomás-Burguera M, Beguería S, Maneta M, Peña-Gallardo M (2018) Global assessment of the standardized evapotranspiration deficit index (SEDI) for drought analysis and monitoring. J Clim 41(14):5371–5393. https://doi.org/10.1175/JCLI-D-17-0775.1

Vicente-Serrano S, Van der Schrier G, Beguería S, Azorin-Molina C, Lopez-Moreno J (2015) Contribution of precipitation and reference evapotranspiration to drought indices under different climates. J Hydrol 526:42–54

Vicente-Serrano SM, Beguería S, López-Moreno L (2010) A multi-scalar drought index sensitive to global warming: the standardized precipitation evapotranspiration index—SPEI. J Clim 23:1696–1718

WB (World Bank) (2021) Climate change portal for development and practitioners and decision makers. https://climateknowledgeportal.worldbank.org/country/saudi-arabia/climate-data-historical

WCG (World Climate Guide) (2024) Temperature data. https://www.climatestotravel.com/temperatures/saudi-arabia

Wehbe Y (2022) Unraveling the spatio-temporal dynamics of satellite-inferred water resources in the Arabian peninsula. In: Shaban A (ed) Chapter in book: satellite monitoring of water resources in the Middle East. Springer Publisher. https://doi.org/10.1007/978-3-031-15549-9_7

Xie P, Chen M, Yang S, Yatagai A, Hayasaka T, Fukushima Y, Liu C (2007) A gauge-based analysis of daily precipitation over East Asia. J Hydrometeorol 8:607–626. https://doi.org/10.1175/JHM583.1

Youssef A, Sefry S, Pradhan B, Abu Alfadail E (2016) Analysis on causes of flash flood in Jeddah city (Kingdom of Saudi Arabia) of 2009 and 2011 using multi-sensor remote sensing data and GIS. Geomat Nat Hazards Risk 7:1018–1042

Zeng J, Zhou T, Qu Y, Benton V, Qi J, Xu Y, Li Y (2023) An improved global vegetation health index dataset in detecting vegetation drought. Sci Data 10:338. https://doi.org/10.1038/s41597-023-02255-3

Zhang X, Li M, Ma Z, Yang Q, Lv M, Clark R (2019) Assessment of an evapotranspiration deficit drought index in relation to impacts on ecosystems. Adv Atmos Sci 36(11):1273–1287. https://doi.org/10.1007/s00376-019-9061-6

Trends and Future Challenges for Regional Safety in the Middle East: The Case of Growing Drought Crisis in Syria

Michael Skaf and Shifa Mathbout

1 Introduction

The increasing impacts of climate change on environmental sustainability and human resources development have become a critical global concern (Chen et al. 2023; Din et al. 2024). The frequency and intensity of concurrent drought and extreme heat events are rising, with severe economic and societal consequences worldwide (Kong et al. 2020; Hao et al. 2022; Suarez-Gutierrez et al. 2023). Dryland ecosystems, particularly under global warming, are experiencing escalating risks and vulnerabilities (Abel et al. 2023; Zhang et al. 2024; Blicharska et al. 2024). In the South-Eastern Mediterranean (SEMED) and the Middle East, the increasing occurrence of droughts and water shortages is inducing substantial shifts in terrestrial ecosystems, exacerbating food insecurity and altering ecosystem structure and function (Hameed et al. 2020). The SEMED region is particularly vulnerable to compound hot-dry events, defined as simultaneous occurrences of extreme heat and drought that intensify socio-economic vulnerabilities (Lazoglou et al. 2024; Santos et al. 2024; Vicente-Serrano et al. 2025a, b). Under ongoing global warming, compound hot and dry events have become more frequent and severe, exerting unprecedented pressure on agriculture, water resources, and economic stability (Furtak and Wolinska 2023; Tabari and Willems 2023; Ying et al. 2024). The Mediterranean region, identified as a global climate change hotspot, is warming at a rate exceeding the global means by

---

M. Skaf
Forestry and Ecology Department, Agriculture Engineering Faculty, Latakia University, Latakia, Syria

S. Mathbout (✉)
Centre of Excellence in Risk and Decision Sciences (CERIDES), European University, Nicosia, Cyprus
e-mail: shifamathbout@yahoo.com; s.mathbout@research.euc.ac.cy

Climate Change and Landscape Ecology Group, Department of Geography, University of Barcelona, Barcelona, Spain

M. M. Al Saud (ed.), *Hydroclimate Changes: Implications of the Middle East Region*,
https://doi.org/10.1007/978-3-032-20053-2_9

20% annually and up to 50% during summer months (Lionello and Scarascia 2018). This warming, coupled with rising atmospheric evaporative demand, is amplifying drought severity (Dai et al. 2018) and prolonging arid conditions (Mathbout et al. 2021; Gimeno-Sotelo et al. 2024). Large-scale teleconnection patterns, such as the North Atlantic Oscillation (NAO), the East Atlantic, West Russia (EAWR) pattern, and the Scandinavian (SCAND) pattern-exert a strong influence on year-to-year climate variability. For instance, positive NAO phases are typically linked to high-pressure systems and decreased rainfall in the Levant, whereas positive EAWR and SCAND phases tend to enhance moisture transport and precipitation over portions of the Eastern Mediterranean (Mathbout et al. 2021). These complex interactions between atmospheric circulation shifts and teleconnection patterns are central to understanding the region's drought dynamics.

Syria, situated at the crossroads of Asia and the Mediterranean, is among the most water-stressed regions globally, experiencing recurrent and intensifying droughts driven by rising temperatures and increasing evaporation rates (Barlow et al. 2016). Over the past four decades, multiple severe drought events have been recorded, notably in 1999, 2010, 2014, 2017, and 2021 (Cook et al. 2016; Barlow et al. 2016; Mathbout et al. 2018; Alsafadi et al. 2022). Long-term drought analysis (1981–2021) reveals an increasing trend in aridity, characterized by prolonged and severe dry periods (Mathbout et al. 2025). This persistent aridification has led to significant declines in agricultural productivity, exacerbating socio-political instability and contributing to regional conflicts (Saleem et al. 2024; Figari 2025). The geographical expansion of arid conditions is evident, with semi-arid regions transitioning into arid zones at a rate of 6.2% over recent decades, driven by sustained arming (Houmsi et al. 2019). This eastward and southward spread of aridity underscores the urgent need for climate adaptation strategies, particularly in Mediterranean countries where socio-economic vulnerabilities are compounded by climate change (Abel et al. 2019; Nastos and Saaroni 2024; Favretto and Stringer 2024). The 2006–2010 multi-year drought in Syria was the most prolonged and intense in the last 900 years, triggering a sharp decline in crop yields and devastating agricultural communities (Cook et al. 2016). This extreme event precipitated the collapse of rural livelihoods, driving mass displacement and heightening the risk of regional conflicts, ultimately contributing to Syria's descent into war (Kelley et al. 2017; Linke and Ruether 2021). The ongoing trend of increasing aridity due to climate change is expected to reshape the structural and functional dynamics of global drylands (Zhang et al. 2024). In Syria, persistent warming has already driven significant geographic shifts in aridity, transforming 6.2% of semi-arid lands into arid zones over recent decades (Houmsi et al. 2019). This expanding aridification, coupled with socio-political vulnerabilities, underscores the necessity for urgent climate resilience measures to mitigate the escalating risks of prolonged droughts and environmental degradation (Mori et al. 2024; Mezieobi et al. 2025).

2 Study Area, Data and Methodology

2.1 Climatic Conditions and Precipitation Patterns in Syria

Syria, located in southwestern Asia north of the Arabian Peninsula, experiences a Mediterranean climate marked by hot, dry summers and mild, rainy winters, with colder and more humid winter conditions occurring further inland (Lionello and Scarascia 2018; Abou Zakhem and Hafez 2010). Precipitation patterns vary significantly across the country, ranging from 300 to 500 mm annually along the coast to as little as 30 mm in the arid southeastern deserts (Mathbout et al. 2018; Homsi et al. 2021). Nearly 90% of Syria receives <200 mm of annual rainfall, with precipitation peaking during the winter months and dropping to minimal levels during the driest periods, especially in arid and semi-arid regions (Fig. 1). Orographic effects, combined with prevailing atmospheric patterns, amplify precipitation in elevated western and southwestern areas (Abou Zakhem and Hafez 2010). The southern region typically receives over 450 mm of rainfall annually, while the western plains and foothills get between 300 and 600 mm. Coastal zones receive the highest precipitation, ranging from 750 to 1000 mm, whereas the northern areas bordering Turkey receive between 400 and 500 mm. Mean annual temperatures vary from approximately 10 °C in the southwestern Anti-Lebanon Mountains to 22 °C in the eastern deserts, with coastal plains generally exceeding 15 °C. Syria receives substantial solar radiation, averaging around 18,720 kJ m^{-2} day^{-1}, with the highest levels recorded in the inland regions.

These areas, characterized by arid and semi-arid climates, experience intense solar exposure, making them ideal for solar energy applications. Additionally, wind speed at height of 10 m exceeds 4 m/s in these inner regions, highlighting strong potential for wind energy development alongside solar power. The country's climate is influenced by three primary air masses: the continental polar air mass from northern and northeastern Europe, the continental air mass from North Africa responsible for "Khamasin depressions," and air masses originating from the Atlantic Ocean and Western Europe, which play a significant role in shaping Syria's rainfall patterns (Al Charideh and Abou Zakhem 2010).

2.2 Climate Datasets

The climate parameters analyzed in this study were obtained from the Climatic Research Unit (CRU) of the University of East Anglia (CRU TS v4.08; Kogan 2007), with a spatial resolution of 0.5° × 0.5°, accessible at (https://crudata.uea.ac.uk/cru/data/hrg/). The CRU dataset is recognized for its reliability and advanced gap-filling techniques, making it suitable for long-term climate analyses (García-León et al. 2019; Morsy et al. 2022; Gorgin Paveh et al. 2022; Liu et al. 2023; Araghi and Martinez 2024; Alemayehu et al. 2024; Wei et al. 2025). Given the scarcity

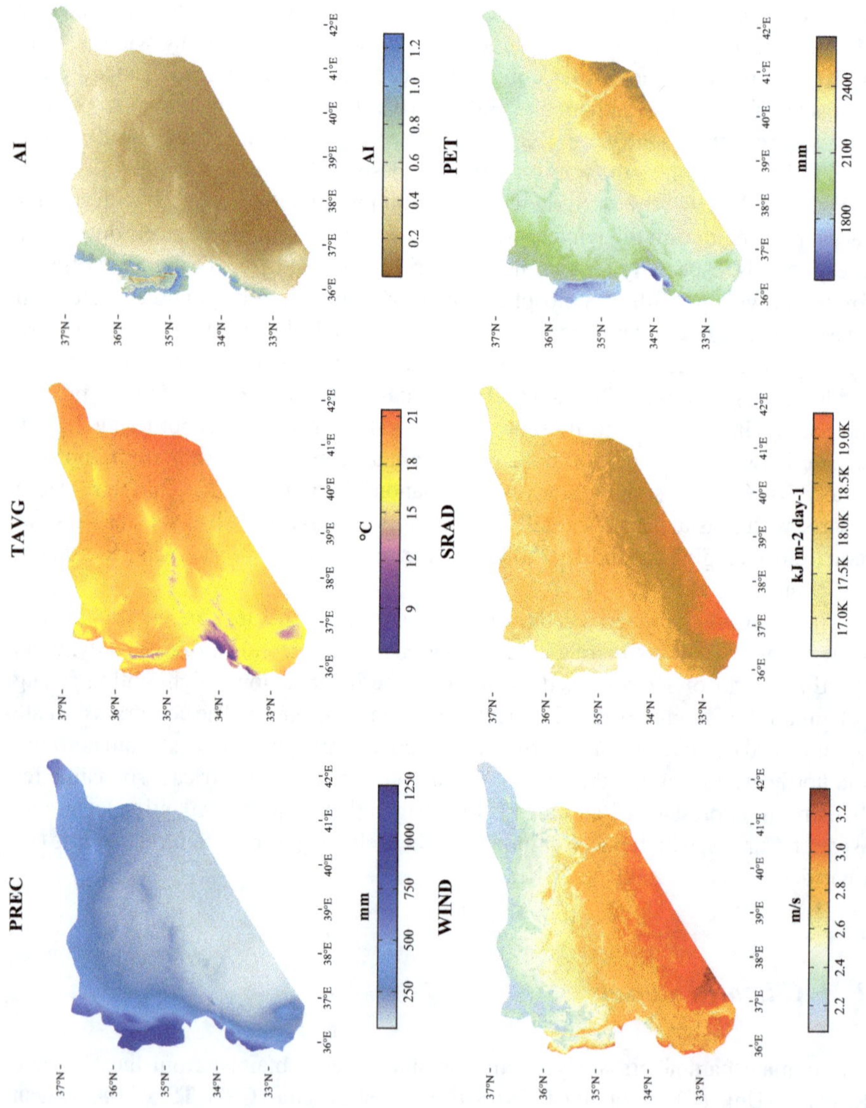

Fig. 1 Spatial distribution of historical total annual precipitation (PREC), mean annual temperature (TAVG), aridity index (AI), wind speed (WIND), solar radiation (SARD), and annual potential evapotranspiration (PET) in Syria during 1961–2023

of meteorological stations and significant gaps in precipitation records, particularly following the Syrian Crisis in 2011, CRU data was selected for its historical consistency and relevance to Syria from 1961 to 2023. The CRU TS v4.08 dataset provides essential climate variables, including precipitation (pre, mm), potential evapotranspiration (pet, mm), and average temperature (Tavg, °C). The accuracy of the CRU TS gridded dataset in Syria has been validated by Mathbout et al. (2023, 2025), exhibiting strong correlation coefficients (CC) when compared to observed temperature and precipitation records. High CC and Nash–Sutcliffe Efficiency (NSE) values were consistently recorded, exceeding 0.70 for precipitation during the rainy season and surpassing 0.91 for temperature across all seasons. Additionally, low Root Mean Square Error (RMSE) values were observed, remaining below 1.0 for temperature and under 3.0 for precipitation. These results highlight the CRU dataset's reliability in accurately representing observed climatic conditions across Syria, particularly for multi-year averaged monthly timescales. To enhance the understanding of drought characteristics and their impacts in Syria, this study also utilises an observed daily precipitation dataset. The original daily series were thoroughly validated and reconstructed using the R package RClimdex (Aguilar and Prohom 2011; Zhang et al. 2011), which enables the detection of duplicate dates, checks for internal coherence, corrects rounding issues, and identifies outliers. Station data were excluded when the number of observations was insufficient for statistical analysis, when the series contained excessive gaps, or when data ended before 2021, except for some Syrian stations that ceased reporting in 2012 due to the political situation. For each year, the number of missing values was calculated, and time series with <5% missing data were deemed suitable for subsequent analysis. Outliers exceeding four standard deviations from the mean were excluded. Consequently, 436 out of the 500 available rainfall series for the period 2000–2021 were retained for analysis, as illustrated in Fig. 2. Syria's land cover is diverse, with vegetation and tree cover primarily concentrated in the west, southwest, and northeast, where precipitation levels are higher. However, in recent years, vegetation cover has been declining due to frequent droughts and wildfires, leading to significant environmental changes (Zubkova et al. 2021; Mathbout et al. 2025). Eight teleconnection patterns were selected for analysis at the annual time scale based on their known influence on regional climate variability. These include the North Atlantic Oscillation (NAO), Mediterranean Oscillation (MO), Western Mediterranean Oscillation (WeMO), Upper-Level Mediterranean Oscillation Index (ULMO), East Atlantic (EA) pattern, East Atlantic/West Russia (EAWR) pattern, Scandinavia (SCAND) pattern, and Southern Oscillation (SO). These indices were incorporated to evaluate their correlation with various DIs across different seasons. For land cover classification, we employed the Global Land Cover 2000 (GLC 2000; Bartholomé and Belward, 2005) dataset, which adheres to the Land Cover Classification System (LCCS) developed by the FAO (Bartholomé and Belward 2005). The LCCS is a globally applicable scheme designed to accommodate diverse data sources and socioeconomic contexts. It relies on a combination of visual interpretation and knowledge-based assessments derived from observable land surface features (Arafat et al. 2014).

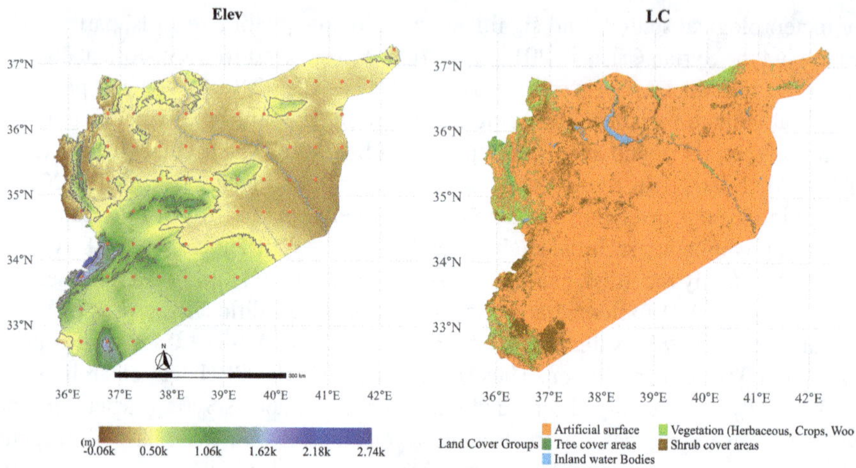

Fig. 2 Hypsometric map (Elev) of Syria alongside the spatial distribution of meteorological stations (left panel) and the dominant land cover types (LC) across the country (right panel)

2.3 *Selection of Drought Indices (DIs): The SPI, SPEI, PDSI, sc-PDSI, and EDI*

DIs are widely used to provide a quantitative assessment of drought characteristics and support risk management strategies (Li et al. 2024; Keune et al. 2025). These indices capture various aspects of drought formation and its impacts by incorporating key climatic variables such as precipitation, temperature, and evapotranspiration (Vicente-Serrano et al. 2025a, b; Saharwardi et al. 2025). Utilizing multiple DIs at different spatial and temporal scales enhances the accuracy of drought detection and provides a clearer picture of drought dynamics, particularly in arid and semi-arid regions (Khosravi and Ouarda 2025; Terzi and Önöz 2025). Among the most widely used meteorological DIs is the SPI, developed by McKee et al. (1993a, b, 1995). The SPI relies on a probabilistic precipitation approach, making it a robust tool for assessing rainfall deficits across multiple time scales. Due to its effectiveness, the World Meteorological Organization (WMO) recommends SPI as a primary index for drought characterization and precipitation deficit quantification (WMO 2012; Kostopoulou et al. 2017). To integrate the effects of temperature along with precipitation, the Standardised Precipitation Evapotranspiration Index (SPEI) is employed. Originally developed by Vicente-Serrano et al. (2010), the SPEI incorporates both water availability and atmospheric evaporative demand, offering a more holistic, multiscale assessment of drought conditions (Peng et al. 2024). Previous systematic reviews (Trenberth et al. 2014; Zhong et al. 2020) highlight the pivotal role of temperature in shaping drought severity and dynamics. In this study, both the Standardised Precipitation Index (SPI) and SPEI were computed over 3- (SPI-3, SPEI-3) and 12-month (SPI-12, SPEI-12) periods to evaluate short-term and long-term drought conditions. The 3-month indices reflect seasonal changes and transient dry events, whereas the 12-month indices capture sustained drought patterns and broader climatic aridity (Mckee et al. 1993a, b). To assess soil moisture stress, the (sc-PDSI; Wells et al. 2004) is employed. Unlike its predecessor, PDSI, the sc-PDSI provides better spatial comparability by adjusting for regional climatic conditions and soil water-holding capacity (Zhong et al. 2020). This index estimates drought conditions based on water supply and demand rather than raw precipitation and PET changes, offering a more precise representation of moisture availability (Zargar et al. 2011). The sc-PDSI is particularly valuable for analyzing long-term moisture trends across diverse climatic regions, including arid and semi-arid zones (Dai 2011; Zhong et al. 2019; Fanta et al. 2025). Additionally, the EDI developed by Byun and Wilhite (1999) is included in this study to capture drought onset, persistence, and recovery dynamics. EDI was initially designed for monitoring drought conditions on a daily time scale (Deo et al. 2017; Habibi et al. 2024). Over time, its application was expanded to include monthly drought monitoring (Deo et al. 2009; Deo and Byun 2014; Guenang et al. 2023). EDI and SPI use a similar classification system for drought severity, allowing for comparable assessments of drought intensity across different time scales. Unlike traditional indices, EDI considers cumulative daily precipitation anomalies, using effective precipitation calculated over a 15-day period. This approach enables real-time monitoring of drought evolution, making it particularly valuable for detecting emerging drought events and assessing their progression across different timescales. Furthermore, SPI and SPEI values are calculated using the SPEI package created by Beguería et al. (2014), while sc-PDSI values are computed through the R algorithm developed by Zhong et al. (2018). All analyses were performed in the R environment (version 4.3.3), with graphical representations and maps generated using tools such as ggplot2.

2.4 Comprehensive Analysis of Drought Event Characteristics and Trends

Drought characteristics, including drought duration (DD), drought intensity (DINT), and maximum drought duration (MDD), are essential for understanding drought behavior and its potential impacts (Dai et al. 2020; Batool et al. 2025; Gironás and Núñez 2025). DD refers to the number of consecutive months during which a drought index remains below a predefined threshold, while DINT represents the average drought index value over the entire drought period (Zhai et al. 2020). MDD, on the other hand, is defined as the longest continuous drought period recorded within a given timeframe, offering insight into extreme and prolonged drought events. These drought characteristics are crucial in assessing the frequency, severity, and persistence of drought conditions, which are influenced by climate variability and change (Mathbout et al. 2023; Matimolane et al. 2024; Ipek et al. 2025). To ensure a consistent classification of drought severity, specific threshold values were applied across different DIs. In this study, SPI and SPEI values of −1 or lower were classified as dry conditions, while PDSI and sc-PDSI values below −2 were considered indicators of drought conditions. These thresholds provide a standardised approach to evaluating drought trends and severity across Syria. To detect trends in drought conditions, the Mann–Kendall test (Hamed 2008) was employed, identifying statistically significant increases or decreases over time. The Sen's slope estimator (Sen 1968) was used to quantify the magnitude of these trends. The Mann–Kendall test was applied at a 95% confidence level, ensuring that only statistically significant trends were considered. The Mann–Kendall trend test and Sen's slope estimator were calculated using the (Kendall V.2.1; McLeod 2025) and (zyp, V. 0.11-1; Bronaugh 2025) packages in R. The Kendall package was applied to perform the Mann–Kendall test, which identifies monotonic trends in time series data. The zyp package was used to calculate Sen's slope, offering an estimate of the trend magnitude. These two packages are widely utilized in trend analysis to assess the significance and direction of changes over time. Additionally, Pearson's correlation coefficient was utilized to evaluate the strength of linear relationships between observed and modeled drought data, improving the reliability of trend assessments. This comprehensive approach provides a robust analysis of drought dynamics, aiding in the assessment of long-term climate variability and its impacts on water resources in Syria.

3 Results

3.1 Changes in Drought Features Across Syria from 1961 to 2023 Based on Monthly Drought Indices

The analysis of drought events in Syria from 1961 to 2023 using multiple DIs reveals significant patterns of drought occurrence and severity. Several drought hotspots have been identified across the country, particularly in the last two decades, with megadrought events recorded in the last decade. These changes are particularly evident when examining key drought features such as DD, DINT, and MDD, which are critical indicators for assessing the evolving drought conditions and their impacts on the region. The spatial patterns of drought in Syria from 1961 to 2023, as assessed using multiple DIs (SPI-3, SPEI-3, SPI-12, SPEI-12, PDSI, and sc-PDSI), reveal significant variations across different regions of the country. While all of Syria has experienced recurrent drought events, the severity and persistence of these events vary spatially (Fig. 3).

The northeastern parts of Syria consistently exhibit the highest drought index values, indicating prolonged and severe drought conditions, which aligns with the region's semi-arid to arid climate and high susceptibility to water deficits. The sc-PDSI highlights extreme drought episodes not only in the northeast but also along central Syria, where prolonged dry spells have intensified over the decades. Additionally, sc-PDSI detects severe drought conditions even in the coastal humid zones, which are traditionally characterized by higher precipitation levels. This suggests a

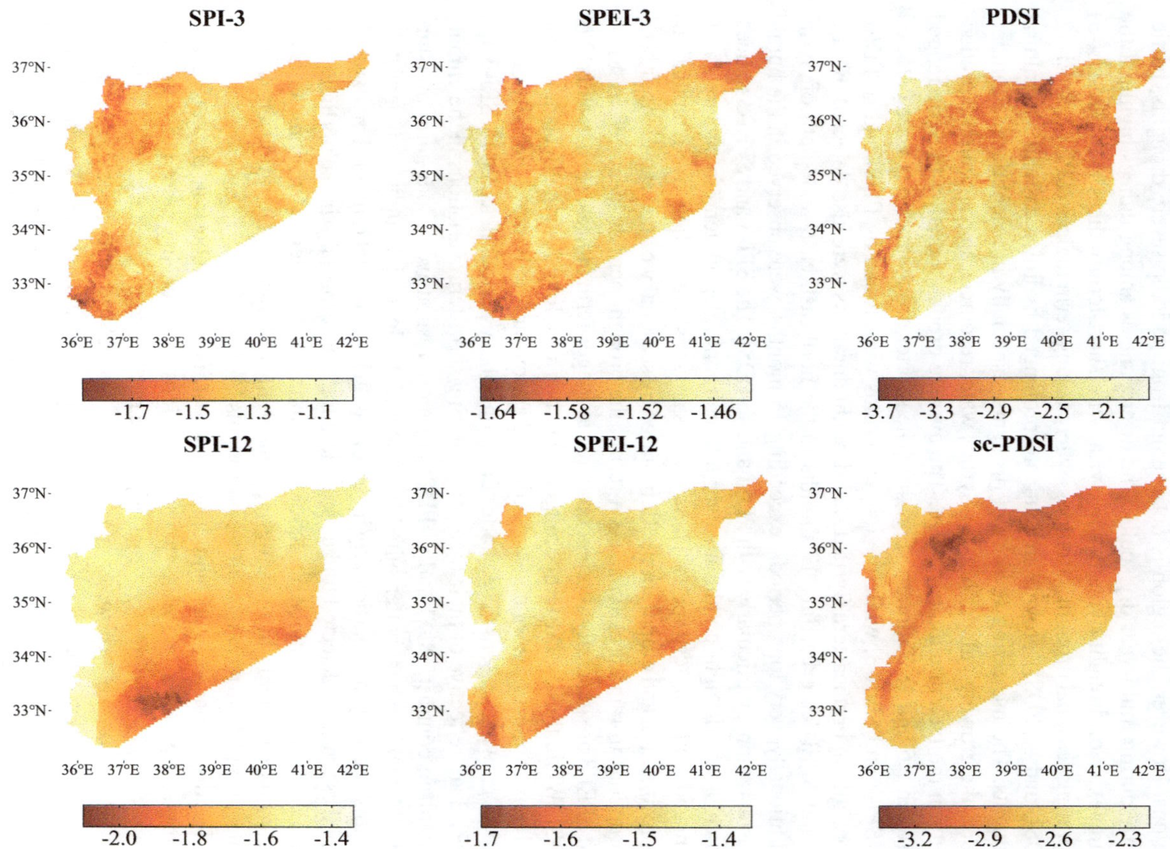

Fig. 3 The spatial extent of drought occurrences in Syria, as assessed using multiple DIs: SPI-3, SPI-12, SPEI-3, SPEI-12, PDSI, and sc-PDSI. This analysis is based on monthly data from the observed CRU dataset, covering the period from 1961 to 2023

significant long-term decline in soil moisture availability in these areas, potentially linked to rising temperatures and increasing evaporative demand. In southwestern Syria, the overall drought trends, as captured by all indices, indicate frequent but relatively less intense drought conditions compared to the northeastern and central regions. However, the region is not exempt from periodic severe droughts, particularly during multi-year dry spells. The discrepancies among indices highlight the complexity of drought dynamics in Syria, with short-term indices (SPI-3, SPEI-3) capturing seasonal drought variations, whereas long-term indices (SPI-12, SPEI-12, PDSI, and sc-PDSI) provide insight into prolonged hydrological and agricultural droughts. The increasing severity of droughts, especially in recent decades, underscores the growing vulnerability of Syria's water resources and agricultural productivity in the face of climate change. The average drought duration in Syria ranged between 2.6 and 18.7 months (Fig. 4a). The shorter drought periods were detected by SPI-3 and SPEI-3, while the longer periods were found by PDSI and sc-PDSI, especially in northeastern Syria. Long drought durations were also detected by SPI-12 and SPEI-12, exceeding 17 months in the northeast, east, and south of Syria. Additionally, very long periods, exceeding 15 months, were observed in the humid coastal regions, particularly with PDSI and sc-PDSI. The SPI-3 and SPEI-3 indices, which assess drought conditions over a three-month period, tend to capture short-term and temporary dry spells. These indices are sensitive to short-lived fluctuations in precipitation and evaporation, often caused by seasonal weather variations or brief disruptions in rainfall patterns. As a result, the drought periods identified by SPI-3 and SPEI-3 are generally shorter compared to longer-term indices, such as SPI-12 or PDSI. While these short-term droughts may not have long-lasting effects, they can still impact immediate water availability and agricultural activities, especially during critical growing seasons. The shorter durations of dry periods detected by SPI-3 and SPEI-3 are reflective of their focus on capturing short-term variations in the climate system. In contrast, PDSI and sc-PDSI are more responsive to long-term moisture deficits, making them better suited for detecting prolonged droughts that persist for months or even years, particularly in regions with less frequent rainfall. This explains why these indices detected longer drought durations, especially in the northeast and coastal areas, where moisture deficits accumulate over extended periods. Overall, these findings highlight the differing sensitivities of various DIs to different time scales, from short-term to long-term droughts.

The discrepancies in drought duration across indices demonstrate the complexity of drought patterns in Syria, with some regions experiencing more persistent and severe drought conditions, particularly in the northeast and coastal areas. The implications are critical for water management, agriculture, and policy planning, as prolonged droughts, especially those lasting more than a year, can severely impact both the environment and local communities. A severe long-term drought period, lasting over two decades from January 2001 to December 2023, was detected using SPEI-12 and sc-PDSI, indicating a prolonged water deficit that significantly affected agriculture, water resources, and ecosystems. This extreme dry period was also captured by short-term DIs such as SPI-3 and SPEI-3, demonstrating their ability to reflect the development and persistence of long-term droughts, while the PDSI index

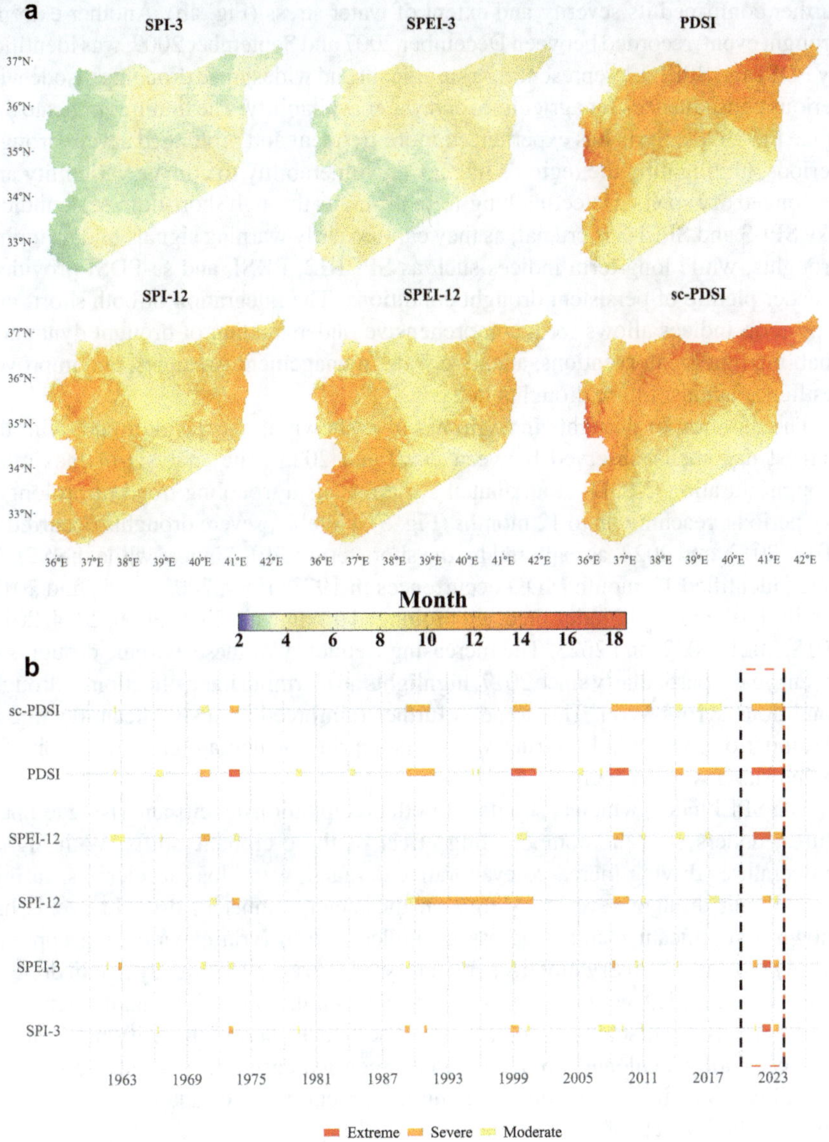

Fig. 4 **a** Spatial distribution of mean drought duration (DD) in Syria (1961–2023) using SPI-3, SPEI-3, SPI-12, SPEI-12, PDSI, and sc-PDSI based on the CRU TS v4.08 dataset. **b** Frequent and prolonged drought events in Syria (1961–2023) ranked by severity across different DIs. The dashed red rectangle represents the extreme mega drought that extended from 2021 until 2023

further confirmed its severity and extent of water stress (Fig. 4b). Another extreme drought event, recorded between December 2007 and September 2009, was identified by PDSI and SPEI-12, representing a significant and widespread drought episode with serious consequences for agriculture and water availability. The findings indicate that since the 1990s, Syria has experienced more frequent and prolonged severe drought periods, highlighting the region's increasing vulnerability to climate variability and prolonged dry spells. Detecting long-term droughts through short-time-scale indices like SPI-3 and SPEI-3 is crucial, as they capture early warning signals of developing droughts, while long-term indices such as SPEI-12, PDSI, and sc-PDSI provide a broader picture of persistent drought conditions. The integration of both short- and long-term indices allows for a comprehensive understanding of drought dynamics, enabling timely interventions, adaptive water management strategies, and improved resilience against future droughts in Syria.

The duration of droughts in Syria has also shown an increasing trend, with the longest dry spells observed between 2020 and 2023. The analysis of maximum drought duration (MDD) at an annual scale reveals a troubling trend of prolonged dry periods, reaching up to 12 months (Fig. 5). Notably, severe droughts occurred in 2017, 2012, and 2022, as captured by the short-term SPEI-3 index, while the SPI-12 index identified 12-month MDD occurrences in 1973, 1999, 2005, 2008, and 2017. Similarly, the SPEI-12 index detected prolonged droughts in 2005, 2009, 2014, 2017, 2018, 2021, 2022, and 2023. The increasing frequency of these extreme droughts in recent years, particularly since 2017, highlights an alarming intensification of drought conditions across Syria. This trend is further reinforced by a significant decline in precipitation, particularly during winter and spring, which are critical seasons for replenishing water resources.

The SPEI index, which accounts for both precipitation deficits and rising temperatures, underscores the compounding effect of these climatic shifts. With higher temperatures driving increased evaporative demand, water loss accelerates, further intensifying drought severity. Moreover, the mean number of drought events has shown a significant increase across all indices, with Syria experiencing approximately 0.6 additional dry months per century on average. This steady rise in drought frequency, coupled with declining seasonal precipitation and escalating temperatures, signals a worsening climate crisis. The persistent and intensifying droughts pose escalating challenges to Syria's water security, agricultural sustainability, and overall ecosystem stability, emphasizing the urgent need for adaptive strategies to mitigate the impacts of climate change.

3.2 *Drought Trends in Syria Over the Past 62 Years*

The analysis of DIs at a 95% confidence level reveals a significant and widespread decline in moisture availability across Syria, with all indices confirming a decreasing trend and an intensifying drying trend across the country, particularly in the last

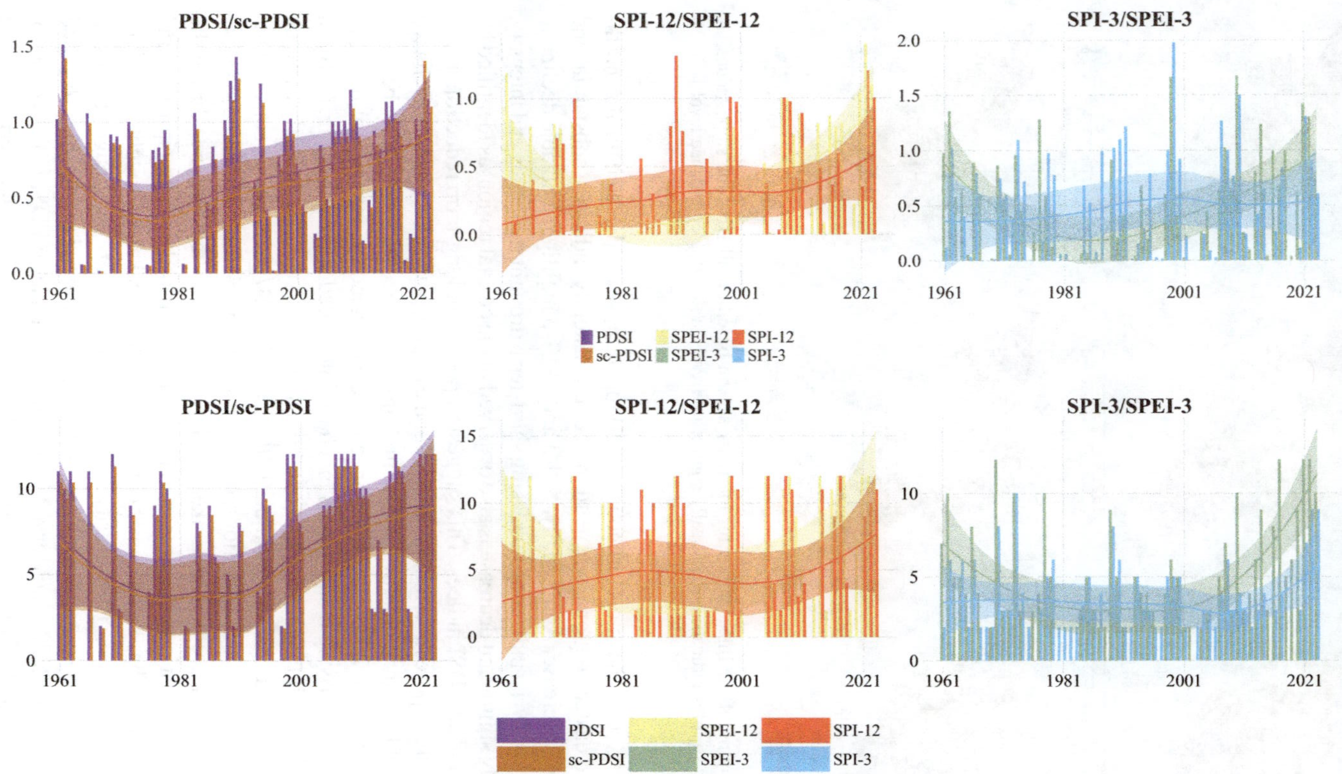

Fig. 5 Evaluation of the mean annual dry spells number (upper panel) and maximum drought duration (MDD) (lower panel) at the country level. The lines represent a smoothed trend of the mean annual drought index values for each year, with the corresponding colour bands displaying the 99% confidence interval around the smoother

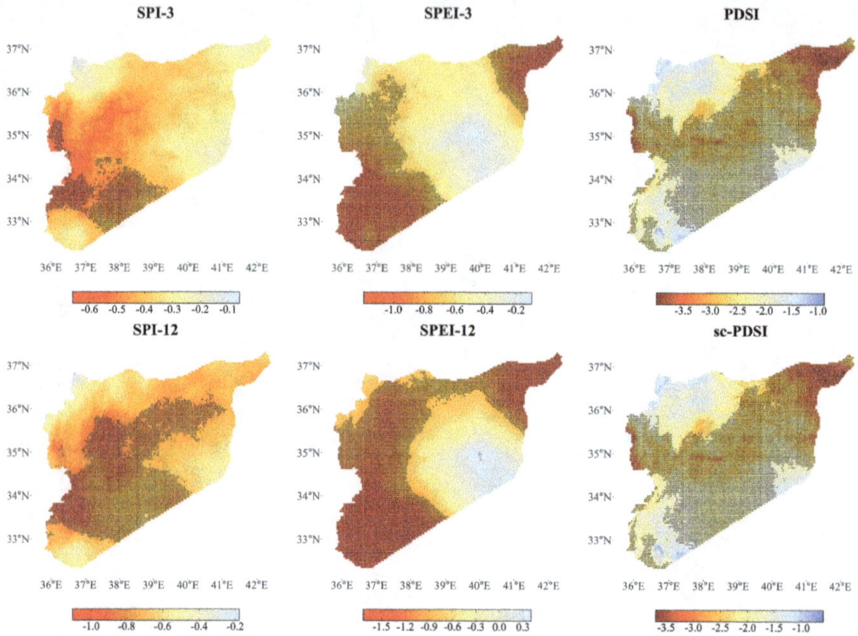

Fig. 6 Mapping of the linear trends for the rainy season based on all DIs. The slope values are shown, with stipples indicating statistically significant trends at the 95% confidence level

decade (Fig. 6). The most pronounced drying trends are detected in the north-eastern and eastern regions, where all indices consistently indicate a substantial and accelerating decrease. Additionally, SPI-3 and SPEI-3 highlight a notable decline in southwestern Syria, indicating increasing short-term drought frequency and intensity.

A highly significant decrease is also observed in the coastal region, as identified by the PDSI and sc-PDSI indices. This suggests a persistent long-term reduction in soil moisture, which poses a serious challenge for the region, particularly given its reliance on stable hydrological conditions. The findings emphasize that, especially between July 2020 and June 2023, climate change has exacerbated the intensity of drought, driven mainly by rising temperatures that have intensified soil drying (Fig. 6). Syria has experienced significant drought conditions in recent years, particularly between July 2020 and June 2023. Both the 2020/2021 and 2021/2022 wet seasons experienced drier-than-average conditions, but these deficits were also compounded by the increasing evaporative demand linked to above-normal air temperatures during the July–September hot and dry season. This combination of lower precipitation and higher evaporation has significantly contributed to the escalating drought conditions observed across Syria. The severity, duration, and intensity of drought events in Syria have increased notably over recent decades, with the highest values recorded between 2020 and 2023, as observed in both the PDSI and sc-PDSI indices (Fig. 7). This period, particularly between 2020 and 2023, has marked the most critical phase in terms of drought severity, with conditions reaching alarming levels. Similarly, drought intensity has been rising steadily from 2000 to 2023, exacerbated by the growing gap between precipitation deficits and rising temperatures, as illustrated in the PDSI and sc-PDSI indices (Fig. 7). The trends observed in Syria reflect a broader regional pattern of increasing drought severity across the Mediterranean, with the period from 2020 to 2023 being one of the most severe. The combination of reduced precipitation and higher evaporation, driven by increasing temperatures, has amplified the drought conditions. As in other Mediterranean regions, Syria is facing a future where the frequency, duration, and intensity of droughts are likely to continue to increase due to ongoing climate change, posing serious risks to agricultural productivity, water resources, and the overall stability of the region.

Moreover, the trend in drought severity has intensified over the last few decades, with a marked increase in the frequency and duration of extreme drought events. The shifts in drought severity have been particularly noticeable in the form of prolonged dry spells, especially in the winter and spring months, as indicated by a growing number of dry spells.

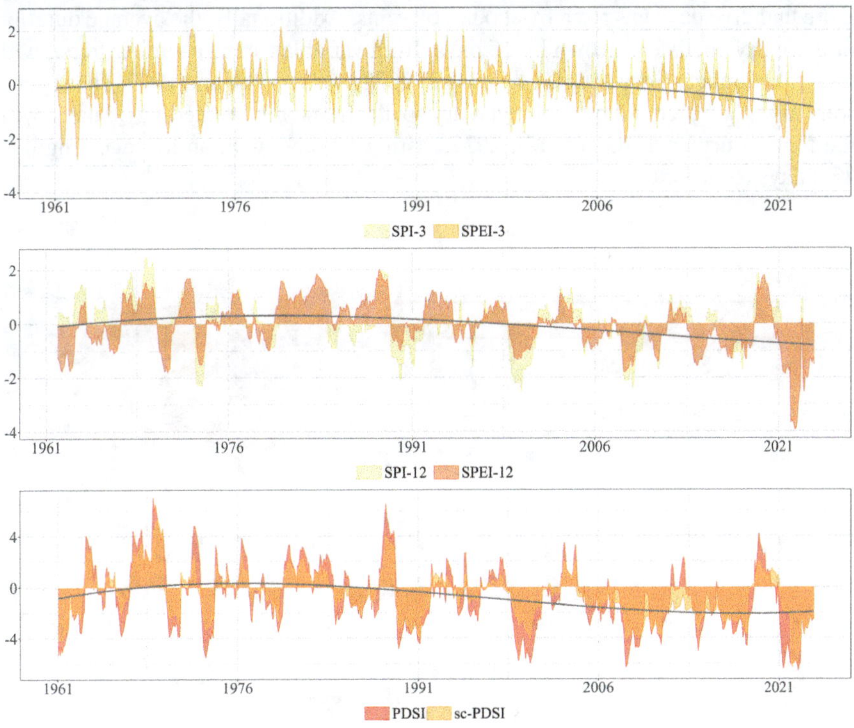

Fig. 7 The evolution of the total monthly drought severity, derived from multiple DIs at the national level from 1961 to 2023, is presented, with a gray regression line used for smoothing. This line represents the average annual values of the DIs for each year

3.3 *Detection of Dry Spells in Syria Based on the Daily Effective Drought Index (EDI)*

To capture the short-term variability and rapid onset of drought events, EDI is employed, offering a daily-scale assessment of drought conditions. Unlike long-term indices such as PDSI or SPEI-12, which primarily detect prolonged drought periods, EDI provides a more responsive measure, making it particularly useful for identifying rapidly developing droughts (flash droughts) and assessing short-term fluctuations in water availability. This section examines drought dynamics in Syria using EDI, highlighting its effectiveness in detecting drought onset, intensity, and recovery phases. The number of dry days during the rainy season (September–May), identified based on the EDI criterion (EDI ≤ -1), has exhibited a significant increasing trend over the study period. On average, the region experiences approximately 246 dry days per rainy season, with a clear tendency toward more frequent and prolonged dry periods in recent decades. Since 1980, the number of dry days per year has increased by approximately 8.7 days. Over the past 44 years, the occurrence of dry spells lasting more than five days has risen by about 2.6 events. Additionally, the average duration of these dry spells has grown by 15 days. Both variables show a similar trend, with the peak number of dry events occurring in 2022, 2021, 2023, and 1999. These years were also characterized by prolonged dry spells, frequently exceeding 50 days, with the longest dry spell recorded in 2022, lasting 62 days with an average length of 47.2 days (Fig. 8).

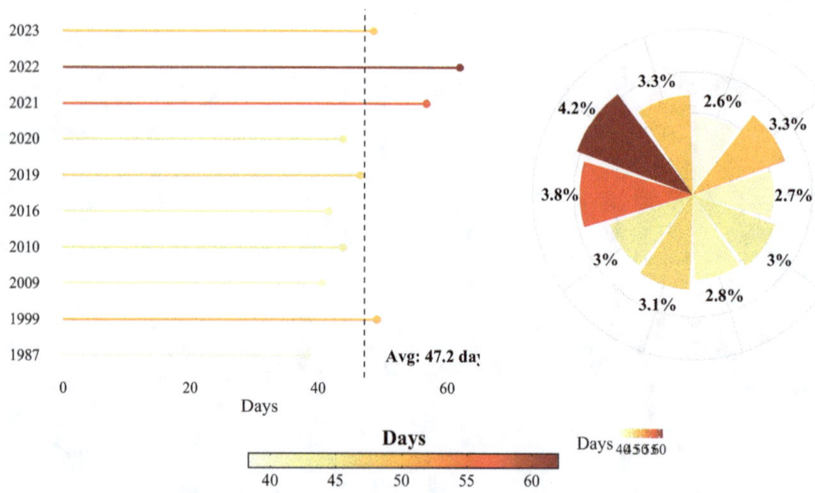

Fig. 8 The ten longest extreme daily dry spells extracted from EDI index at country level (Left) and the percentage of these longest spells relative to total drought duration over the period 1980–2023 (Right). The black vertical dashed line represents the average duration of dry periods for the extreme 10 Events

Over the past 3 years, Syria has experienced significant dry spells, with the longest occurring in 2022, representing 4.8% of the total drought duration recorded between 1980 and 2023. In comparison, the longest dry spells in 2021 and 2023 accounted for approximately 3.8 and 3.3% of the total drought duration over the same period, respectively. These prolonged dry spells are indicative of a significant precipitation deficiency, which has been further exacerbated by rising temperatures. DIs reveal a clear trend of increasing drought severity, reflecting not only the decline in precipitation but also the intensifying effects of higher temperatures. This combination of reduced rainfall and elevated temperatures underscores the growing challenges of drought conditions in Syria in recent years. Additionally, Syria has seen a notable rise in the mean intensity of dry spells, with the highest values occurring in the last three years. The mean intensity of dry spells is defined as the average cumulative drought magnitude of all individual dry spells within a given period. The mean intensity of dry spells was 120 in 2020, 130 in 2021, and 140 in 2022. These figures reflect a clear trend of both longer and more intense dry spells. This simultaneous rise in dry spells and prolonged dry spells signals a shift towards more persistent drought conditions, which could have significant implications for regional water resources and ecosystem stability. To facilitate direct comparison, normalized values are employed, which standardize different variables on a common scale, typically ranging from 0 to 1, while retaining their relative differences.

Figure 9 illustrates the normalized trends of the number of dry events and their mean duration using Min–Max scaling, allowing for direct comparison despite differences in their original ranges. In this scaling, a value of 1 represents the highest recorded occurrence, while 0 indicates the lowest, providing a standardised basis for analysis. The dumbbell plot in Fig. 8 effectively captures the evolving relationship between these two drought indicators, revealing a strong alignment over time. Notably, both the frequency of dry events and their duration have reached their highest recorded levels in the years from 2019 until 2023. During this period, dry events have been more frequent, with prolonged dry spells averaging 52 days in duration. These findings highlight the increasing severity of dry conditions, reinforcing the strong correlation between drought occurrence and duration. This persistent trend underscores the need for further research into the underlying climatic drivers and the potential consequences for water resources and agriculture in drought-prone regions. The Effective EDI demonstrates a strong correlation with short-term DIs, exhibiting a highly significant Pearson correlation of $r = 0.93$ with SPI-3 and $r = 0.76$ with SPEI-3 at a 0.05 confidence level. This indicates that EDI is highly responsive to short-duration drought conditions and effectively captures their variability (Fig. 10).

This confirms EDI as a reliable tool for real-time drought monitoring and early warning systems, making it particularly useful in regions prone to rapid moisture fluctuations. Its alignment with both precipitation-based (SPI-3) and temperature-sensitive (SPEI-3) indices suggests that EDI effectively integrates multiple climatic factors. Given its responsiveness, EDI can enhance drought forecasting and risk management, especially when combined with long-term DIs for a comprehensive assessment. The Mean Absolute Error (MAE) in our analysis ranges from 0.2 to 1.8, indicating varying levels of agreement between EDI and the other DIs. The lowest

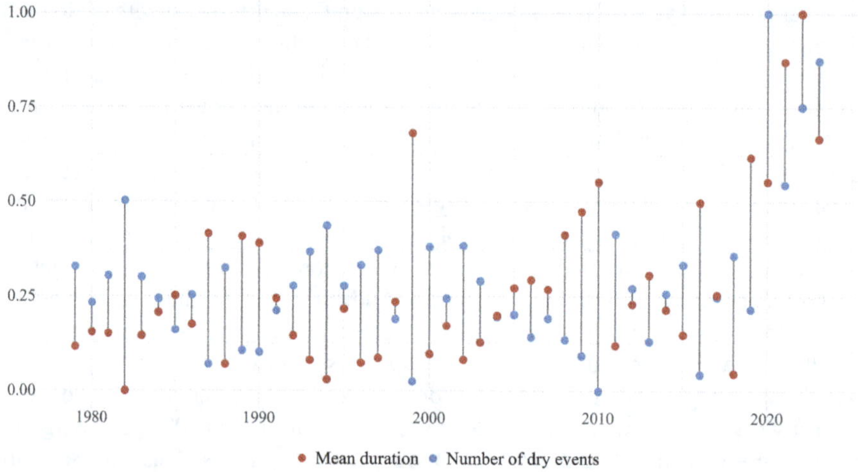

Fig. 9 Normalized trends of dry spells number and their mean duration using Min–Max scaling. The dumbbell plot presents the Min–Max normalized values of dry events (blue dots) and mean dry spell duration (red dots, with a minimum threshold of 5 days). Gray lines connect the two metrics for each year, highlighting their relative differences and concurrent trends over time

MAE (0.2) is observed with SPI-3, which also corresponds to the highest Pearson correlation (r = 0.93), highlighting a strong alignment between EDI and short-term drought conditions. On the other hand, the highest MAE values (up to 1.8) are observed with PDSI and sc-PDSI, suggesting greater discrepancies. This reflects differences in how EDI compares to long-term DIs, which incorporate soil moisture conditions and cumulative hydrological stress, whereas EDI is more responsive to short-term variations in precipitation.

3.4 *Teleconnection Patterns (TPs) and Their Influence on Droughts in Syria*

This study identifies statistically significant negative correlations between the Southern Oscillation (SO) and all examined DIs across Syria, indicating that positive SO phases are consistently linked to drier conditions, while negative SO phases correspond to wetter anomalies (Fig. 11). This inverse relationship is particularly evident during the winter season, with the strongest signal occurring in December and January, underscoring the critical role of SO in modulating winter precipitation across the Eastern Mediterranean. These results are consistent with earlier findings (e.g., Mathbout et al. 2021), which highlight the sensitivity of regional hydroclimate to large-scale tropical Pacific variability. In parallel, a significant positive correlation is observed between the EATL/WRUS teleconnection pattern and DIs in Syria,

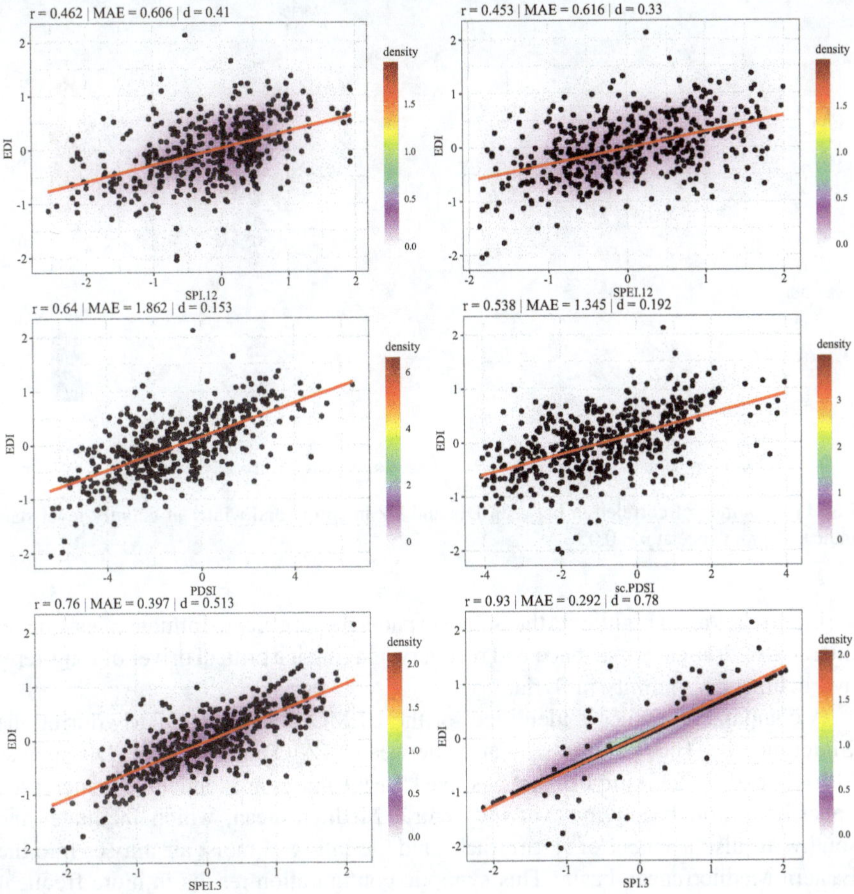

Fig. 10 Density-based comparison between EDI and various dis: agreement and error statistics (d, MAE, Pearson's r) with high-disagreement points highlighted

with correlation coefficients ranging from r = 0.25–0.35. This suggests that positive EATL/WRUS phases, often characterized by enhanced cyclonic activity over the Eastern Mediterranean, promote increased moisture transport and higher winter precipitation, leading to a noticeable reduction in drought severity. These findings further support the key role of mid-latitude atmospheric patterns in shaping the regional drought climatology, as corroborated by previous studies focused on the Eastern Mediterranean domain. Beyond the influence of SO and EATL/WRUS, this study also reveals a pronounced impact of the Scandinavian (SCAND) pattern on long-term drought conditions, particularly during the spring season. Strong positive correlations were detected between the SCAND index and both the SPEI-12 and the sc-PDSI. This indicates that positive SCAND phases, often associated with high-pressure anomalies over northern and eastern Europe, enhance the southward advection of moisture toward the Eastern Mediterranean. The resulting increase in

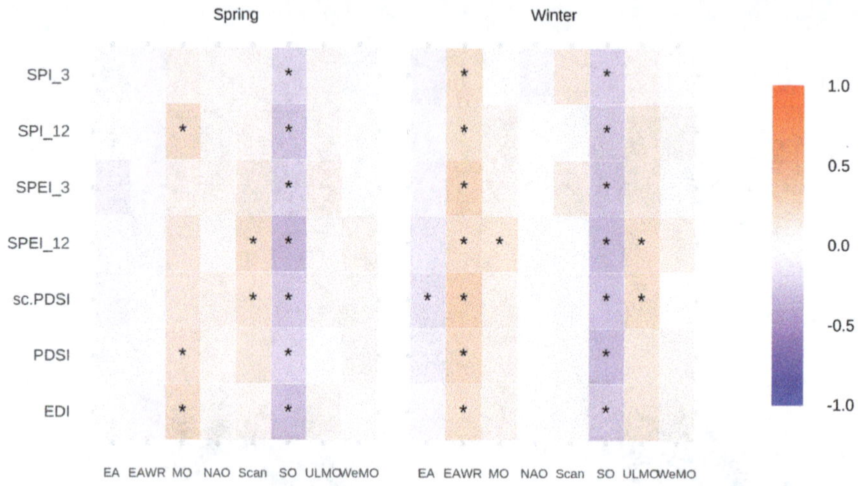

Fig. 11 Heatmap of correlation between DIs and TP in spring and winter at country level stars indicate significance at $p < 0.05$

spring precipitation highlights the SCAND pattern's capacity to influence seasonal to interannual drought persistence and recovery, making it a critical driver of long-term hydroclimatic variability in Syria.

A comparable impact is identified for the ULMO index, particularly during the winter months. The positive association between ULMO and long-term DIs such as SPEI-12 and sc-PDSI implies that positive ULMO phases are conducive to increased upper-level trough activity over the central Mediterranean, which facilitates the southward displacement of storm tracks and the influx of moist air masses into the Eastern Mediterranean basin. This synoptic configuration results in more frequent and intense winter precipitation events, which are vital for recharging soil moisture and water resources in Syria's predominantly semi-arid environment. The robust correlations with long-term indices confirm that ULMO significantly contributes to shaping the seasonal hydrological regime, particularly during the climatologically wettest period of the year. These results are consistent with Mathbout et al. (2021), who reported strong positive correlations between ULMO and precipitation over Israel and Greece, highlighting ULMO as a dominant driver of Mediterranean precipitation variability, more influential than NAO and EATL/WRUS. Furthermore, they emphasized that the combined effect of ULMO and NAO exerts a stronger influence on annual precipitation over the southern Levant than the NAO alone.

3.5 *Socioeconomic and Political Implications of Drought in Syria*

The intensification of drought conditions in Syria, as projected by CMIP6-based future climate scenarios, has profound socioeconomic and political consequences (Mathbout et al. 2023). The worsening drought severity, frequency, and duration will not only impact natural ecosystems but also exacerbate water scarcity, agricultural productivity declines, economic instability, and regional security concerns (Eklund et al. 2022; Ingrao et al. 2023). This section explores the broader implications of increasing drought risks in Syria, integrating the latest scientific findings with real-world consequences. Agriculture, a vital sector in Syria, is severely threatened by declining precipitation and increasing temperatures, leading to reduced crop yields, soil degradation, and heightened food insecurity (Gleick 2014; Kelley et al. 2015). NDVI and VHI indices indicate worsening vegetation health, confirming that drought events are increasingly damaging plant cover, reducing pastureland, and accelerating desertification (Mathbout et al. 2025). Water scarcity, particularly in the Euphrates basin and semi-arid regions, is reaching critical levels due to over-extraction of groundwater and declining rainfall, which CMIP6 projections suggest will worsen (Morante-Carballo et al. 2022). As a result, migration pressures are expected to increase, leading to overcrowded urban areas, resource competition, and heightened social tensions (Abel et al. 2019; Ash 2023). The drought-induced decline in cropland productivity has been particularly severe in recent years, with croplands exhibiting greater vulnerability than non-agricultural lands. The Normalized Difference Vegetation Index (NDVI) and the Vegetation Health Index (VHI) analyses have demonstrated that agricultural areas are increasingly exposed to prolonged dry spells, particularly in northeastern Syria, a region that has experienced significant desertification and reduced agricultural output. The loss of fertile land has exacerbated rural-to-urban migration, contributing to demographic shifts and socio-economic instability (Mathbout et al. 2025). The economic impact extends beyond agriculture, affecting employment, GDP, and market stability, while increasing the likelihood of protests and unrest due to inflation and reduced income opportunities (Gleick 2014). Policy and governance challenges further complicate the situation, as Syria lacks a comprehensive drought management strategy, and international cooperation on transboundary water resources remains a contentious issue (Deribe 2024). Climate change serves as a risk multiplier for conflict, with evidence suggesting that past droughts contributed to rural displacement and economic hardships, ultimately fuelling grievances that escalated into instability (Kelley et al. 2015). Future droughts are likely to intensify such risks, exacerbating localized conflicts over water access and agricultural resources (Abel et al. 2019). Additionally, worsening conditions may create broader regional security concerns, as climate migrants move into neighbouring countries, straining diplomatic relations and necessitating international intervention. As CMIP6 projections indicate increasing drought severity in Syria (Mathbout et al. 2023), urgent action is required to enhance resilience through policy reforms, climate adaptation strategies, and regional cooperation (Dinc and Eklund

2023; Granata and Nunno 2025). Addressing the impacts on agriculture, water security, and migration requires a multi-faceted approach that integrates scientific research with sustainable development planning. Without immediate intervention, Syria faces a future of heightened water stress, economic instability, and increased vulnerability to social and political unrest (Selby et al. 2017; Sukkar et al. 2024).

4 Discussion

4.1 Spatial and Temporal Analysis of Drought in Syria (1961–2023): Intensification, Impacts, and Regional Vulnerability

This study focuses on assessing the spatial and temporal characteristics of drought in Syria over the period 1961–2023, using a suite of meteorological DIs including SPI, SPEI, PDSI, sc-PDSI, and EDI. By analysing long-term climatic data and drought trends, the research highlights the intensification of dry spells, rising drought severity, and their growing impact on water resources, agriculture, and socio-economic stability. The study also explores the influence of large-scale teleconnection patterns on regional drought variability, offering critical insights into the drivers of Syria's escalating climate vulnerability. The results of this study reveal a significant intensification of drought conditions in Syria over the past six decades, with a pronounced acceleration in severity and frequency observed since 2000. The increasing number of dry events, coupled with longer drought durations and rising mean intensity of dry spells, underscores the profound impacts of climate change on the region. The period between 2020 and 2023 exhibited the most severe drought characteristics, with extreme and prolonged dry spells exacerbating water scarcity, agricultural decline, and ecosystem degradation. These findings align with previous studies that highlight the Mediterranean region as a climate change hotspot, where warming trends are exceeding global averages and intensifying aridification processes (Lionello and Scarascia 2018; Lazoglou et al. 2024). One of the most alarming findings of this study is the rising trend in the mean intensity of dry spells, which reached 120 in 2020, 130 in 2021, and 142 in 2022. These values surpass previous records, except for the extreme drought event of 1999 (142.2). The simultaneous increase in the frequency and duration of dry spells further indicates a shift towards more persistent drought conditions, which is consistent with global observations of prolonged dry spells linked to increasing temperatures and declining precipitation (Dai et al. 2018). This study reveals a clear and concerning intensification of drought conditions across Syria over the past six decades, with a sharp escalation in frequency, duration, and severity observed since the early 2000s. The years between 2020 and 2023 were particularly severe, marked by record-high values in dry spell intensity, prolonged water deficits, and expanding spatial extent of drought-affected areas. These patterns align with broader regional findings and are consistent with recent work (Houmsi et al. 2019; Eklund et al. 2022; Mathbout et al. 2023), which confirmed similar trends using Meteorological DIs and highlighted increasing vulnerability even in typically more humid zones. Spatial analysis confirms that northeastern and eastern Syria remain the most drought-prone regions, reflecting their inherently arid climate and limited water resources. However, the emergence of significant drought signals in coastal and southwestern areas suggests that traditionally less-vulnerable regions are also experiencing long-term hydrological stress. The integration of multiple DIs in this study—ranging from short-term (SPI-3, SPEI-3) to long-term (SPI-12, SPEI-12, PDSI, sc-PDSI)—enabled a detailed understanding of both seasonal and chronic drought conditions. Notably, rising temperatures and increased evaporative demand are intensifying these effects, contributing to widespread and persistent soil moisture deficits. The agricultural sector has been particularly impacted. Recurrent and extended dry spells have undermined crop productivity, reduced soil quality, and strained irrigation systems. During multi-year droughts such as the 2007–2018 event, large swaths of agricultural land, over 45% by some estimates, were exposed to moderate to severe drought stress. Subsequent years, especially 2021 and 2022, saw even more severe impacts in key production zones, reflecting the growing intensity and persistence of climate stressors. These developments have significant socio-economic implications. The compounding effects of climate-induced water scarcity and agricultural disruption have exacerbated rural vulnerability, driven migration, and heightened the risk of social and political instability. As prolonged droughts increasingly overlap with critical growing periods, the resilience of both livelihoods and food systems is being undermined.

4.2 *Physical Interpretation of Drought Intensification in Syria and the Eastern Mediterranean*

The intensification of drought conditions observed in Syria and the broader Eastern Mediterranean region is driven by a complex interplay of atmospheric and oceanic processes influenced by anthropogenic climate change. One of the primary physical drivers is the regional amplification of warming, where surface temperatures are rising at rates significantly higher than the global average, particularly during the summer months (Lionello and Scarascia 2018; Zittis et al. 2022). This enhanced warming increases evaporative demand, intensifies soil moisture depletion, and reduces effective precipitation, thereby prolonging and deepening drought conditions. Moreover, the Mediterranean region is a well-documented climate change hotspot (Lazoglou et al. 2024), where changes in large-scale atmospheric circulation patterns have contributed to reduced rainfall and shifting storm tracks. The northward displacement of mid-latitude westerlies and a weakening of the Mediterranean storm track have led to fewer and less intense winter precipitation events, which are crucial for recharging soil moisture and surface water resources (Vázquez et al. 2023; Sandler et al. 2024). Anthropogenic forcing is widely believed to disrupt large-scale atmospheric circulation mechanisms, particularly through the expansion of the Hadley circulation cell and a northward shift of the subtropical high-pressure belts. This shift results in the suppression of convection in the subtropical regions and promotes the prevalence of anticyclonic conditions, especially over the Mediterranean basin. One of the most significant consequences of these changes is a reduction in the occurrence of extratropical storms that typically bring winter precipitation to the region. As a result, the Mediterranean experiences a decline in seasonal rainfall, which in turn exacerbates the frequency and severity of meteorological droughts (Gandham et al. 2025). Teleconnection patterns, particularly the NAO, EAWR, SCAND, and the ULMO, also play a critical role in modulating interannual drought variability in the region. Positive phases of the NAO are typically associated with reduced rainfall over the Levant due to enhanced anticyclonic activity and subsidence over the Eastern Mediterranean basin (Mathbout et al. 2021). In Syria specifically, significant negative correlations have been identified between the Southern Oscillation (SO) index and all DIs especially during the winter season, suggesting that positive SO phases lead to intensified drought conditions through reduced precipitation. At the oceanic scale, warming of the Mediterranean Sea and adjacent marine regions affects atmospheric moisture transport and regional convection patterns, further inhibiting precipitation. The synergy between sea surface temperature anomalies and altered upper-level atmospheric dynamics reinforces a feedback loop of drying and warming, particularly in inland and semi-arid zones (Liu et al. 2025). Together, these physical and teleconnection drivers explain the increasing frequency, severity, and persistence of droughts in the Eastern Mediterranean and justify the observed geographical expansion of aridity into regions previously considered climatically stable. Without significant mitigation and adaptation efforts, continued warming and drying trends are likely to intensify, posing long-term risks to water security, agriculture, and ecosystem functioning in Syria and surrounding areas.

4.3 Toward Resilience: Strategic Responses to Intensifying Drought in Syria and the Eastern Mediterranean

The increasing occurrence of multi-year droughts in Syria, most notably the prolonged dry period from 2001 to 2023, signals a fundamental shift toward more extreme and persistent drought regimes (Dinc and Eklund 2023). This intensification raises serious concerns about the long-term sustainability of the country's water resources. When coupled with ongoing socio-political instability, the impacts of climate-induced drought become even more severe, compounding the risks of food insecurity, displacement, and economic decline (Kelley et al. 2017; Abel et al. 2019). Addressing this crisis requires a comprehensive suite of short- and long-term adaptation and mitigation strategies. Central to this is the promotion of sustainable water use through improved irrigation efficiency, the expansion of rainwater harvesting, desalination, and wastewater reuse, all of which can reduce reliance on depleted groundwater reserves (Xing and Wang 2024). Simultaneously, scaling up drought-resilient agricultural practices, such as the use of climate-adapted crop varieties, soil conservation, and crop rotation, can support food security under changing climatic conditions (Wang and Ren 2025). Afforestation and reforestation initiatives are also critical. These strategies enhance soil moisture retention, improve local microclimates, and reduce the risks of desertification (Mao et al. 2024; Cho et al. 2025). Equally important is the development of Integrated Water Resource Management (IWRM) frameworks that balance water allocation across sectors and enforce controls on unsustainable withdrawals, particularly in arid regions (Koop et al. 2022). Policy and governance reforms must be reinforced by financial incentives and technical support for farmers adopting sustainable practices (Rosário et al. 2022). Moreover, community-level adaptation, including drought monitoring networks, public awareness campaigns, and participatory planning, can significantly boost local resilience (Moridi et al. 2025). Given Syria's geopolitical landscape, regional cooperation is essential. Coordinated management of transboundary water resources and the exchange of adaptation technologies can strengthen climate resilience at a broader scale (Piemontese et al. 2024). Leveraging international support through mechanisms such as the Green Climate Fund (GCF) and UN-backed climate adaptation programs will be crucial in financing and implementing large-scale, climate-resilient development projects. Ultimately, the findings of this study point to an urgent need for an integrated and proactive approach, one that combines scientific innovation, policy coherence, ecosystem restoration, and inclusive governance. Without such action, Syria and its neighbours in the Eastern Mediterranean face a growing threat from escalating water stress, agricultural collapse, and socio-economic disruption. Future research should prioritize high-resolution climate modeling, early warning systems, and adaptive governance frameworks tailored to regional vulnerabilities.

5 Conclusion

This study provides a comprehensive look at how climate change—especially in the form of intensifying drought, is reshaping Syria and the wider Middle East. Over six decades of data (1961–2023), using a range of drought indicators like SPI, SPEI, PDSI, and others, clearly shows a troubling pattern: droughts are becoming more frequent, lasting longer, and having a deeper impact than ever before.

One of the most striking findings is the presence of a prolonged drought from 2001 to 2023, a megadrought that now ranks among the longest and most severe in the region's recent history. What makes it especially alarming is how it's expanded beyond traditionally dry zones, affecting Syria's once-humid coastal and agricultural areas. As a result, the country is shifting from facing occasional droughts to enduring a more permanent climate stress.

Global climate systems, like the NAO and others, also play a key role in this shift. These large-scale patterns are becoming more erratic with global warming and are increasing the unpredictability of Syria's weather. This adds another layer of difficulty for farmers, water managers, and policymakers. The impacts are already being felt: crops are failing, rural communities are shrinking, and key water sources like the Euphrates River are under pressure. Drought in Syria is no longer just an environmental concern, it is becoming a serious challenge to national security, food systems, and human survival.

At the local level, farming systems must adapt through more water-efficient practices, better irrigation, and land conservation. Nationally, stronger drought monitoring, early warning systems, and coordinated emergency responses are essential. On a regional scale, Syria must work closely with neighbouring countries on managing shared water resources. And globally, support from international climate adaptation funds and organisations will be vital to help Syria cope and rebuild with resilience.

The drought crisis in Syria is not a distant possibility, it's a current and growing reality. But with informed decisions, proactive strategies, and collaboration across borders, there's still a chance to reduce its impacts and protect the lives and livelihoods of millions.

References

Abel G, Brottrager M, Crespo Cuaresma G et al (2019) Climate, conflict and forced migration. Glob Environ Change 54:239–249. https://doi.org/10.1016/j.gloenvcha.2018.12.003

Abel C, Abdi AM, Tagesson T et al (2023) Contrasting ecosystem vegetation response in global drylands under drying and wetting conditions. Glob Chang Biol 29(14):3954–3969. https://doi.org/10.1111/gcb.16745

Abou-Zakhem A, Hafez R (2010) Climatic factors controlling chemical and isotopic characteristics of precipitation in Syria. Hydrol Process 24(18):2641–2654. https://doi.org/10.1002/hyp.7646

Aguilar E, Prohom M (2011) RClimDex-extraQC (EXTRAQC quality control software) user manual. In: Centre for climate change. University Rovira i Virgili, Tarragona. http://www.c3.urv.cat/data/Manual_rclimdex_extraQC.r.pdf. Accessed 26 May 2025

Al-Charideh AR, Abou-Zakhem B (2010) Distribution of tritium and stable isotopes in precipitation in Syria. Hydrol Sci J 55(5):832–843. https://doi.org/10.1080/02626667.2010.487977

Alemayehu S, Olago D, Alfred O et al (2024) Spatiotemporal analysis of rainfall and temperature variability and trends for a mixed crop-livestock production system: its implications for developing adaptation strategies. Int J Clim Change Strateg Manag 17(1):268–290. https://doi.org/10.1108/IJCCSM-08-2024-0133

Alsafadi K, Al-Ansari N, Mokhtar A et al (2022) An evapotranspiration deficit-based drought index to detect variability of terrestrial carbon productivity in the Middle East. Environ Res Lett 17:014051. https://doi.org/10.1088/1748-9326/ac4765

Arafat SM, Saleh NS, Aboelghar M et al (2014) Mapping of North Sinai land cover according to FAO-LCCS. Egypt J Remote Sens Space Sci 17(1):29–39. https://doi.org/10.1016/j.ejrs.2014.03.002

Araghi A, Martinez CJ (2024) Evaluation of CRU-JRA gridded meteorological dataset for modeling of wheat production systems in Iran. Int J Biometeorol 68:1201–1211. https://doi.org/10.1007/s00484-024-02659-9

Ash K (2023) Fallow agriculture and climatic stress independently predict migration during Syria's 2006–2010 drought. Reg Environ Change 23:119. https://doi.org/10.1007/s10113-023-02115-1

Barlow M, Zaitchik B, Paz S et al (2016) Review of drought in the middle east and southwest Asia. J Clim 29:8547–8574. https://doi.org/10.1175/JCLI-D-13-00692.1

Bartholomé E, Belward AS (2005) GLC2000: A new approach to global land cover mapping from Earth observation data. Int J Remote Sens 26(9):1959–1977. https://doi.org/10.1080/01431160412331291297

Batool A, Kartal V, Ali Z (2025) Development of trivariate Multiscalar–standardized drought index (TMSDI) for assessing drought characteristics. Environ Monit Assess 197:268. https://doi.org/10.1007/s10661-025-13742-y

Beguería S, Vicente-Serrano SM, Reig F et al (2014) Standardized precipitation evapotranspiration index (SPEI) revisited: parameter fitting, evapotranspiration models, tools, datasets and drought monitoring. Int J Clim 34:3001–3023. https://doi.org/10.1002/joc.3887

Blicharska M, Haddad F, Riccardi T et al (2024) Global presumed drylands: drivers, pressures, state, impacts, responses. J Environ Plan Manag 12:1–28. https://doi.org/10.1080/09640568.2024.2351424

Bronaugh D (2025). Zhang + Yue-Pilon trends package. Version 0.11-1. https://cran.r-project.org/web/packages/zyp/zyp.pdf

Byun HR, Wilhite DA (1999) Objective quantification of drought severity and duration. J Clim 12(9):2747–2756

Chen XM, Sharma A, Liu H (2023) The impact of climate change on environmental sustainability and human mortality. Environments 10(10):1650. https://doi.org/10.3390/environments10100165

Cho S, Baral S, Burlakoti D (2025) Afforestation/reforestation and avoided conversion carbon projects in the United States. Forests 16:115. https://doi.org/10.3390/f16010115

Cook BI, Anchukaitis KJ, Touchan R et al (2016) Spatiotemporal drought variability in the Mediterranean over the last 900 years. J Geophys Res 121(5):2060–2074. https://doi.org/10.1002/2015JD023929

Dai A (2011) Characteristics and trends in various forms of the palmer drought severity index during 1900–2008. J Geophys Res Atmos 116:D12115. https://doi.org/10.1029/2010JD015541

Dai A, Zhao T, Chen J (2018) Climate change and drought: a precipitation and evaporation perspective. Curr Clim Change Rep 4:301–312. https://doi.org/10.1007/s40641-018-0101-6

Dai M, Huang S, Huang Q et al (2020) Assessing agricultural drought risk and its dynamic evolution characteristics. Agric Water Manag 231:106003. https://doi.org/10.1016/j.agwat.2020.106003

Deo R, Byun HR (2014) Quantifying the severity, duration and intensity of drought in Australia using the effective drought index. J Hydrol 12(9):2747–2756. https://doi.org/10.1175/1520-0442(1999)012%3c2747:OQODSA%3e2.0.CO;2

Deo RC, Syktus J, McAlpine C et al (2009) Impact of historical land cover change on daily indices of climate extremes including droughts in eastern Australia. Geophys Res Lett 36:L08705. https://doi.org/10.1029/2009GL037666

Deo RC, Byun HR, Adamowski JF et al (2017) Application of effective drought index for quantification of meteorological drought events: a case study in Australia. Theor Appl Climatol 128:359–379. https://doi.org/10.1007/s00704-015-1706-5

Deribe H (2024) Review on effects of drought stress on maize growth, yield and its management strategies. Commun Soil Sci Plant Anal 56(1):1–21. https://doi.org/10.1080/00103624.2024.2404663

Din A, Yang Y, Yan R et al (2024) Growing success with sustainability: the influence of green HRM, innovation, and competitive advantage on environmental performance in the manufacturing industry. Heliyon 10(10):e30855. https://doi.org/10.1016/j.heliyon.2024.e30855

Dinc P, Eklund L (2023) Syrian farmers in the midst of drought and conflict: the causes, patterns, and aftermath of land abandonment and migration. Clim Dev 16(5):349–362. https://doi.org/10.1080/17565529.2023.2223600

Eklund L, Theisen OM, Baumann M et al (2022) Societal drought vulnerability and the Syrian climate-conflict nexus are better explained by agriculture than meteorology. Commun Earth Environ 3:85. https://doi.org/10.1038/s43247-022-00405-w

Fanta SS, Yesuf MB, Merkos MT et al (2025) The spatio-temporal trend of climate and characterization of drought in Borana Zone. Ethiopia. Sci Rep 15:4743. https://doi.org/10.1038/s41598-025-87345-5

Favretto N, Stringer LC (2024) Climate resilient development in vulnerable geographies. Mitig Adapt Strateg Glob Change 29:90. https://doi.org/10.1007/s11027-024-10187-5

Figari S (2025) Climate change response: input adjustment in agriculture. J Dev Econ 175:103472. https://doi.org/10.1016/j.jdeveco.2025.103472

Furtak K, Wolińska A (2023) The impact of extreme weather events as a consequence of climate change on the soil moisture and on the quality of the soil environment and agriculture a review. CATENA 231:107378. https://doi.org/10.1016/j.catena.2023.107378

Gandham H, Dasari HP, Luong TM et al (2025) Declining summer circulation over the Eastern Mediterranean and Middle East. NPJ Clim Atmos Sci 8:177. https://doi.org/10.1038/s41612-025-01072-2

García-León D, Contreras S, Hunink J (2019) Comparison of meteorological and satellite-based drought indices as yield predictors of Spanish cereals. Agr Water Manag 213:388–396. https://doi.org/10.1016/j.atmosres.2025.107929

Gimeno-Sotelo L, Fernández-Alvarez JC, Nieto R et al (2024) The increasing influence of atmospheric moisture transport on hydrometeorological extremes in the Euro Mediterranean region with global warming. Commun Earth Environ 5:604. https://doi.org/10.1038/s43247-024-01787-9

Gironás J, Núñez M (2025) Definition and characterization of drought. In: Rivera D, Donoso G (eds) Droughts in Chile. Global issues in water policy, vol 31. Springer, Cham

Gleick P (2014) Water, drought, climate change and conflict in Syria. Weather Clim Soc 6(3):331–340. https://doi.org/10.1175/WCAS-D-13-00059.1

Gorgin-Paveh F, Ramezani-Etedali H, Collins B (2022) Evaluation of CRU TS, GPCC, AgMERRA, and AgCFSR meteorological datasets for estimating climate and crop variables: a case study of maize in Qazvin Province. Iran. J Arid Land 14:1361–1376. https://doi.org/10.1007/s40333-022-0108-7

Granata F, Di Nunno F (2025) Pathways for hydrological resilience: strategies for adaptation in a changing climate. Earth Syst Environ 10:203–231. https://doi.org/10.1007/s41748-024-00567-x

Guenang GM, Komkoua-Mbienda AJ, Djiotang-Tchotchou LA et al (2023) Comparative study of two drought description models in Central-Africa: the revisited effective drought index and the standardized precipitation index. Model Earth Syst Environ 9:1775–1792. https://doi.org/10.1007/s40808-022-01584-3

Habibi B, Meddi M, Emre T et al (2024) Drought assessment and characterization using SPI, EDI and DEPI indices in northern Algeria. Nat Hazards 120:5201–5203. https://doi.org/10.1007/s11069-024-06408-4

Hamed K (2008) Trend detection in hydrologic data: the Mann-Kendall trend test under the scaling hypothesis. J Hydrol 349:350–363. https://doi.org/10.1016/j.jhydrol.2007.11.009

Hameed M, Ahmadalipour A, Moradkhani H (2020) Drought and food security in the middle east: an analytical framework. Agric for Meteorol 281:107816. https://doi.org/10.1016/j.agrformet.2019.107816

Hao Z, Hao F, Xia Y et al (2022) Compound droughts and hot extremes: characteristics, drivers, changes, and impacts. Earth Sci Rev 235:10424. https://doi.org/10.1016/j.earscirev.2022.104241

Homsi R, Shahid S, Iqbal Z et al (2021) Historical trends in crop water demand over semiarid region of Syria. Theor Appl Climatol 146:555–566. https://doi.org/10.1007/s00704-021-03751-5

Houmsi MR, Shiru MS, Nashwan MS et al (2019) Spatial shift of aridity and its impact on land use of Syria. Sustainability 11:1–18. https://doi.org/10.3390/su11247047

Ingrao C, Strippoli R, Lagioia G et al (2023) Water scarcity in agriculture: an overview of causes, impacts and approaches for reducing the risks. Heliyon 9(8):e18507. https://doi.org/10.1016/j.heliyon.2023.e18

Ipek AF, Altas A, Serencam U et al (2025) Spatio-temporal drought assessment by using the innovative probability of drought severity (PDS) method. Water Resour Manag 39:5135–5150. https://doi.org/10.1007/s11269-025-04199-2

Kelley CP, Mohtadi S, Cane MA et al (2015) Climate change in the fertile crescent and implications of the recent Syrian drought. Proc Natl Acad Sci 112(11):3241–3246. https://doi.org/10.1073/pnas.1421533112

Kelley C, Mohtadi S, Cane M et al (2017) Commentary on the Syria case: climate as a contributing factor. Polit Geogr 60:245–247. https://doi.org/10.1016/j.polgeo.2017.06.013

Keune J, Di Giuseppe F, Barnard C et al (2025) ERA5—drought: global drought indices based on ECMWF reanalysis. Sci Data 12:616. https://doi.org/10.1038/s41597-025-04896-y

Khosravi Y, Ouarda TBMJ (2025) Drought risks are projected to increase in the future in central and southern regions of the Middle East. Commun Earth Environ 6:384. https://doi.org/10.1038/s43247-025-02359-1

Kogan FN (2007) Remote sensing of weather impacts on vegetation in nonhomogeneous areas. Int J Rem Sens 11:1405–1419. https://doi.org/10.1080/01431169008955102

Kong Q, Guerreiro SB, Blenkinsop S et al (2020) Increases in summertime concurrent drought and heatwave in Eastern China. Weather Clim Extrem 28:100242. https://doi.org/10.1016/j.wace.2019.100242

Koop SHA, Grison C, Eisenreich SJ (2022) Integrated water resources management in cities in the world: global solutions. Sustain Cit Soc 68:104137. https://doi.org/10.1016/j.scs.2022.104137

Kostopoulou E, Giannakopoulos C, Krapsiti D et al (2017) Temporal and spatial trends of the standardized precipitation index (SPI) in Greece using observations and output from regional climate models. In: Karacostas T, Bais A, Nastos P (eds) Perspectives on atmospheric sciences. Springer atmospheric sciences. Springer, Cham

Lazoglou G, Papadopoulos-Zachos A, Georgiades P et al (2024) Identification of climate change hotspots in the Mediterranean. Sci Rep 14:29817. https://doi.org/10.1038/s41598-024-80139-1

Li Y, Yuan F, Zhou Q et al (2024) Drought indices performance for predicting agricultural drought. In: Spatiotemporal dynamics of meteorological and agricultural drought in China. Springer, Singapore. https://doi.org/10.1007/978-981-97-4214-1_12

Linke AM, Ruether B (2021) Weather, wheat, and war: security implications of climate variability for conflict in Syria. J Peace Res 58(1):114–134. https://doi.org/10.1177/0022343320973070

Lionello P, Scarascia L (2018) The relation between climate change in the Mediterranean region and global warming. Reg Environ Change 18:1481–1493. https://doi.org/10.1007/s10113-018-1290-1

Liu H, Liu Y, Chen Y et al (2023) Dynamics of global dryland vegetation were more sensitive to soil moisture: evidence from multiple vegetation indices. Agric for Meteorol 331:109327. https://doi.org/10.1016/j.agrformet.2023.109327

Liu T, Yu L, Yan Z et al (2025) Enhanced climate mitigation feedbacks by wetland vegetation in semi-arid compared to humid regions. Geophys Res Lett 25(9):e2025GL115242. https://doi.org/10.1029/2025GL115242

Mao L, Miao Y, Ge Y et al (2024) Effects of different afforestation years on soil moisture and nutrient content in Maxian Mountain of the Loess Plateau. Sci Rep 14:16194. https://doi.org/10.1038/s41598-024-66408-z

Mathbout S, Lopez-Bustins JA, Martin-Vide J et al (2018) Spatial and temporal analysis of drought variability at several time scales in Syria during 1961–2012. Atmos Res 200(1):153–168. https://doi.org/10.1016/j.atmosres.2017.09.016

Mathbout S, Lopez-Bustins JA, Royé D et al (2021) Mediterranean-scale drought: regional datasets for exceptional meteorological drought events during 1975–2019. Atmosphere 12(8):941. https://doi.org/10.3390/atmos12080941

Mathbout S, Martin-Vide S, Lopez-Bustins JA (2023) Drought characteristics projections based on CMIP6 climate change scenarios in Syria. J Hydrol Reg Stud 50:101581. https://doi.org/10.1016/j.ejrh.2023.101581

Mathbout S, Boustras G, Papazoglou P et al (2025) Integrating climate indices and land use practices for comprehensive drought monitoring in Syria: impacts and implications. Environ Sustain Indic 26:100631. https://doi.org/10.1016/j.indic.2025.100631

Matimolane S, Strydom S, Mathivha FI et al (2024) Evaluating the spatiotemporal patterns of drought characteristics in a semi-arid region of Limpopo Province. South Africa. Environ Monit Assess 196:1062. https://doi.org/10.1007/s10661-024-13217-6

McKee TB, Doesken NJ, Kleist J (1995) Drought monitoring with multiple time scales. Proceeding of the ninth conference on applied climatology. American Meteorological Society, Dallas, pp 233–236

Mckee TB, Doesken NY, Kleist Y (1993) The relationship of drought frequency and duration to time scales. In: Proceedings of the 8th conference on applied climatology, Anaheim, CA, pp 179–184

Mckee TB, Doesken NJ, Kleist J (1993) The relationship of drought frequency and duration to time scales. In: Proceedings of the eighth conference on applied climatology. American Meteorological Society, pp 179–184

McLeod AI (2025) Kendall rank correlationand Mann-Kendall trend tTest. Version 2.2.1. https://cran.r-project.org/web/packages/Kendall/Kendall.pdf

Mezieobi KC, Alum EU, Egwu CK et al (2025) Strategies for resilience: mitigating the effects of climate change on hunger and mental health. J Agric Food Res 24:102023. https://doi.org/10.1016/j.jafr.2025.102023

Morante-Carballo F, Montalván-Burbano N, Quiñonez-Barzola X et al (2022) What do we know about water scarcity in semi-arid zones? A global analysis and research trends. Water 14:2685. https://doi.org/10.3390/w14172685

Mori AS, Gonzalez A, Seidl R et al (2024) Urgent climate action is needed to ensure effectiveness of protected areas for biodiversity benefits. One Earth 7(10):1874–1885. https://doi.org/10.1016/j.oneear.2024.08.003

Moridi M, Rahimian M, Movahed RG et al (2025) Policy insights for drought adaptation: farmers' behavior and sustainable agricultural development. Environ Sustain Indic 12:100603. https://doi.org/10.1016/j.indic.2025.100603

Morsy M, Moursy FI, Sayad T et al (2022) Climatological study of SPEI drought index using observed and CRU gridded dataset over Ethiopia. Pure Appl Geophys 179:3055–3073. https://doi.org/10.1007/s00024-022-03091-z

Nastos P, Saaroni H (2024) Living in Mediterranean cities in the context of climate change: a review. Int J Climatol 44(10):3169–3190. https://doi.org/10.1002/joc.8546

Peng L, Sheffield J, Wei Z et al (2024) An enhanced standardized precipitation–evapotranspiration index (SPEI) drought-monitoring method integrating land surface characteristics. Earth Syst Dynam 15:1277–1300. https://doi.org/10.5194/esd-15-1277-2024

Piemontese L, Terzi S, Di Baldassarre G et al (2024) Over-reliance on water infrastructure can hinder climate resilience in pastoral drylands. Nat Clim Chang 14:267–274. https://doi.org/10.1038/s41558-024-01929-z

Rosário J, Madureira L, Marques C et al (2022) Understanding farmers' adoption of sustainable agriculture innovations: a systematic literature review. Agronomy 12:2879. https://doi.org/10.3390/agronomy12112879

Saharwardi MS, Dasari HP, Hassan WU et al (2025) Projected increase in droughts over the Arabian Peninsula and associated uncertainties. Sci Rep 15:1711. https://doi.org/10.1038/s41598-025-85863-w

Saleem A, Anwar S, Nawaz T et al (2024) Securing a sustainable future: the climate change threat to agriculture, food security, and sustainable development goals. J Umm Al-Qura Univ Appl Sci 94:177. https://doi.org/10.1007/s43994-024-00177-3

Sandler D, Saaroni H, Ziv B et al (2024) The connection between North Atlantic storm track regimes and eastern Mediterranean cyclonic activity. Weather Clim Dynam 5:1103–1116. https://doi.org/10.5194/wcd-5-1103-2024

Santos R, Russo A, Gouveia CM (2024) Co-occurrence of marine and atmospheric heatwaves with drought conditions and fire activity in the Mediterranean region. Sci Rep 14:19233. https://doi.org/10.1038/s41598-024-69691-y

Selby J, Dahi O, Fröhlich C et al (2017) Climate change and the Syrian civil war revisited. Polit Geogr 60:232–244. https://doi.org/10.1016/j.polgeo.2017.05.007

Sen PK (1968) Estimates of the regression coefficient based on Kendall's Tau. J Am Stat Assoc 63(324):1379–1389. https://doi.org/10.1080/01621459.1968.10480934

Suarez-Gutierrez L, Müller WA, Marotzke J (2023) Extreme heat and drought typical of an end-of-century climate could occur over Europe soon and repeatedly. Commun Earth Environ 4:415. https://doi.org/10.1038/s43247-023-01075-y

Sukkar A, Abulibdeh A, Essoussi S et al (2024) Investigating the impacts of climate variations and armed conflict on drought and vegetation cover in Northeast Syria (2000–2023). J Arid Environ 225:105278. https://doi.org/10.1016/j.jaridenv.2024.105278

Tabari H, Willems P (2023) Global risk assessment of compound hot-dry events in the context of future climate change and socioeconomic factors. NPJ Clim Atmos Sci 6:74. https://doi.org/10.1038/s41612-023-00401-7

Terzi TB, Önöz B (2025) Drought analysis based on nonparametric multivariate standardized drought index in the Seyhan River Basin. Nat Hazards 121:11051–11078. https://doi.org/10.1007/s11069-025-07234-y

Trenberth KE, Dai A, Van der Schrier S et al (2014) Global warming and changes in drought. Nat Clim Change 4(1):17–22. https://doi.org/10.1038/nclimate2067

Vázquez M, Nieto R, Liberato MLR et al (2023) Influence of teleconnection patterns on global moisture transport during peak precipitation month. Int J Climatol 43(2):932–949. https://doi.org/10.1002/joc.7843

Vicente-Serrano SM, Begueria S, Lopez-Moreno JI et al (2010) A new global 0.58 gridded dataset (1901–2006) of a multiscalar drought index: comparison with current drought index datasets based on the Palmer drought severity index. J Hydrometeorol 11:1033–1043. https://doi.org/10.1175/2010JHM1224.1

Vicente-Serrano SM, Domínguez-Castro F, Beguería S et al (2025a) Atmospheric drought indices in future projections. Nat Water 3:374–387. https://doi.org/10.1038/s44221-025-00416-9

Vicente-Serrano SM, Tramblay Y, Reig F et al (2025b) High temporal variability not trend dominates Mediterranean precipitation. Nature 639:658–666. https://doi.org/10.1038/s41586-024-08576-6

Wang L, Ren W (2025) Drought in agriculture and climate-smart mitigation strategies. Cell Rep 24:100386. https://doi.org/10.1016/j.crsus.2025.100386

Wei L, Jiang S, Ren L et al (2025) Detection of the 2022 extreme drought over the Yangtze River basin using two satellite-gauge precipitation products. Atmos Res 315:10792. https://doi.org/10.1016/j.atmosres.2025.107929

Wells N, Goddard S, Hayes MJ (2004) A self-calibrating palmer drought severity index. J Clim 17(12):2335–2351. https://doi.org/10.1175/1520-0442

World Meteorological Organization (2012) Guidelines on the selection of reference climatological stations (RCSs) from the existing climatological station network. 1986. WMO/TD-No. 130. World Climate Programme, WMO, Geneva, Switzerland. World Meteorological Organization. Standardized Precipitation Index User Guide. (M. Svoboda, M. Hayes and D. Wood). WMO-No. 1090. Geneva, p 16

Xing Y, Wang X (2024) Precision agriculture and water conservation strategies for sustainable crop production in arid regions. Plants 13:3184. https://doi.org/10.3390/plants13223184

Ying Y, Bojie F, Yanxu L et al (2024) Compound hot–dry events greatly prolong the recovery time of dryland ecosystems. Natl Sci Rev 11(10):nwae274. https://doi.org/10.1093/nsr/nwae274

Zargar A, Sadiq R, Bahman N et al (2011) A review of drought indices. Environ Rev 19:333–349. https://doi.org/10.1139/a11-013

Zhai J, Mondal SK, Fischer T et al (2020) Future drought characteristics through a multi-model ensemble from CMIP6 over South Asia. Atmos Res 246:105111. https://doi.org/10.1016/j.atmosres.2020.105111

Zhang X, Alexander L, Hegerl GC et al (2011) Indices for monitoring changes in extremes based on daily temperature and precipitation data. Wiley Interdiscip Rev Clim Change 2(6):851–870. https://doi.org/10.1002/wcc.147

Zhang X, Evans JP, Burrell AL (2024) Less than 4% of dryland areas are projected to desertify despite increased aridity under climate change. Commun Earth Environ 5:300. https://doi.org/10.1038/s43247-024-01463-y

Zhong Z, He B, Guo L et al (2019) Performance of various forms of the palmer drought severity index in China from 1961 to 2013. J Hydrometeorol 20(9):1867–1885. https://doi.org/10.1175/JHM-D-18-0247.1

Zhong L, Hua L, Yan Z (2020) Datasets of meteorological drought events and risks for the developing countries in Eurasia. Earth Syst Sci Data 4:191–223. https://doi.org/10.1080/20964471.2019.1710383

Zhong R, Chen X, Wang Z et al (2018) scPDSI: calculation of the conventional and self-calibrating Palmer drought severity index, R package version 0.1.3. https://github.com/Sibada/scPDSI. Accessed 28 May 2025

Zittis G, Almazroui M, Alpert P et al (2022) Climate change and weather extremes in the eastern Mediterranean and Middle East. Rev Geophys 60(3):762. https://doi.org/10.1029/2021RG000762

Zubkova M, Giglio L, Humber ML et al (2021) Conflict and climate: drivers of fire activity in Syria in the twenty-first century. Earth Interact 25(1):119–135. https://doi.org/10.1175/EI-D-21-0009.1

Assessing Climate Vulnerability, the Impacts of Future Climate Scenarios and Coastal Resilience in the Tuban Delta, Yemen

Khaldoon A. Mourad, Joris Oele, Waleed Yacoob, Julie Greenwalt, Mohammed Zain, Abdulraqeb Al-Okaishi, and Alaa Aulaiah

1 Introduction

1.1 Climatic Scenarios

Yemen faces climate change and among its most vulnerable regions is the Tuban Delta, which is formed by the seasonal Wadi Tuban, a critical watercourse that supports agriculture and settlements in Lahij Governorate. This area, historically an agricultural hub, is now confronting compounding climate risks that threaten livelihoods, ecosystems, and regional stability.

Over 70% of Yemen's population relies on agriculture (FAO 2020), and the Tuban Delta's degradation directly threatens food security for millions. The delta's fertile soils and proximity to the port city of Aden have made it a vital region for crop production, including sorghum, vegetables, and cash crops like qat.

The Tuban Delta relies on seasonal rains and groundwater, but rising temperatures (projected to increase by 1.5–3 °C by 2050) and erratic rainfall have reduced recharge rates. Prolonged droughts, such as those in 2018–2021, have diminished agricultural yields, forcing farmers to abandon lands or over-pump groundwater, accelerating aquifer depletion.

The Tuban Delta's low-lying coastal zones are highly susceptible to sea-level rise, which the IPCC estimates could reach 0.5–1 m by 2100. Saltwater intrusion has already contaminated aquifers and soils, rendering farmland uncultivable. Storm surges from cyclones in the Arabian Sea (Cyclones Chapala and Mekunu in 2015

K. A. Mourad (✉) · J. Oele · W. Yacoob · J. Greenwalt · M. Zain · A. Al-Okaishi · A. Aulaiah
United Nations Human Settlements Programme Aden, Aden, Yemen
e-mail: khaldoonmourad@yahoo.com

K. A. Mourad
The Center for Sustainable Visions, Lund, Sweden

M. M. Al Saud (ed.), *Hydroclimate Changes: Implications of the Middle East Region*,
https://doi.org/10.1007/978-3-032-20053-2_10

and 2018) have caused catastrophic flooding, displacing communities and destroying infrastructure.

The following are the main climate risks facing the study area and Yemen in General:

(a) Water scarcity and drought

Yemen is one of the most water-stressed countries globally, with per capita freshwater availability at just 2% of the global average (World Bank 2017). The Tuban Delta relies on seasonal rains and groundwater, but rising temperatures—projected to increase by 1.5–3 °C by 2050 (IPCC 2021)—and erratic rainfall have reduced recharge rates. Prolonged droughts, such as those in 2018–2021, have diminished agricultural yields, forcing farmers to abandon lands or over-pump groundwater, accelerating aquifer depletion (Al-Sakkaf 2016).

(b) Sea-level rise and coastal flooding

The Tuban Delta's low-lying coastal zones are highly susceptible to sea-level rise (SLR), which the IPCC estimates could reach 0.5–1 m by 2100 (IPCC 2021). Saltwater intrusion has already contaminated aquifers and soils, rendering farmland uncultivable. Storm surges from cyclones in the Arabian Sea (e.g., Cyclones Chapala and Mekunu in 2015 and 2018) have caused catastrophic flooding, displacing communities and destroying infrastructure (CARE International 2021).

(c) Extreme weather events

Climate models predict increased intensity of cyclones and flash floods in Yemen's coastal regions (IPCC 2021). Sudden floods overwhelm Wadi Tuban's ephemeral channels, eroding topsoil and damaging irrigation systems. In 2020, floods affected 300,000 people nationwide, with the Tuban Delta among the hardest-hit regions (UNDP 2019).

(d) Socioeconomic and conflict-related vulnerabilities

Yemen's ongoing conflict has decimated institutional capacity to address climate risks (UNDP 2019). The Tuban Delta lacks adaptive infrastructure, such as flood barriers or desalination plants, due to funding shortages and political instability. Competition over dwindling water resources has also fueled local conflicts, particularly between farmers and pastoralists (Al-Sakkaf 2016). Displacement from climate shocks further strains urban centers like Aden, where 40% of Yemen's internally displaced persons (IDPs) reside (CARE International 2021).

(e) Ecological and health impacts

Salinization and desertification have reduced the delta's arable land by an estimated 30–40% since 2000 (Ministry of Water and Environment, Yemen 2018). Mangrove ecosystems, which buffer against coastal erosion, are in decline due to pollution and overharvesting. Contaminated water supplies have heightened cholera and malnutrition rates, with Yemen enduring the world's worst cholera outbreak in 2017–2019 (CARE International 2021).

1.2 Climatic Scenarios

The climate modeling community has created four Representative Concentration Pathways (RCPs) based on CO_2 concentrations that result in global warming of 2.6, 4.5, 6 or 8.5 W/m^2 under the Coupled Model Inter-comparison Project 5 (CMIP5) as follows:

- RCP 2.6: Carbon dioxide (CO_2) emissions must start to decline by 2020 and reach zero by 2100. In this situation, carbon capture, utilization, and storage (CCUS) need to be strengthened.
- RCP 4.5, emissions peak will be in 2040, then CO_2 concentration will start to decline.
- RCP 6, after a peak around 2080, emissions start to drop.
- RCP 8.5 is the worst-case scenario, by which emissions will increase throughout the twenty-first century.

On the other hand, the Coupled Model Inter-comparison Projects 6 (CMIP6) considered Shared Socioeconomic Pathways (SSPs) to project socioeconomic global trends up to 2100. The SSPs are built on five storylines that describe various socioeconomic processes based on five narratives describing alternative socio-economic developments, including sustainable development (SSP1), regional rivalry (SSP3), inequality (SSP4), fossil-fueled development (SSP5), and middle-of-the-road development (SSP2) (Riahi et al. 2017), (Fig. 1).

SSPs based scenarios further refine the RCPs. Thus, couplings SSPs and RCPs is used to describe possible alternative conditions of human society and nature at

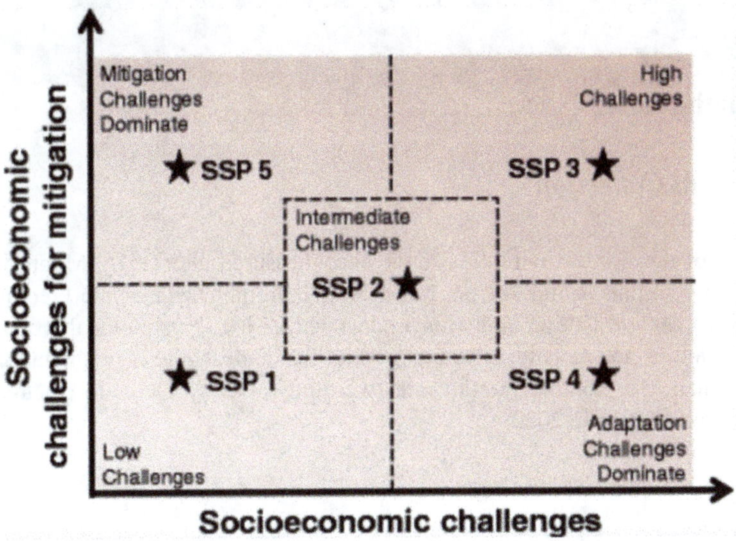

Fig. 1 Combinations of challenges to mitigation and adaptation (O'Neill et al. 2014)

the macroscale (Frame et al. 2018) that could shape future society as the following (Hausfather 2018):

- SSP1: a world of sustainability-focused growth and equality based on the Sustainable Development Goals (SDGs) (RCP 1.9 and RCP 2.6).
- SSP2: "middle of the road" world where trends broadly follow their historical patterns RCP 4.5.
- SSP3: a fragmented world of regional rivalry and resurgent nationalism RCP7.
- SSP4: a world of ever-increasing inequality; and
- SSP5: (fossil fuel development): a world of rapid and unconstrained growth in economic output and energy use. RCP 8.5.

SSP3 and SSP4 show low investments in health and education in developing nations, a rapidly expanding population, and rising inequality.

SSP1 and SSP5 are indicators of effective institutions, wise investments in health and education, and quick economic expansion. SSP1 is based on sustainable practices, whereas SSP5 is based on a fossil fuel-driven economy (Bassetti 2022).

Examining how well CMIP5 and CMIP6 Global Climate Models (GCMs) performed in various nations revealed that CMIP6 GCMs significantly improved at reproducing temperature and rainfall. Except for minimum temperature at various timeframes, CMIP5 GCMs perform better at replicating geographical distribution but less well at simulating spatial variability of most climatic variables (Kamruzzaman et al. 2021). As a result, the Meteorological Research Institute Earth System Model Version 2.0 (MRI-ESM 2.0) will be considered under CMIP6. It is anticipated that MRI-ESM 2.0 will perform better in many of the experiments scheduled for CMIP6 than earlier models that took part in the fifth phase of the Coupled Model Intercomparison Project (CMIP5) (Yukimoto et al. 2019a).

2 Methodology

2.1 Data Collection

Because of the war, no rainfall records were measured after 2015. Monthly rainfall rates were thus acquired from Multi-Source Weighted-Ensemble Precipitation (MSWEP), a worldwide precipitation product with a 3 h 0.1° resolution that combines gauge, satellite, and reanalysis data to produce the highest quality precipitation estimates. Then, the monthly rainfall rates were processed in QGIS to calculate the annual precipitation in 2022.

2.2 Climate Scenarios

Based on the Coupled Model Inter-comparison Project 6 (CMIP6), which supports the IPCC's Sixth Assessment Report, the future projections in this report have been performed under to climatic scenarios SSP3 and SSP5, due to the increase of conflicts and the traditional socioeconomic activities all over the world:

- SSP3 is coupled to RCP7, which suggests high obstacles in adaptation and mitigation due to conflicts and instabilities, and
- SSP5 is coupled to RCP8.5, which indicates high challenges in mitigation leading to the increase of fusel fuel consumption.

In general, SSP3 and SSP5 represent bad scenarios. However, the situation during the last 13 years is not optimistic to consider better scenarios such as SSP2.

Some of climate projections up to 2100 were obtained by the support of UNESCWA. Other climate projections were brained from different climate services platforms such as Climate Change Knowledge Portal (CCKP), which was created by the World Bank for development practitioners and policy makers to solve a need. https://climateknowledgeportal.worldbank.org. Moreover, the time frames for the extreme events scenarios have been modified from 2010–2039; 2035–206; 2060–2089; and 2070–2099 to be 2023–2040; 2041–2060; 2061–2080; and 2081–2100.

2.3 Runoff

Runoff water can be calculated by subtracting evaporation and infiltration from total rainfall (Dingman 2015). The rational method, Cook's method, or curve number are three basic runoff estimation methods that can be used to estimate runoff and water availability (Juniati et al. 2021). The rational method was utilized in this paper to estimate the runoff in the Tuban Delta using the following equations (Juniati et al. 2021):

$$\mathrm{C} = \sum (\mathrm{ci} \times \mathrm{Ai}) / \sum \mathrm{Ai} \tag{1}$$

$$\mathrm{R} = \sum \mathrm{Ri}/\mathrm{m} \tag{2}$$

$$\mathrm{WA} = 10 \times \mathrm{C} \times \mathrm{R} \times \mathrm{A} \tag{3}$$

where, WA = water availability (m^3/year), C = weighted runoff coefficient, Ci = Land-use coefficient, Ai = Land Area (ha), R = average yearly rainfall data (mm/year), Ri = Rainfall in station i, m = number of rainfall stations, A = Total Area (ha), 10 = conversion factor from ha to m^3.

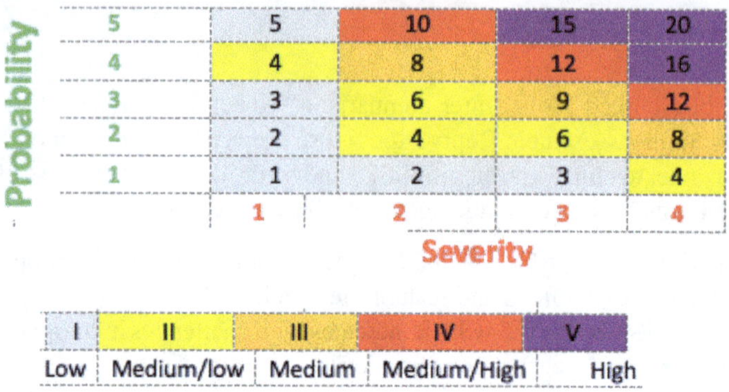

Fig. 2 Risk assessment values and categories

2.4 Hazard Assessment

Hazard assessment values are determined by multiplying the scores for severity values and the probability together for each recognized risk. Severity is the amount of damage or harm a hazard could create and it is often ranked on a four-point scale: 4—Catastrophic; 3—Critical; 2—Marginal; and 1—Negligible. Probability is the likelihood of the hazard occurring and it is often ranked on a five-point scale: 5—Frequent; 4—Probable; 3—Occasional; 2—Remote; and 1—Improbable. Then hazard values can be categorized into four categories (I) Low; (II) Medium/Low; (III) Medium; (IV) Medium/High; and (V) High as (Fig. 2).

3 Results and Discussion

3.1 Heat Waves

According to the climate model MRI-ESM2-0, there will be an increase in the number of hot days per year might increase starting from 2050 to reach in 2100 40 days/year under RCP 7, and 70 days/year under RCP 8.5, (Fig. 3). This increase of temperature will increase evapotranspiration rates and water scarcity in the Tuban Delta, thus reduce agriculture production (food security) and domestic water supply. The severity of heat waves, from 1 to 4 scale, is shown in Table 1.

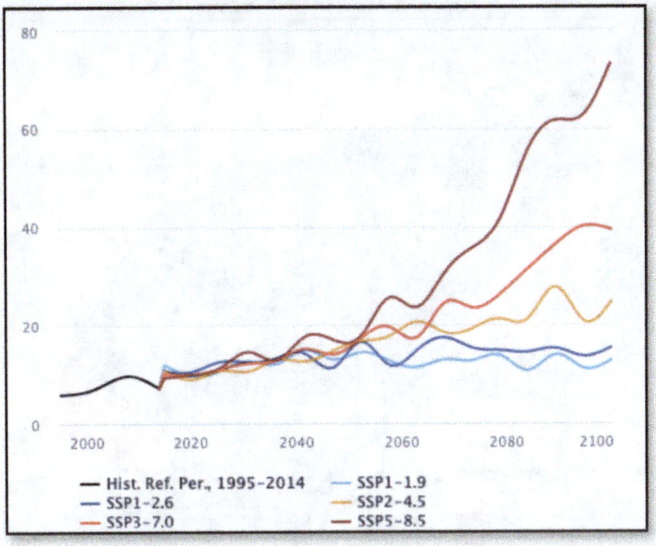

Fig. 3 The number of hot days >40 °C in Lahj

Table 1 The severity of heat wave risks

Years/Regions	SSP3 (RCP 7)			SSP5 (RCP 8.5)		
	UR	MR	LR	UR	MR	LR
2023–2040	1	1	2	1	1	2
2041–2060	2	2	3	2	2	3
2061–2080	3	3	4	3	4	4
2081–2100	3	4	4	3	4	4

3.2 Saltwater Intrusion

The salinity of groundwater in Tuban Delta was estimated based on the Electric Conductivity (EC). In general, EC concentrations in Tuban Delta have increased near the coast due to overexploitation and saltwater intrusion. Moreover, the aquifer type near the coast is semi-confined (Squarespace n.d.), which allows saltwater intrusion and the movement of the salty water from the brackish aquifer.

On the other hand, there are medium salinity levels in the MR and the UR due to the characteristics of the rocks, (Fig. 4).

According to Yemini drinking water standards, EC for drinking water is preferred to be <1000 μS/cm and the maximum allowed limit can reach 2500 μS/cm (Saleh et al. 2017a, 2017b). WHO recommends EC in drinking water to be about 400–800 μS/cm, while water with EC from 800 to 2500 μS/cm can be used for irrigation and livestock (Al-Khashman and Jaradat 2014). Figure 5 shows Electric Conductivity

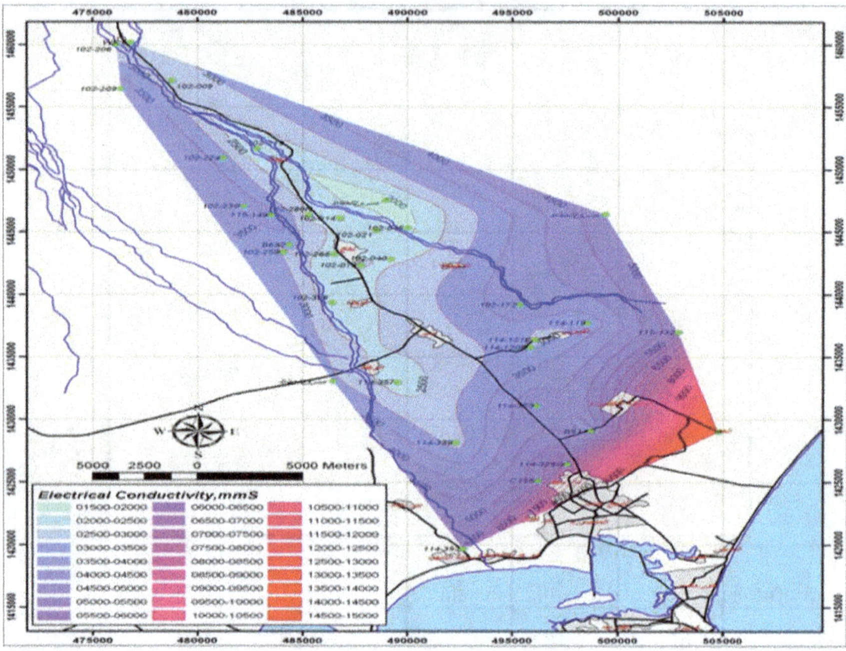

Fig. 4 EC concentration lines in Tuban Delta

(EC) concentrations in some wells in the Tuban Delta. It shows that all wells that are located near the coast have high EC concentrations especially those that are located at a brackish aquifer.

In 2023, based on NWSA data of 2019, many wells of Dar Almanasira well field that is located 15 km northern the Eastern costal line have high EC concentration (more than 2500 μS/cm), which means this discharge brackish water, which is about 20 m below sea level. Taking an average annual drop in water table of 5 m all wells of Dar Almanasira well field will become brackish by 2040. Thus, discharged water won't be drinkable without a proper purification/treatment via Reverse Osmosis system (RO), for example. However, it can be used to irrigate salinity tolerant crops. In this regard, biosaline agriculture, which is the production and growth of plants in saline rich groundwater and/or soils, is important because it allows using low quality of water to irrigate some species and to adapt to local climatic conditions (Oumara and El Youssfi 2022). For example, in Morocco, some crop species were tested and showed very high tolerance to salinity, such as Quinoa, Pearl millet, Barley, and Panicum Blue (Hirich et al. 2021).

Only few wells outside of the Bir Ahmed well field that is located 8–14 km from the coast, have EC <2000 μS/cm. These wells are located at low productive aquifer and above the brackish water, which is about 65–70 m below the surface, which means these wells are vulnerable to drought and to sea-level rise. The well field is located at 40 m above sea level. Thus, in case of drought, people might think of

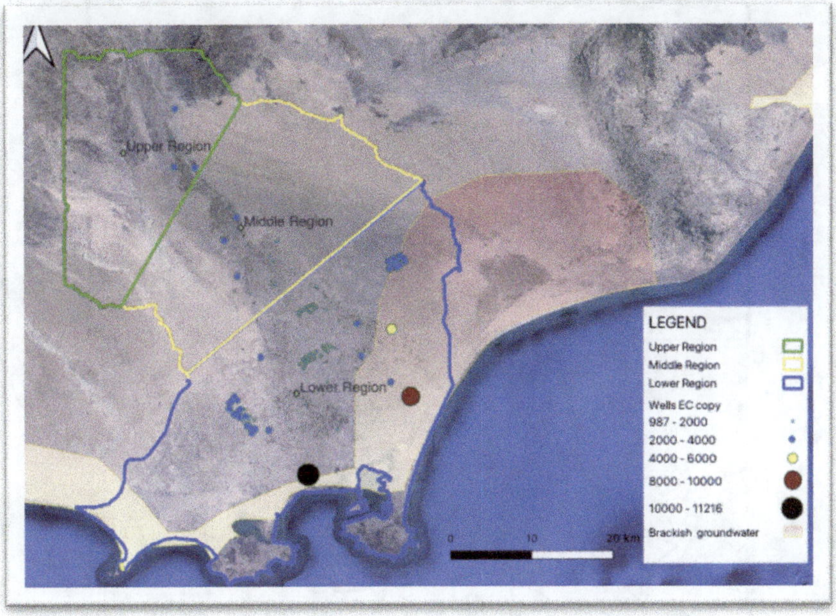

Fig. 5 EC concentrations in some well in the Tuban Delta

deepen these wells ignoring the possibility of reaching the brackish aquifer. Thus, in case of drought and continuous over abstraction, these well will become brackish by 2060. Abubakr et al. (2012) argued that wells that have a depth of 65 m or more will get brackish.

Bir Naser well field that is located about 16 km from the coast has fair water quality, based on the Yemeni standards, EC range 1500–1800 μS/cm. However, these wells are vulnerable to sea-level rise and overexploitation. Thus, EC concentrations will increase. S.W.L of the wells are about 20–30 m below sea level. The water will become brackish when S.W.L reaches 80 m below sea level. Thus, in case of drought and continuous over abstraction, these well will become brackish by 2060.

If these wells (Fig. 6), which are the main drinking water resource, become brackish, people in the Lower Region, mainly Aden, will face water scarcity as domestic water needs will be affected. Moreover, other legal and illegal wells that are used for irrigation at the coastal area will be affected as well. Therefore, new allocation strategies will be needed and looking for new domestic water resource will be a must. The National Authority of Water and Sanitation in Aden suggested drilling new wells with a depth of 300–350 m to cover the increasing demand. However, it is not recommended to deepen wells in the coastal area as it may reach the brackish water. Thus, this might not be a sustainable solution. Constructing solar-powered desalination plants can be a sustainable option to cover the increasing demand in Aden.

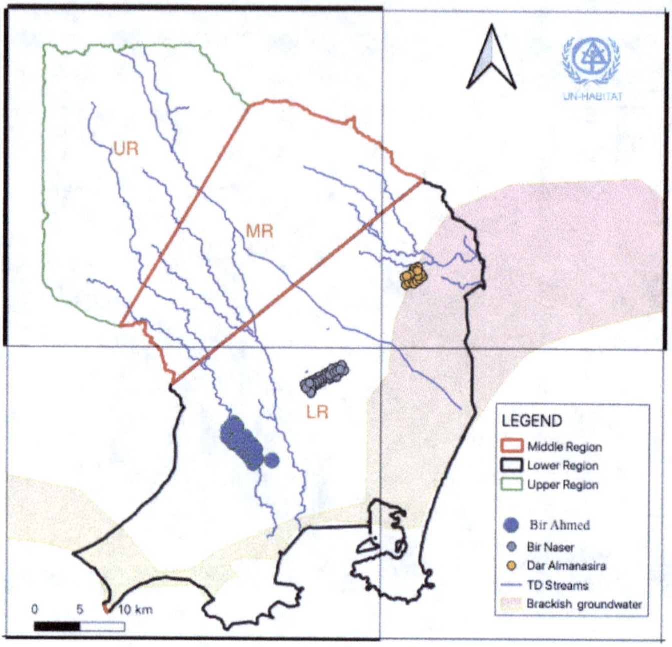

Fig. 6 Main well fields in the lower region

Saltwater intrusion will keep increasing due to increase water demand and drought years between 2023 and 2040. After 2040, the probability for the years to be wetter is high however, population increase, and unmonitored pumping might not help in reducing saltwater intrusion. Table 2 shows saltwater intrusion risks at the main water supply well fields in Aden taking into consideration the continues illegal and unmonitored groundwater withdrawal.

Table 2 The risk of saltwater intrusion in the coastal area of the Tuban Delta

Years/Well fields	Saltwater intrusion		
	Dar Almanasira	Bir Ahmed	Bir Naser
2023–2040	High risk	Medium risk	Medium risk
2041–2060	Brackish	High risk	High risk
2061–2080	Brackish	Brackish	Brackish
2081–2100	Brackish	Brackish	Brackish

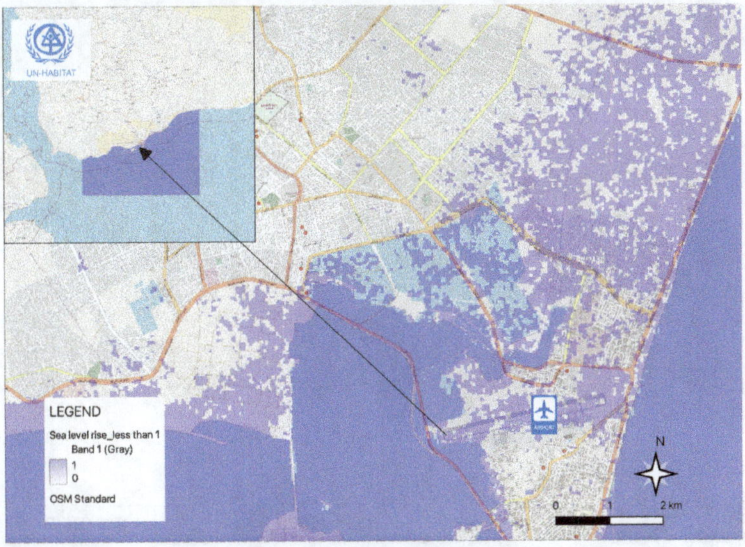

Fig. 7 The impacts of sea-level rise in the coastal areas of Aden

3.3 Sea-Level Rise

Mapping Seal Level Rise

QGIS has been used to map the affected areas if sea-level rises by 1 m. The simulations show that most of the coastal areas in Aden will be affected especially the airport (Fig. 7).

In this regard, based on Climate Change Knowledge Portal (CCKP), (Fig. 8), shows seal-level rise from 2008 to 2100 under RCP2.6, RCP4.5 and RCP8.5. While based on the CMIP6, Table 3 shows projected seal-level rise under SSP5 and SSP4 (Climate Central 2023).

The Impacts of Sea-Level Rise

Sea-level rise might affect most of the building near the coast that has an elevation of zero or below sea level. Mapping sea-level rise under RCP 8.5 shows that the risk will be low in 2040 in the coastal area compared with 2100, by which the airport will be affected if no measures are taken (Fig. 9).

In this regard, the airport is located at elevation range between 0 and −10 m below sea level, (Fig. 10), which means seal-level rise will be a real threat to the airport and its facilities.

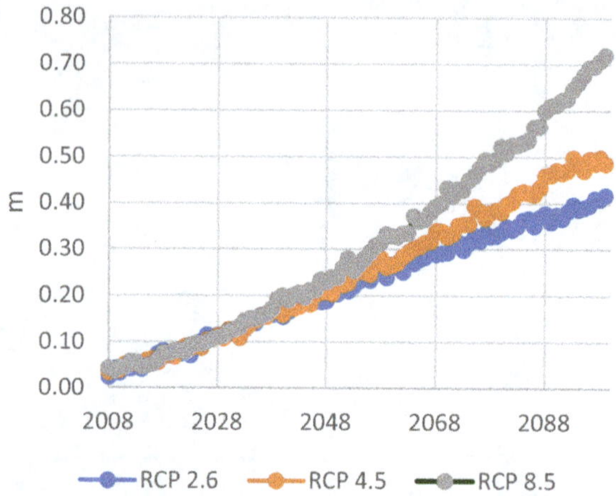

Fig. 8 Future projections of sea-level of coastal Yemen

Table 3 Projected sea-level rise (m) under CMIP6

Year	SSP5 (RCP 8.5)	SSP3 (RCP 7)
2020	0.05	0.05
2040	0.16	0.16
2060	0.31	0.28
2080	0.51	0.46
2100	0.77	0.67

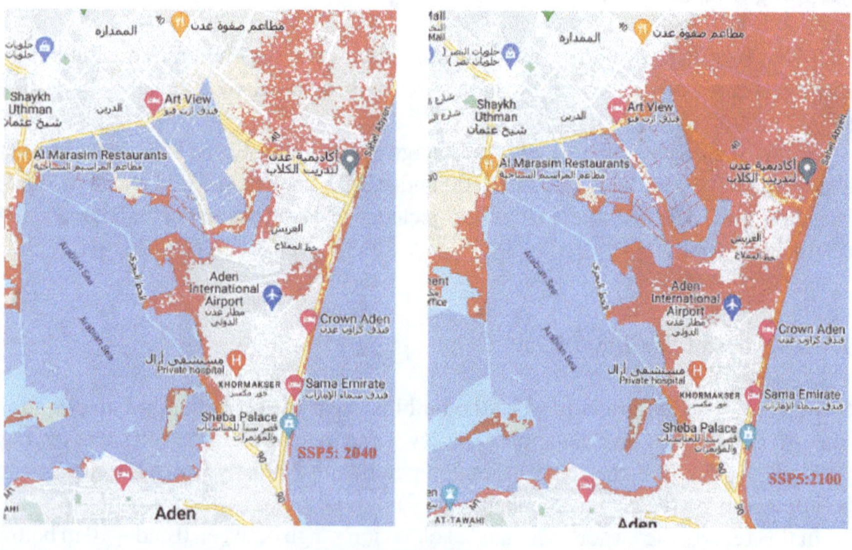

Fig. 9 Sea-level rise in 2040 and 2100 under RCP8.5

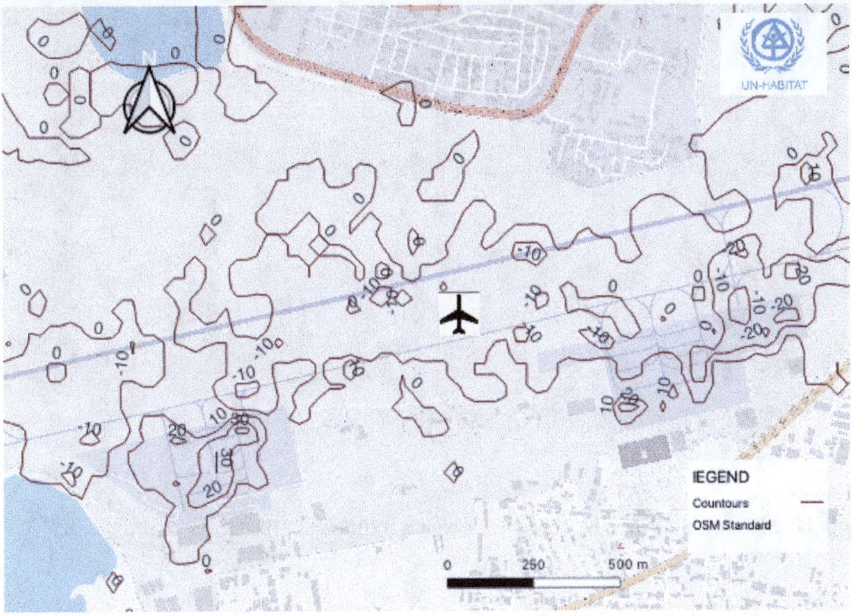

Fig. 10 The contours at the airport's area

On the other hand, sea-level rise will increase saltwater intrusion into many wells in Aden, especially the remaining wells of Bir Ahmed well field, which will affect domestic water supply in Aden.

3.4 Flooding

Flooding is likely to increase due to due to the increase in heavy rainfall/rainstorm. Using the MRI-ESM2.0 climate model (Yukimoto et al. 2019b), the projections show extreme rainfall in some months that may reach, 80–100 mm, which might occur between 2038 and 2040 under both RCPs and after 2060–2063 and 2072–2077 under SSP3 (RCP 7), (Fig. 11) and between 2033 and 2037 under SSP5 (RCP 8.5), resulting flooding, (Fig. 12).

Quantifying the Runoff

Quantifying runoff due the largest one-day precipitation gives an idea about flood's significance. However, to estimate the risk of the flash flood, the main Tuban watershed will be considered. As shown in (Fig. 13), the water runoff will come first from 1730, 990, and 1550 Km^2 of Al Dhali, Ibb and Ta'izz, respectively through 1400 of

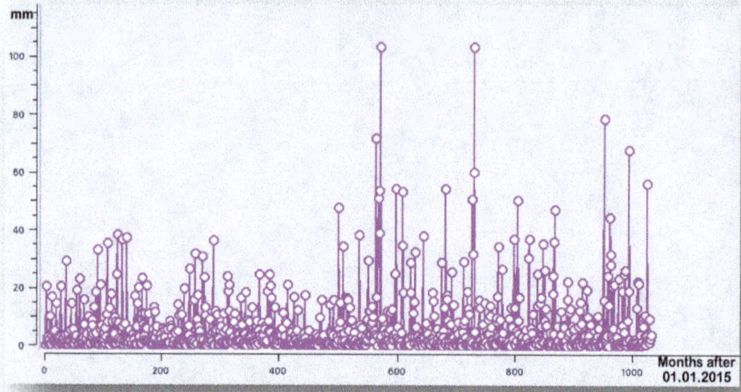

Fig. 11 Monthly rainfall in the Tuban Delta between 2015 and 2100, SSP3

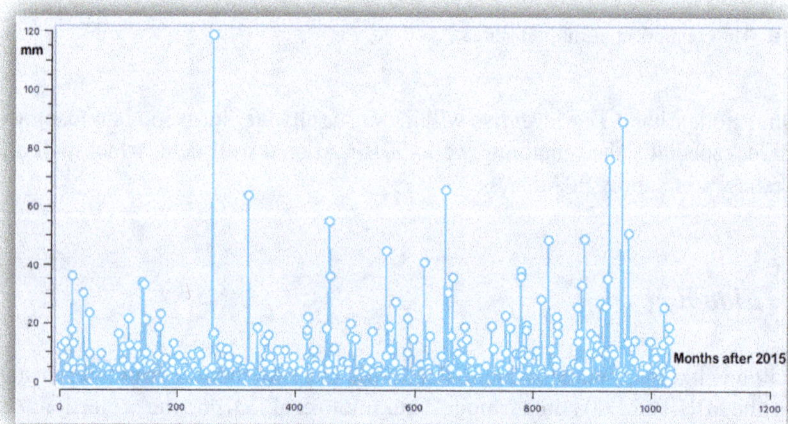

Fig. 12 Monthly rainfall in the Tuban from 2015 to 2100 under SSP5

km^2 Lahj to Dukeim at the top of Tuban Delta to the 392 Km^2 of the UR, 136 Km^2 of the MR, and 84 Km^2 of the LR, then to the ocean in Aden.

The severity of floods can be quantified based on the largest one-day rainfall rate in each area, Table 4.

Then the resulted peak flow (runoff) can be estimated using the rational method at Tuban Delta, Table 5.

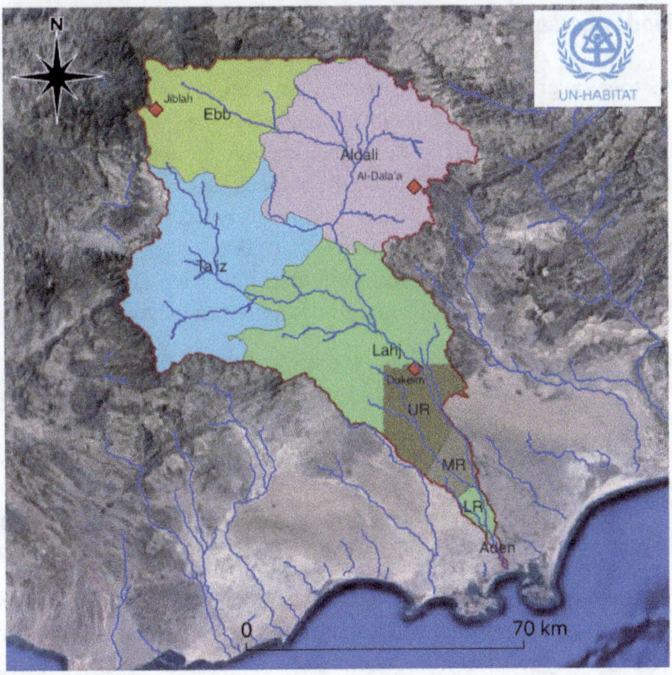

Fig. 13 The boundaries of the Tuban watershed that contributes to flash flood in Tuban Delta

Table 4 Largest one-day precipitation (mm)

Regions	SSP3 (RCP 7)							SSP3 (RCP 8.5)						
	Taiz	Ebb	Aldali	Lahj	UR	MR	LR	Taiz	Ebb	Aldali	Lahj	UR	MR	LR
2023–2040	9	11	10	8	8	8	1.5	1	7	6	2	2	2	2
2041–2060	13	14	16	10	10	10	4	9	14	15	11	11	11	5
2061–2080	23	19	16	19	19	19	24	27	28	24	20	20	20	20
2081–2100	30	39	29	23	23	23	18	33	30	26	30	30	30	41

Table 5 Runoff peak flow due to the largest one-day precipitation (m^3/s)

Years	SSP3			SSP5		
	Dukeim/UR	MR	LR/Aden	Dukeim/UR	MR	LR/Aden
2023–2040	310	329	336	126	131	133
2041–2060	441	463	473	403	428	439
2061–2080	632	675	702	809	854	880
2081–2100	974	1026	1053	976	1044	1087

Urban Flood Assessment

In general, the lower region faces four main climate hazards namely flooding, drought, sea-level rise, and saltwater intrusion. While the middle region faces mainly flooding and drought. Comparing the actual average annual rainfall rates that were 45, 54 and 62 mm in the LR, MR, and UR, respectively; with the projected rainfall rates, the percentages of these changes can be estimated, Table 6. The table shows general increase in annual rainfall partners after 2040.

The severity of urban floods can be divided into four categories based on the increase percentages of the annual rainfall compared to the average values:

- Negligible: Low severity when the increase in rainfall patterns is <51%,
- Marginal: Medium severity when the increase in rainfall patterns is between 51 and 99%,
- Critical: Serious severity when the increase in rainfall patterns is between 100 and 200%,
- Catastrophic: High severity when the increase in rainfall patterns is more than 200%.

Table 7 shows urban flood severity assessment under RCP 7 and RCP 8.5.

The probability of urban flood occurrence has been estimated based on probability of the largest one-day rainfall that generate serious or high severity in each region, as show in Table 8.

- Improbable (<20%):
- Remote (between 20 and 39%)

Table 6 Future change in rainfall rates in Tuban Delta

Years/Regions	Annual change (%) /RCP 7			Annual change (%) 7 RCP 8.5		
	LR	MR	UR	LR	MR	UR
2023–2040	56	29	23	84	49	42
2041–2060	71	27	21	100	68	60
2061–2080	147	98	88	74	40	33
2081–2100	125	77	68	117	57	50

Table 7 Urban flood severity for Tuban Delta

Years/Regions	RCP 7			RCP 8.5		
	UR	MR	LR	UR	MR	LR
2023–2040	1	1	2	1	1	2
2041–2060	1	1	2	2	2	3
2061–2080	2	2	3	1	1	2
2081–2100	2	2	3	2	2	3

Table 8 The probability of urban flood occurrence

Years/Regions	RCP 7			RCP 8.5		
	UR	MR	LR	UR	MR	LR
2023–2040	1	1	1	1	1	1
2041–2060	2	1	1	2	2	2
2061–2080	4	3	3	1	1	1
2081–2100	3	2	3	3	3	3

Table 9 Urban flood hazard assessment under SSP3 and SSP5

Years/Regions	SSP3 (RCP 7)			SSP5 (RCP 8.5)		
	UR	MR	LR	UR	MR	LR
2023–2040	I	I	I	I	I	I
2041–2060	I	I	I	II	II	II
2061–2080	III	II	III	I	I	I
2081–2100	II	II	III	II	II	III

- Occasional (between 40 and 59%)
- Probable (between 60 and 80%)
- Frequent: more than 80%

Thus, urban flood hazards assessment values can be estimated and categorized into five categories (from I to V); low, low/medium, medium, medium/high, or high, Table 9.

Flash Flood Assessment

Flash flood occurs due to the runoff flow that comes from the highland before the Tuban Delta (at Dukeim). The average annual water flow at Dukeim was 110 MCM between 1980 and 2014. The largest monthly flow was about 31 MCM in September while the lowest was 0.25 MCM in March. However, in March 1982 the flow reached 225 MCM resulting a catastrophic flood. Considering the flood is the result of five days of rainfall, the average flow at Dukeim would be about 521 m^3/s. Thus, the severity of flash floods can be assessed based on the peak flow at Dukiem and Aden as the following:

- Low severity when the peak flow <200 m^3/s,
- Medium severity when the peak flow is between 200 and 350 m^3/s,
- Serious severity when the 350 and 500 m^3/s
- High severity peak flow in more than 500 m^3/s.

Table 10 presents flash flood severity projections up to 2100 in the three regions.

Table 10 Flash flood severity in the Tuban Delta

Years	SSP3			SSP5		
	Dukeim/UR	MR	LR/Aden	Dukeim/UR	MR	LR/Aden
2023–2040	2	2	2	1	1	1
2041–2060	3	3	3	3	3	3
2061–2080	4	4	4	4	4	4
2081–2100	4	4	4	4	4	4

Table 11 Flash flood hazard assessment

Years	SSP3			SSP5		
	Dukein/UR	MR	LR/Aden	Dukein/UR	MR	LR/Aden
2023–2040	I	I	I	I	I	I
2041–2060	I	I	I	I	I	I
2061–2080	II	II	II	II	II	II
2081–2100	II	II	II	II	II	II

Then the flash flood hazard assessment, Table 11, can be categorized base on severity and considering a low probability, (<20%).

The Impacts of Flooding

There are more than 200 dams and water barrier in the Tuban basin mainly in the upper region, which have a capacity of 5 MCM. These dams are 70% full. Thus, they can absorb only 1.5 MCM. Thus, based on the extreme rainfall rates that are presented Table 7, large amounts of flood water will flow downstream in the streams and may pose big risks to the infrastructure and agriculture lands near the streams and at the coastal area. For example, in the Upper Region two villages (namely Al shqa'ah and Zaedah) and some agricultural lands near the streams might be affected due to the flooding from the Wadi Tuban, (Fig. 14).

In this regard, based on FGDs, an extreme flooding occurred in 1982 due to an extreme rainfall rate that led to an extreme water flow, 350 225 MCM in March 1982, at Dukeim (at the top of Tuban Delta).

Taking into consideration the risk of sea-level rise after 2080, flood risk that might occur after 2060 and the location (distance to the streams/ocean) (Fig. 15), Table 12 presents the main infrastructure in the Tuban delta and their associated climate risks based on their locations.

Based on the Camp Coordination and Camp Management Cluster (CCCM) National Flood Hazard Analysis for the Internally Displaced People (IDP) sites in Yemen (REACH 2023), Table 13.

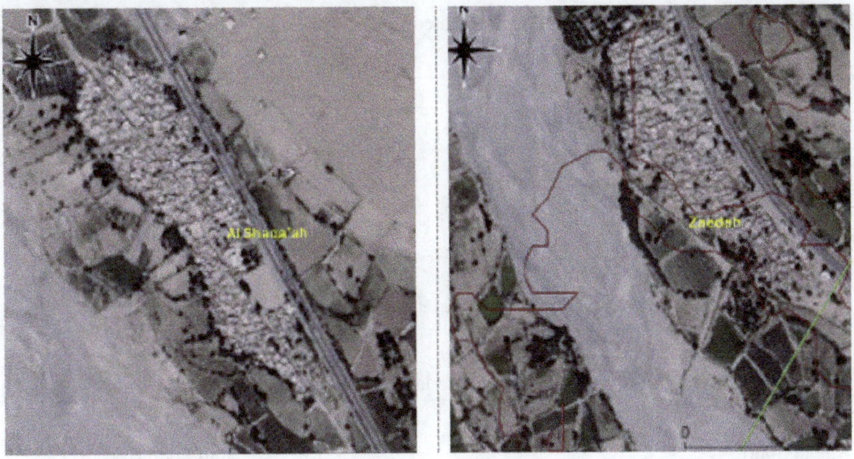

Fig. 14 Two villages in the upper regions might be affected in case of flooding in Wadi Tuban

Fig. 15 Main streams and infrastructure in the Tuban Delta

Table 12 Main infrastructure and their associated risks in the Tuban Delta

Name	Region	Climate risks
Al-Anad water supply station	UR	High flood risk
Dar Al-Araes Palace	UR	High flood risk
Ibn Khldoon hospital	MR	Low risk
Abass power station	MR	Low flood risk
Al-Hawtah water supply station	MR	Low flood risk
Al-Rawda palace	MR	Low flood risk
Al-Hawtah wastewater treatment plant	MR	Low flood risk
Al-Khudad IDP camp	MR	Low flood risk
The natural channel from Byzag Weir to Al-Hadarm Weir	MR	Medium flood risk
Al-Qomondan palace	MR	Low risk
Aden airport	LR	High flood risks
Al-Waht hospital	LR	Medium flood risk
Aden hospital	LR	Medium flood risk
Al-Gamhoria hospital	LR	Low risk
Al-Sadakh hospital	LR	Low risk
Refinary hospital	LR	Low risk
22 May hospital	LR	Low risk
Al-Haswah Termoelctric plant	LR	Low risk
Al-Mansorah power station	LR	Low risk
Khormksr power station	LR	Low risk
Hugaif power station	LR	Low risk
Chihnaz power station	LR	Medium flood risk
Bir Naser water supply station	LR	Low flood risk
Al-Barzakh water supply station	LR	Medium flood risk
Bir Ahmed water supply station	LR	Low flood risk
Saber wastewater treatment plant	LR	Low flood risk
Al-Mansorah wastewater treatment plant	LR	Low flood risk
Al-Areesh wastewater treatment plant	LR	Low flood risk
Salah Addin wastewater treatment plant	LR	High flood risk
Al-Fashlah waste landfills	LR	Low flood risk
Al-Haswah waste landfills	LR	Medium sea-level rise risk
Al-What Al-Rugaa road	LR	High flood risk
Al-What Al-Rugaa road	LR	High flood risk
Aden National museum	LR	Low risk
Al-Heswah wetlands reserve	LR	Medium sea-level rise risk
Nature reserve of the Swans	LR	High sea-level rise risk
Al-Heswah wetlands	LR	Medium sea-level rise risk

(continued)

Table 12 (continued)

Name	Region	Climate risks
Al-Feuosh IDP camp	LR	High flood risk
Al-Rugaa IDP camp	LR	Low risk
Al-Rubat IDP camp	LR	High flood risk
Madinat Al-Shaab IDP camp 1	LR	Low risk
Madinat Al-Shaab IDP camp 2	LR	Low risk

Table 13 IDP under flood risk in the Tuban Delta

Governorates	District name	Site name EN	Regions	Flood risk
Aden	Al Burayqah	Al Burayqah	LR	High risk
Aden	Dar Sad	Dar Sad	LR	High risk
Aden	Dar Sad	Dar Sad	LR	High risk
Aden	Dar Sad	Dar Sad	LR	High risk
Aden	Dar Sad	Dar Sad	LR	High risk
Lahj	Tuban	Al Hawtah, Tuban	LR	High risk
Lahj	Tuban	Al Hawtah, Tuban	MR	High risk
Lahj	Tuban	Al Hawtah, Tuban	MR	High risk
Lahj	Tuban	Al Hawtah, Tuban	LR	High risk
Lahj	Radfan	Al Habilin	LR	High risk
Lahj	Radfan	Al Habilin	UR	High risk
Lahj	Al Malah	Al Malah	LR	High risk
Lahj	Radfan	Al Habilin	LR	High risk

3.5 Drought

Drought is based on the degree of dryness and the duration of the dry period. Drought has four categories:

- Meteorological drought, which represents the change in metrological measurement in the metrological stations (rainfall, temperature, humidity, …. etc.)
- Agricultural drought, which occur due to the decrease of soil humidity that affect agriculture production.
- Hydrological drought due to law rainfall rates that may affect runoff and water storage.
- Socioeconomic drought due to the huge difference between supply and demand. This category is influenced manly by population increase and the lack of local water supply management.

Drought indices have been developed over the past few decades to assimilate data on rainfall, snowpack, streamflow, and other water supply indicators into a comprehensible big picture. These indices include the Standardized Precipitation Index (SPI), Palmer Drought Severity Index (PDSI), Standardized Runoff Index (SRI), rainfall anomaly index (RAI), the Deciles Index (DI), drought area index (DAI), standardized precipitation evaporation index (SPEI), surface humidity index, and streamflow drought index (SDI) (Eslamian et al. 2017).

Analysing precipitation data is the most common method to study drought in each area because precipitation is the most common variable which directly impact the level of soil moisture content, steam flows and groundwater resources (Eslamian et al. 2017). Therefore, in this report, SPI and SPEI are good options to assess the temporal variability of rainfall in the Tuban Delta.

In this report two drought categories will be considered in the assessment namely the metrological drought and the socioeconomic drought.

Metrological Drought

Metrological drought can be assessed by SPI and SPEI. SPI is mainly caused by a deficiency of precipitation. A long-term precipitation record is needed to calculate SPI. After the statistical fitting and transformation of the long-term precipitation data, region-specific deviations are mostly minimized. SPI is a probability-based index, so the heaviness or lowness of a precipitation event in the SPI is relative to the rainfall characteristics of that area. SPEI is an extension of the SPI. It considers both precipitation and potential evapotranspiration (PET) in determining drought. Thus, it uses the monthly (or weekly) difference between precipitation and PET. SPI and SPEI have a very straightforward classification of different drought severities starting from extremely dry to extremely wet, Table 14.

To assess drought conditions based on SPI in 2022, monthly SPI maps have been downloaded from the Emergency Management Service. The maps show that the SPI for the upper and the middle regions is near normal (0), while it is extremely dry (−2) in the Lower Region, (Fig. 16).

For future projections, Under SSP3, high annual rainfall might occur in Aden and Lahij between 2060 and 2065 resulting wet years (Fig. 17). However, limited rainfall

Table 14 Drought classification based on SPI and SPEI indices

SPI classification	Condition
+2.0	Extremely wet
1.5 to 1.99	Very wet
1.0 to 1.49	Moderately wet
−0.99 to 0.99	Near normal
−1 to −1.49	Moderately dry
−1.5 to −1.99	Severely dry
−2 and less	Extremely dry

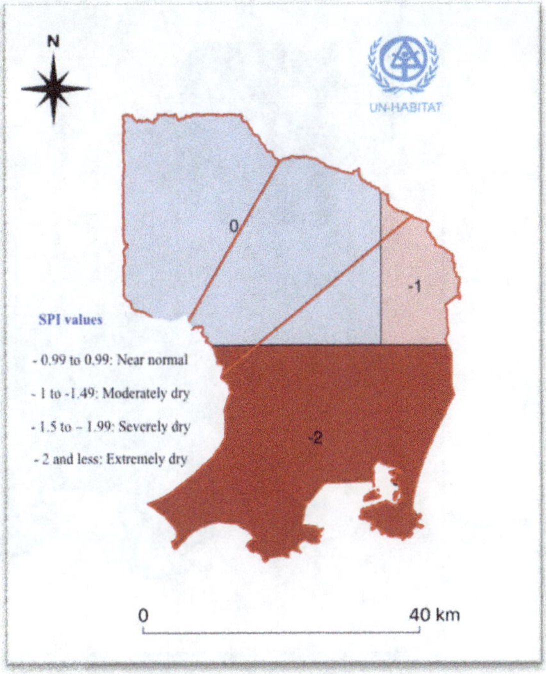

Fig. 16 SPI drought index in the Tuban Delta in 2022

resulting drought (<50 mm) may occur in Aden and Lahij in 2030, 2031 and 2033, (Fig. 18).

Under RCP 8.5, the drought might occur between 2070 and 2073, where annual precipitation will be <50 mm, (Fig. 19). However, high rainfall rates might occur between 2090 and 2100, (Fig. 20).

The severity of drought can be divided into four categories based on the increase's percentages of the annual rainfall compared to the average values:

- Negligible: Low severity when the decrease in rainfall patterns is <25%,
- Marginal: Medium severity when the decrease in rainfall patterns is between 25 and 50%,
- Critical: Serious severity when the decrease in rainfall patterns is between 51 and 75%,
- Catastrophic: High severity when the decrease in rainfall patterns is more than 75%.

The probability of drought occurrence has been estimated based on probability of the low precipitation rates that generate serious or high severity.

- Improbable (<20%):
- Remote (between 20 and 39%)
- Occasional (between 40 and 59%)
- Probable (between 60 and 80%)

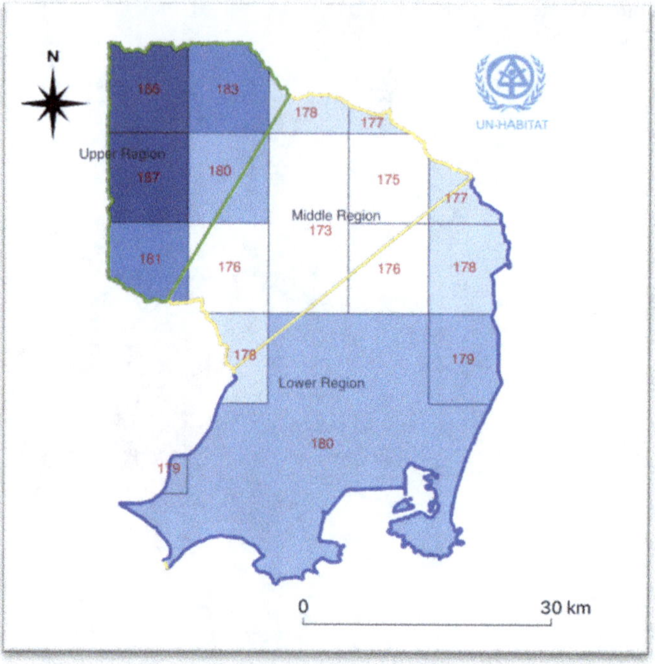

Fig. 17 Annual rainfall distribution (2060–2063) in Tuban Delta (SSP3)

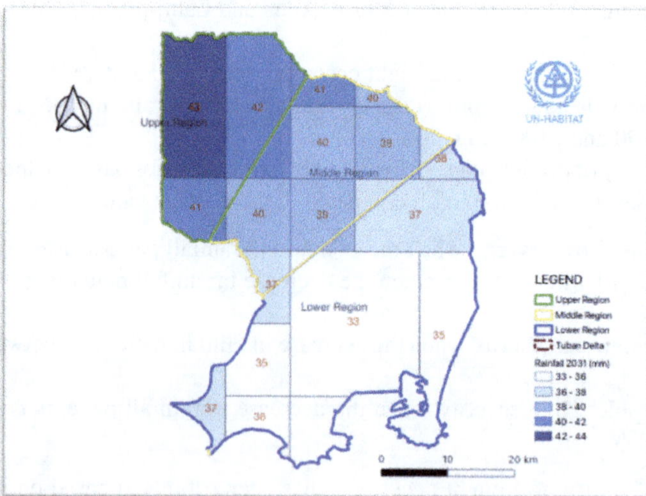

Fig. 18 Annual rainfall distribution (2030–2033) in Tuban Delta (SSP3)

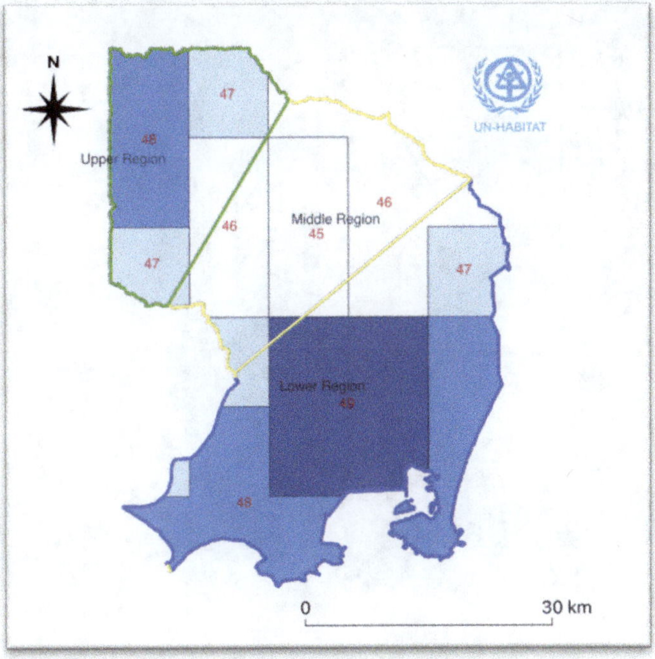

Fig. 19 Average annual rainfall distribution in (2070–2073) in Tuban Delta (SSP5)

- Frequent: more than 80%

Following the same flooding risk assessment procedure, Table 15 presets drought risk assessment under SSP3 and SSP5. The table shows that the Lower region might have more drought periods compared to the other regions.

Socioeconomic Drought

For the socioeconomic drought, future water balance projections show water shortage under the three demand scenarios especially in the lower region under SSP3 and SSP5 as seen in Tables 16 and 17. The most optimistic scenario is No. 2, which needs raising awareness and actual development in modern irrigation, and domestic supply systems.

The hazard of socioeconomic drought has been categorized into five categories under SSP3 and SSP 5, Tables 18 and 19, based on the water balances that have been estimated in Tables 16 and 17:

- Low, if annual water shortage is between 1 and 10 MCM,
- Low/Medium, if annual water shortage is between 10.1 and 40 MCM,
- Medium, if annual water shortage is between 41 and 100 MCM,

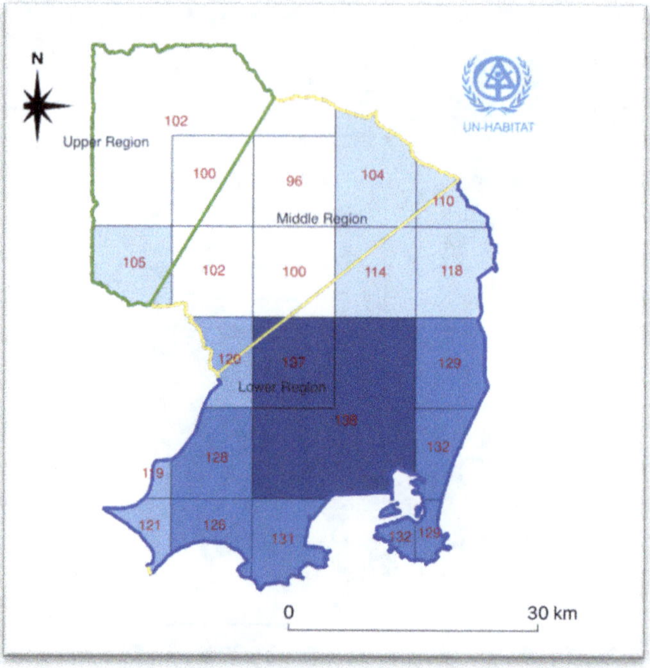

Fig. 20 Average annual rainfall distribution (2095–2100) in Tuban Delta (SSP5)

Table 15 Drought risk assessment under SSP3 and SSP5

Years/Regions	SSP3 (RCP 7)			SSP5 (RCP 8.5)		
	UR	MR	LR	UR	MR	LR
2023–2040	II	II	III	II	II	III
2041–2060	I	II	II	I	I	II
2061–2080	I	I	I	III	III	III
2081–2100	I	I	I	I	I	I

Table 16 Water balance projections in MCM under SSP3

Scenarios	Demand scenario 1			Demand scenario 2			Demand scenario 3		
Regions	UR	MR	LR	UR	MR	LR	UR	MR	LR
2022	14	2.9	−19	14	3	−19	14	2.9	−19.3
2040	0.3	−7	−24	0.3	−6.6	−24	0	−10	−39
2060	37.3	42	3.7	37	43	11	30	26	−30
2080	9.7	6	−112	11	8	−87	3	−13	−144
2100	7.7	2	−223	10	7	−156	1	−17	−228

Table 17 Water balance projections in MCM under SSP5

Scenarios	Demand scenario 1			Demand scenario 2			Demand scenario 3		
Regions	UR	MR	LR	UR	MR	LR	UR	MR	LR
2022	14	2.9	−19	14	2.9	−19	14	2.9	−19.3
2040	4	−1.9	−18	4	−1.9	−18	3	−5.4	−33
2060	20	20	−30	20	21.2	−23	13	4.3	−64
2080	13	10	−111	14	12.3	−86	6	−8	−143
2100	11.5	6.7	−206	23	−27	−143	5	−12	−212

Table 18 Socioeconomic drought hazard assessment under SSP3

Scenarios	Demand scenario 1			Demand scenario 2			Demand scenario 3		
Regions	UR	MR	LR	UR	MR	LR	UR	MR	LR
2022	–	–	II	–	–	II	–	–	II
2040	–	I	II	–	I	II	–	I	II
2060	–	–	–	–	–	II	–	–	II
2080	–	–	IV	–	–	III	–	II	IV
2100	–	–	V	–	–	V	–	II	V

Table 19 Socioeconomic drought hazard assessment under SSP5

Scenarios	Demand scenario 1			Demand scenario 2			Demand scenario 3		
Regions	UR	MR	LR	UR	MR	LR	UR	MR	LR
2022	–	–	II	–	–	II	–	–	II
2040	–	I	II	–	I	II	–	I	II
2060	–	–	II	–	–	II	–	–	III
2080	–	–	IV	–	–	III	–	I	IV
2100	–	–	V	–	II	IV	–	II	V

- Medium/High, if annual water shortage is between 101 and 150 MCM,
- High, if annual water shortage is more than 150 MCM.

The Impacts of Drought

Drought will increase water insecurity in the Tuban Delta, mainly in the Lower region leading to low recharge rates and overexploitation of groundwater resources, which will increase groundwater depletion and saltwater intrusion. Thus, many wells in the lower region will go dry, especially the wells of Bir Ahmed well field affecting the main domestic water source of Aden.

The thickness of groundwater aquifers in the Tuban Delta increases from 30 to 120–170 m (Squarespace n.d.). Due to drought, groundwater recharge rates will be limited, and it might reach zero. In the current conditions, ground water levels drop by about 3–7 m/year (UNDP 2021) due to overexploitation. Taking 5 m drop of groundwater level and a thickness of 170 m for groundwater aquifers, fresh groundwater resources in the lower region will not last more than 40 years if no urgent actions are taken. The same estimation can be reached based on the water balance of the Lower Region. Water shortage of 7.6 MCM in 2022 lowering groundwater table by 5 m every year. Thus, all wells will go dry by 2080 as the estimated water shortage based on the three demand scenarios were 80–100 MCM.

4 Conclusions

The climate model MRI-ESM2-0, has been used to analyse potential climate impacts on the Tuban Delta. The findings indicated that the Tuban Delta will have high annual rainfall, resulting in wet years between 2039 and 2042, and **low** annual rainfall, resulting in drought between 2070 and 2073. High rainfall rates may cause flooding in some villages and agricultural areas near streams, while local severe rainfall rates may cause flooding in cities and airports if rainwater systems are not maintained and planned properly. Furthermore, the model predicts a rise in the number of hot days per year beginning in 2050, reaching 40 days per year under SSP3, and 70 days per year under SSP5 in 2100. Moreover, Modelling seal-level rise using QGIS showed that most of the coastal areas will be affected mainly the airport by 2100. Therefore, adaptation measures are needed to prepare for such kind of extreme events. Moreover, new versions of water law and a water policy are needed considering climate change as these frameworks remain inactive and have failed to effectively address the challenges in the water sector.

Acknowledgements This study is part of the Green Climate Fund (GCF) Readiness project, which is implemented by UN-Habitat in coordination with the Environmental Protection Authority (EPA) of Yemen.

Competing Interests The authors have no conflicts of interest to declare that are relevant to the content of this chapter. Selecting the "Competing Interests" option will conveniently add the heading "Competing Interests" for you.

Ethics Approval Not applicable.

References

Abdurabu WA, Saleh MA, Ramli AT et al (2016) Occurrence of natural radioactivity and corresponding health risk in groundwater with an elevated radiation background in Juban District. Yemen. Environ Earth Sci 75:1360. https://doi.org/10.1007/s12665-016-6142-z

Abubakr MM, Al Saafani MA, Nagi HM, Hajer A, Alhababy AM (2012) Coastal zone vulnerability and adaptation assessment, Aden Governorate, Republic of Yemen

Al-Gheethi A et al (2014) Effectiveness of selected wastewater treatment plants in Yemen for reduction of faecal indicators and pathogenic bacteria in secondary effluents and sludge. Water Pract Technol 9(3):293–306

Al-Khashman OA, Jaradat AQ (2014) Assessment of groundwater quality and its suitability for drinking and agricultural uses in arid environment. Stoch Environ Res Risk Assess 28:743–753. https://doi.org/10.1007/s00477-013-0787-x

Al-Sabri A, Halim MK (2012) Status and new developments on the use of brackish water for agricultural production in the Near East. In: Yemen country report. United Nations Food and Agriculture Organization Regional Office for the Near East (RNE), Cairo. Accessed 22 March 2023

Al-Sakkaf A (2016) Water scarcity and conflict in Yemen: the role of climate change. Middle East J 70(3):363–380

Bassetti F (2022) Shared socioeconomic pathways. In: Foresight: the CMCC observatory on climate policies and future. Accessed 06 May 2023

Bellafiore D, Ferrarin C, Maicu F, Manfè G, Lorenzetti G, Umgiesser G et al (2021) Saltwater intrusion in a Mediterranean delta under a changing climate. J Geophys Res Oceans 126:e2020JC016437. https://doi.org/10.1029/2020JC016437

CARE International (2021) Climate change and conflict in Yemen: a vicious cycle

Climate Central (2023) Coastal risk screening tool. https://coastal.climatecentral.org. Accessed 05 May 2023

Dingman SL (2015) Physical hydrology, 3rd edn. Waveland Press, Inc

EPA (2013) Yemen's second national communication under the United Nations framework convention on climate change. Accessed 20 March 2023

Eslamian S, Ostad-ali-askari K, Singh VP, Dalezios NR (2017) A review of drought indices. Int J Constr Res Civil Eng 3(4):48–66

FAO (2020) Yemen: agricultural livelihoods and food security in the context of conflict. Food and Agriculture Organization

Frame B, Lawrence J, Ausseil A-G, Reisinger A, Daigneault A (2018) Adapting global shared socio-economic pathways for national and local scenarios. Clim Risk Manag 21:39–51. https://doi.org/10.1016/j.crm.2018.05.001

Girgirah AA, Maktari MS, Sattar HA, Mohammed MF, Abbas HH, Shoubihi HM (n.d.) Wadi development for agriculture in PDR Yemen. https://floodbased.org/wp-content/uploads/2021/05/Wadi-development-for-agriculture-in-PDR-Yemen.pdf

Haidera M, Noaman A (2008) Application of decision support tools for water resources management in coastal arid areas (case study: Aden, Yemen). In: The 3rd international conference on water resources and arid environments and the 1st Arab water forum

Harb S, El-Kamel AH, Zahran AH, Abbady AA, Ahmed FA (2013) Natural radioactivity of ground water in some areas in Aden governorate South of Yemen region. Radiat Protect Environ 36(3):115–121. https://doi.org/10.4103/0972-0464.137476

Hausfather Z (2018) Explainer: how 'shared socioeconomic pathways' explore future climate change. Carbon Brief. Accessed 28 March 2023

Hirich A, Choukr-Allah R, Ezzaiar R, Shabbir SA, Lyamani A (2021) Introduction of alternative crops as a solution to groundwater and soil salinization in the Laayoune area, South, Morocco. Euro-Mediterr J Environ Integr 6:52

Huntjens P et al (2014) The political economy of water management in Yemen: conflict analysis and recommendations. In: Technical report, The Hague Institute for Global Justice

IPCC (2021) Climate change 2021: the physical science basis. In: Contribution of working group I to the sixth assessment report

Juniati AT, Kusratmoko E, Sutjiningsih D (2021) Estimation of potential water availability and water resources carrying capacity for Bogor City spatial plan. J Geogr Trop Environ 5(1):4

Kamruzzaman M, Shahid S, Islam AT, Hwang S, Cho J, Zaman MA, Ahmed M, Rahman M, Hossain B (2021) Comparison of CMIP6 and CMIP5 model performance in simulating historical precipitation and temperature in Bangladesh: a preliminary study. Theor Appl Climatol 145:1385–1406

Kazem M (2015) Challenges and difficulties of living in river deltas. A review of the major river deltas in Asia and Africa—a review of the major river deltas in Asia and Africa, Munich, GRIN Verlag. https://www.grin.com/document/306464

Komex (2001) Water resources management studies in the tuban-abyan region. Komex International Ltd

Marifa TA et al (2021) Ekstraksi Parameter Topografi sebagai Survei Pragmatik untuk Analisis Potensi Longsoran Lereng. Jurnal Teknologi Sumberdaya Mineral 2(1):5412

Ministry of Water and Environment, Yemen (2018) National adaptation plan of action

Mourad KA, Alshihabi O (2016) Assessment of future Syrian water resources supply and demand by WEAP model. Hydrol Sci J 61(2):393–401. https://doi.org/10.1080/02626667.2014.999779

Mourad KA, Berndtsson JC, Berndtsson R (2011) Potential fresh water saving using greywater in toilet flushing in Syria. J Environ Manag 92(10):2447–2453

Naouel D (2021) Planning and management of water resources in the context of economic development and climate change in Algerian highlands by the WEAP model, case of the Gareat el Tarf basin (northwestern Algeria). Faculty of Natural and Life Sciences, Laboratory of Biotechnology, Water, Environment and Health, Khenchela University, Algeria

O'Neill BC, Kriegler E, Riahi K et al (2014) A new scenario framework for climate change research: the concept of shared socioeconomic pathways. Clim Change 122:387–400. https://doi.org/10.1007/s10584-013-0905-2

Oumara NGA, El-Youssfi L (2022) Salinization of soils and aquifers in Morocco and the alternatives of response. Environ Sci Proc 16:65. https://doi.org/10.3390/environsciproc2022016065

Rajosoa AS, Abdelbaki C, Mourad KA (2022) Assessing the impact of climate change on the Medjerda River Basin. Arab J Geosci 15:1052. https://doi.org/10.1007/s12517-022-10288-y

REACH (2023) 2023 REACH—CCCM national flood hazard analysis of IDP sites in Yemen (February 2023). Accessed 02 May 2023

ReliefWeb (2021) Yemen water project. https://reliefweb.int/report/yemen/yemen-water-project. Accessed 28 March 2023

Riahi K et al (2017) The shared socioeconomic pathways and their energy, land use, and greenhouse gas emissions implications: an overview. Global Environ Change 42:153–168. https://doi.org/10.1016/j.gloenvcha.2016.05.009

Saleh SMK, Al-Sallami AMA (2022) Assessment of the level of physicochemical and microbiological contamination of groundwater in parts of Bir Nasser and Bir Ahmed water fields in Tuban Delta in Aden and Lahej Governorates—Yemen. Electr J Univ Aden Basic Appl Sci EJUA-BA 3(2):158

Saleh EE, El-Kassas HI, El-Fiki SAE, Diab HA, Al-Nagashee AAS (2017a) Evaluation of environmental hazards resulted from human activities and natural radioactivity in Tuban Delta in Yemen. J Anal Bioanal Techn 8:374. https://doi.org/10.4172/2155-9872.1000374

Saleh S et al (2017b) Evaluation of groundwater quality and its suitability for drinking and agricultural use of rural areas for Zabid Directorate-Wadi Zabid, Hodiedah. Yemen. J Sci Eng Res 4(7):10–24

Shrestha AB, Ezee GC, Adhikary RP, Rai SK (2012) Resource manual on flash flood risk management. In: International centre for integrated mountain development (ICIMOD), Kathmandu

Squarespace (n.d.) Aquifer system in Yemen. Accessed 20 March 2023

Suresh R (1997) Soil and water conservation. Engineering Standard Publishers Distributors, New Delhi

Tabacchi E, Lambs L, Guilloy H, Planty-Tabacchi AM, Muller E, Decamps H (2000) Impacts of riparian vegetation on hydrological processes. Hydrol Process 14:2959–2976

UNDP (2019) Yemen climate change adaptation strategy. United Nations Development Programme

UNDP (2021) Water availability in Yemen: literature review of the current and future water resources and water demand in Yemen

Vector solutions (2018) Risk matrix calculations: severity, probability, and risk assessment. Accessed 27 April 2023

World Bank (2010) Assessing the impacts of climate change and variability on the water and agricultural sectors and the policy implications. Report No. 54196-YE

World Bank (2017) Beyond scarcity: water security in the Middle East and North Africa

Yukimoto S et al (2019a) The meteorological research institute earth system model version 2.0, MRI-ESM2.0: description and basic evaluation of the physical component. J Meteorol Soc Jpn II 97(5):931–965. https://doi.org/10.2151/jmsj.2019-051

Yukimoto S et al (2019b) MRI MRI-ESM2.0 model output prepared for CMIP6 CMIP. Earth Syst Grid Feder 12:621

Hydroclimatic Changes Along the Eastern Mediterranean: Case of Lebanon

Amin Shaban

1 Climate Characterization

Climate is often characterized by precipitation and temperature as the major two meteorological variables, while other variables (*e.g.*, humidity, solar radiation, etc.), are sometimes used and this depends on data availability from ground-based stations. However, components on terrain surface must be also characterized; especially they have a role in reacting with meteorological conditions at different levels; and hence, governing the regime of water flow and storage. These two physical pillars (meteorology and terrain surface components) must be well identified in order to achieve optimal water resources management. Thus, the relatively humid climate is associated with the rugged topography and both control water volume, including water flow, infiltration, and storage in the substratum and even they govern water loss to the sea. The atmospheric regime and the characteristics of terrain are principal generators for the entire water cycle like the case in Lebanon. The climate of Lebanon is known by wet periods that are pronounced by the rainfall and the snow cover for a considerable number of months over the year (Shaban 2020). The terrain surface and its components has different responding level and aspects. This includes accelerating water flow energy, uniform water recharge into the underlying rock layers, and the flow into the karstic conduits.

There are two principal climatic variables often considered while applying climatic assessment and projections. These are the precipitation and temperature, which can fulfill performing of climatic trend analysis. The use of other meteorological variables (*e.g.* humidity, wind, etc.) would be more favorable. However, this is usually dependent on data availability whether from ground measurements or from space-based techniques.

A. Shaban (✉)
National Council for Scientific Research, Beirut, Lebanon
e-mail: geoamin@gmail.com

M. M. Al Saud (ed.), *Hydroclimate Changes: Implications of the Middle East Region*,
https://doi.org/10.1007/978-3-032-20053-2_11

Climatic data in Lebanon are almost available for few localities; however, complete time series data often do not exist; especially many meteorological stations are out of order. There is always a debate on measuring climatic variables and this can be attributed to:

1. The rugged topography and non-uniformity of Lebanon's terrain, and this requires increased number of meteorological stations to have typical representation of climate datasets.
2. Data inaccuracy is a matter notably that many meteorological stations need maintenance and some of them are totally damaged.
3. The weak collaboration between different institutions who own climatic data series, is also a problem that needs to be addressed on a national level.

Lebanon has diverse climatic regime resulted from its rugged topography; and thus, precipitation is unevenly distributed resulting microclimatic zones between various topographies. This is well pronounced because rain storms blow from several directions and with warm, cold and humid weathers (Arkadan 2008). In addition, meteorological turbulences exist and act in the blowing air; therefore, rain shadow impact occurs among the topography of the Lebanese mountains. According to Shaban and Houhou (2015), cold air masses rising up on the mountains from west and descending air warms and picks up available moisture to the east of the mountain ridges resulting eventually climatic turbulences.

Many assumptions have been raised to describe the climate of Lebanon, while it is well known that Lebanon is a Middle East Region and it is characterized by semi-arid climate, but this has not been evidenced. FAO (2009) reported that the Lebanon's climate is typically Mediterranean climate with heavy rains in the winter season (November to May) and dry and arid conditions in the remaining months of the year. According to the Lebanon's Second National Communication (SNC 2011), Lebanon is a semi-arid area during the dry season to humid to sub-humid in the wet season. CAS (2015) stated that Lebanon is related to the northern temperate zone, while, CAL (1982) described Lebanon with an oceanic climate during winter and a sub-tropical climate during summer.

According to Shaban (2020), in many studies and projects Lebanon is always included among wide geographic regions (*e.g.* MENA, Arab, ME, ESCWA, etc.); especially when regional climatic trends, projections and scenarios are applied; and this is erroneous, because Lebanon with its distinguished mountainous topography must studied separately from surrounding regions. Therefore, Lebanon is characterized by a typical Mediterranean climate, which is moderately cold and rainy in winter, warm in summer, mild in spring and autumn (Shaban 2003).

1.1 Data Sources

There was about 70 ground-based stations for climatic data in Lebanon until 1990s, and 50 of them were in the western part of the country, and more specifically in the coastal zone and the adjacent occidental mountain. Most of these stations are mandated by the Ministry of Public Works and Transport (MoPWT) where it provides meteorological data through the Climatic Atlas of Lebanon (CAL 1971, 1982). In addition, the General Directorate of Civil Aviation GDCA (1999) was also concerned in producing climatic data, but almost for Beirut region. However, these stations have been quit during the conflicts and a few ones remain functioning. Recently, there are about 125 operational meteorological stations in Lebanon fixed/ or operated by different institutions, such as: Lebanese Agriculture Research Institute, Lebanon (LARI 2017) who has more than 60 ground-based climatic stations; while some data are delivered by researchers such as Ghaddar (1995) and Na'ameh (1995).

Due to the lack in sufficient datasets, satellite-based platforms have been raised as a source of climatic data in many regions. Thus, climate data became accessible from remote sensing products, such as the Tropical Rainfall Mapping Mission (TRMM) (2015) which was extended by NASA since 1998, Climate Hazards Group Infrared Precipitation with Stations (CHIRPS 2017); and from NOAA climatic data system—National Oceanographic Data Center (NOAA 2013).

1.2 Rainfall

Precipitation represents all meteoric water in various states including mainly rainfall and snow, and it is the major source of renewable water in Lebanon. There is obvious diversity in rainfall distribution resulted from the topography of Lebanon which generally ranges between 650 and 1500 mm on the coastal zone and the elevated areas; respectively, but anomalous rates are often recorded in the inner zone of Lebanon. Figure 1 shows the rainfall distribution over the three geomorphologic units of Lebanon (Mount Lebanon, Bekaa Plain and Anti-Lebanon). While, acute changes in rainfall rate may occur between along short distances, such as between Al-Arez (more than 1400 mm) and Hermel (<250 mm).

For solid precipitation (snow), it is considered as a principal water-feeding source for surface water and groundwater in Lebanon. In the Middle East Region, Lebanon is the only geographic region where snow cover remains for few months on the mountain chains and covers between 2100 and 2500 km^2 (about 22% of Lebanon surface area).

The calculated average rate of rainfall in Lebanon is still undefined and contradictory estimates have been recorded by various institutes and researchers. For examples, rainfall rate was reported as: 982, 940, 820, 873, 823, and 800 mm according to Rey (1954); UNDP (1970); Plassard (1971); CAL (1982); WB (2003); MoEW and UNPD (2016); respectively.

In this respect, the National Council for Scientific Research in Lebanon (CNRS-L 2015) calculated and mapped a bulk of datasets on rainfall for the period between 1950 and 2015 depending on serval data sources as well as from the satellite-based platforms. Therefore, the produced rainfall maps have also been lately updated by the author with recent data until 2024 (Fig. 2). The updated map shows an average

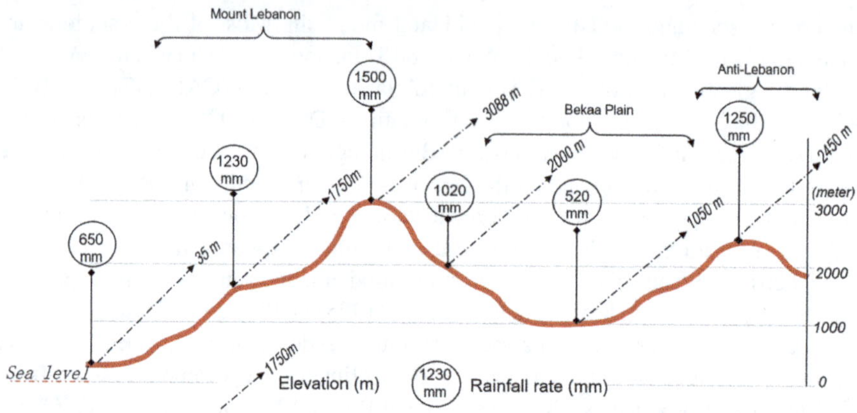

Fig. 1 Schematic representation of rainfall distribution in Lebanese. Adapted from Shaban (2020)

rainfall rate of about 858 mm/year, where it ranges between 363 and 1452 mm/year. In the Mount Lebanon, the geographic distribution of rainfall is completely controlled by altitude, but this is not the case in the inner zone where the climatic turbulences exist by the blowing between mountains and resulting in descending air warms and picks up available moisture the mountain ridges, and that is why lowest rainfall rate (<250 mm/year) is reported in the northern part of the Bekaa Plain.

The area of each rainfall interval was calculated in order to achieve the water volume resulted in each region represented by defined rainfall interval (Table 1); and thus, the total water for the entire Lebanon was calculated at 8973 million m^3. If the evapotranspiration (about 52%) is subtracted from this volume; therefore, about 4307 million m^3 of water remains on land surface and part of it percolates into the underlying rock layers. This volume assures about 1077 m^3/capita/year.

Rainfall data recorded from ground stations and from TRMM platform showed a number rainfall peaks which often exist between November and February where the average monthly rate exceeds to about 230 mm/year, while the lowest rates of rainfall exist between May and August months.

1.3 Snow

As a distinguished water resources in Lebanon, snow is considered the main source of water, notably it has a significant role in the replenishment of rivers, springs and groundwater aquifers. According to Shaban (2020), Lebanon without snow will lose the largest part of its water, and snow contribution is estimated to more than 60%. Moreover, it was estimated that more than 60% of water in Ibrahim River, one of the largest coastal rivers in Lebanon, is derived from snowmelt (Abd EL-Al 1953). While, Mount-Lebanon receives about 1100 million m^3 of water from the melting snow and this is also converted to about 425 mm/year (Shaban et al.

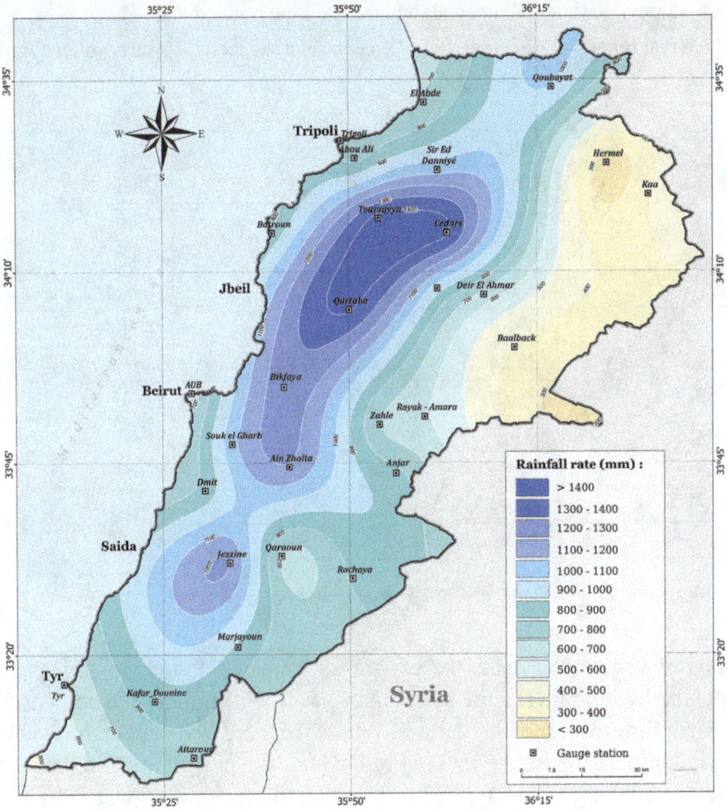

Fig. 2 Updated rainfall distribution map of Lebanon

2004). Recently, the exacerbated challenges on water supply makes it necessary to highlight the significance of snowpack, and snow accumulation/melting regime has been investigated. Many studies have been performed particularly for monitoring snow cover, field measurements, modeling snow cover and reservoirs' capacity, as well as the impact of climate change on snowpack. In this respect, satellite images in combination with the ground stations and in-situ measuring have been utilized.

Snowfall in Lebanon is a yearly meteorological event that resulted from regional weather by wet storms derived either from the European, Scandinavian or Siberian areas. Hence, snow cover over mountains above 1500 m is common, and snowpack remains between 4 and 8 months over the year, and sometime, snowpack of a year is covered by snow of the consequent year, while snowmelt often starts in late March. In addition, the number of snowing days is distributed as: <1, 2–10, 10–20, 20–40 and >40 days for the altitudes below 250 m, 250–750 m, 750–1500 m, 1500–2250 m and >2250 m; respectively (Shaban et al. 2014b).

Recently, snowpack is under threat due to human encroachments and it becomes under deterioration and many natural and man-made impacts are pronounced on the

Table 1 Area of various rainfall intervals and the resulting water volume

Rainfall interval (mm/year)	Area (km^2)	Percentage of the total area (%)	Water volume (million m^3)
<300	158.5	1.5	47.6
300–400	920.2	8.8	322.0
400–500	507.5	5.8	228.4
500–600	307.3	3.9	169.0
600–700	654.7	7.2	425.5
700–800	1460.5	14.9	1095.4
800–900	1681.9	16.0	1429.6
900–1000	1332.4	13.0	1265.8
1000–1100	1190.2	9.7	1190
1100–1200	790.7	7.5	909.5
1200–1300	792.7	5.8	990.9
1300–1400	360.4	3.5	486.5
>1400	294.8	2.8	412.7
Total			8973

mountains of Lebanon. Hence, the expand of human activities to snowpack areas interrupts the hydrologic regime by reducing water infiltration rate and destruct the path of water flow from snowmelt flows. While, the increased temperature with raised sunlight radiation increase the melting rate which was not an issue in the past.

There are several studies applied to calculate the volume of water derived from snow in Lebanon and to evaluate the role of snow contribution in the water budget (examples: Shaban et al. 2004; Corbane et al. 2005; Mhawej et al. 2014, etc.). Most of these studies depended only on measuring snow depth for different time periods and for selected localities. However, satellite images with different spatial and temporal resolutions were processed to make the cartography of snow cover areas, and thematic data was integrated in the geo-information systems. This enables have periodical snow cover calculations which were supported by field verifications to calculate snow depth and snow-water-equivalent (SWE). In this regard, ground instrumentations to measure snow characteristics and all the relevant meteorological data were lacking until CNRS-L, in support of IRD-CESBIO (Institut de Recherche pour le Développement—Centre d'Etudes Spatiales de la Biosphère), installed three snow stations. These stations are installed in Cedars (2800 m), Laqlouq (1850 m) and El-Mazar (2300 m). They are able to measure, in addition to the known climatic variables, the snow depth and snow/water equivalent.

In this respect, MODIS images were the most common satellite type used to study snow cover. These images were retrieved from Goddard Space Flight Center "DISC" website of NASA Agency. The products MODIS images were delivered on the basis of 8-days at 250, 500 m and 1 km spatial resolution; and they belong to MODIS-Terra (MOD10-A2) and MODIS-Aqua (MYD10-A2) Snow Cover 8-Day L3 Global

Fig. 3 MODIS satellite images processed with various band ordering

500 m Grid data sets. The visibility and band ordering of these images are illustrated as (examples in Fig. 3):

- MODIS-Terra: true color, corrected reflectance
- MODIS-Terra: 7–2–1, corrected reflectance
- MODIS-Terra: 3–6–7, corrected reflectance
- Suomi NPP, VIIRS, corrected reflectance.

In addition, high resolution satellite images were also processed, with a special emphasis to Spot-4, Veg (1 km resolution) and Landsat 7 ETM$^+$ (30 m spatial resolution), plus datasets from the Google Earth Engine (GEE). These images provide narrow time period observations as well as to compare snow cover area with those obtained from MODIS images for more characterization of the snowpack at different altitudes and localities (Shaban et al. 2004). The mean monthly snow cover area was also calculated on the Lebanese mountains using MODIS images. The calculated snow cover area for the period 2000–2023, ranges between 1760 and 2890 km^2, with an average of about 2230 km^2.

Oscillations between the maximum and minimum snow cover area as well as the peak values were obviously noted (Fig. 4).

According to Telesca et al. (2018), there are three aspects patterns for the existing of snow accumulation peaks, as follows:

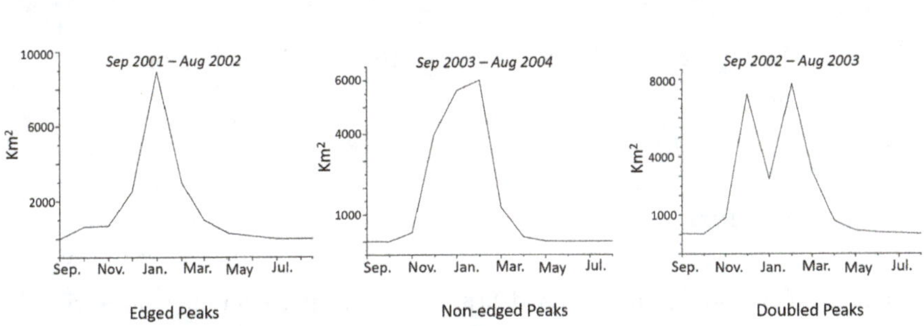

Fig. 4 Examples of peaks for snow cover area in Lebanon (Telesca et al. 2018)

1. Edged peaks: This is a sharp spike as a function of regular snow accumulation and melting processes and it show almost symmetry between snowpack accumulation and melting rate. Thus, edged peak refers to one big snow storm.
2. Non-edged peaks: This shows that the snow accumulation process is slower than the snow melting.
3. Double peaks: These are snow peaks with two or more spikes, and these show immediate change in the snow cover areas that belongs to consequent snow storms (e.g. tropical storm, regional, etc.).

For the volume of water derived from snowmelt, SWE was the most critical variable to be calculated. In this regard, UNDP (1970) estimated the average annual SWE in Lebanon was calculated between 1.82×10^6 and 2.57×10^6 million m^3 for the years between 2008 and 2012, which is equivalent to 174 and 246 mm/year. While, Mhawej et al. (2014) estimated SWE is about 2.42×10^6 million m^3 for the years between 2002 and 2011.

The study obtained by CNRS-L (2015) calculated SWE using MODIS satellite images (250 m spatial resolution) accompanied with wide field coverage for Lebanon, including the selection of 275 different sites where the ground measures where ground-based measurements taken on weekly basis for the years between 2010 and 2014. Calculating SWE includes measuring the volume of water melted from snowpack, and this requires calculating snow area, snow thickness and snow density. Thus, SWE was calculated according to the formula number 1 (Shaban et al. 2014b):

$$SWE = \left(A_{sp} \times T_{sp} \times D_{sp}\right) \tag{1}$$

where:

A_{sp} is the area of snowpack.

T_{sp} is the thickness (depth) of snowpack.

D_{sp} is the density of snowpack.

The calculated average snowpack cover area (A_{sp}) for the investigated years was about 2585 km^2; which is equivalent to about 25% of the Lebanese surface area. The average snowpack thickness (T_{sp}) was 0.82 m. While the calculated (*i.e.* consequent measurements) snowpack density (D_{sp}) was about 63%. Therefore, SWE for the entire Lebanon will be:

$$\text{SWE} = 2585 \times 10^6 \times 0.82 \times 63/100 = 1335 \text{ million m}^3$$

1.4 Temperature

Similar to precipitation, temperature data are also unavailable for long time series and the uniform geographic coverage, as well as, temperature is also altitude-controlled meteorological variable and substantially affected by the rugged topography of

Lebanon. The lack, availability and functioning of meteorological satiations is also the same for that of precipitation.

The temperature of Lebanon is considered as moderate and cannot be described as hot or very cold even during the extreme weather; however, the temperature rate has been obviously increased in Lebanon and it affected several sectors, notably the agriculture one. Hence, there is contradictory in temperature estimations, but the average annual temperature ranges between 11 °C in winter to 27 °C in summer in the coastal zone, and thus averaging 18.37 °C. In the mountainous regions, the average annual temperature is below 12 °C and it ranges between 2 and 18 °C in summer and winter; respectively. While, the hottest temperature rates in Lebanon is between June and August, while the coldest months are December and January and sometimes extending to February CNRS-L (2015).

Recently, trend analysis for temperature has been performed using long time series for the entire Lebanon depending on various data records (Shaban and Hamzé 2017). Results show that the lower calculated average temperature ranges between 5 and 10 °C in regions above 1800 m altitude; around 15 °C on areas higher than 1100 m; and it is about 20 °C along the coastal zone. According to Shaban (2020), the distribution of average annual temperatures with different altitudes can be summarized as follows:

- Coastal zone (<250 m) = 24 °C
- Plateau (250–750) = 22.5 °C
- Moderately elevated areas = (750–1500) = 19.5 °C
- Heights (1500–2500) = 18.5 °C
- Crests (>2500) = 14 °C.

1.5 Evapotranspiration

Comprehensive data evapotranspiration (Et) in Lebanon is still lacking and inadequate to build Et estimated over different geographic areas. Measuring lysimeter are few enough and the old ones are not functioning. However, many estimations were made by various traditional methods; and thus, diverse results were obtianed and estimated Et between 48 and 58% of the precipitated water volume. The Et estimates were put as: 48, 50, 52 and 53% by Fawaz (1992); MoE/ECODIT (2002); MoEW (2010) and Shaban (2003); respectively. While, FAO (2009), calculated the mean annual potential evapotranspiration between 1100 mm in the coastal zone and 1200 mm in the inner zone (*i.e.*, Bekaa Plain). Besides, Karam et al. (2003) calculated the annual potential evapotranspiration at 936 mm for the entire Lebanon.

Lately, two Bowen ratio stations have been installed in the Bekaa Plain, the agricultural land of Lebanon, by the CNRS-L to measure crop yield and Et. They compute daily average ET, which is based on FAO Penman Monteith equation and uses accurate net radiation and soil heat flux measured by Bowen Ratio. Results applied showed a daily average potential Et at 6.44 mm/day; while the total annual potential Et was 1140 mm/year. While, the actual Et, as it was calculated from satellite-based data for northern Bekaa Plain (Al Assi River Basin), was 0.53 mm/day and averaging about

190 mm/year. This indicates that actual ET is about 63% from the precipitated water in the driest region of Lebanon.

1.6 Aridity

Due to the discrepancy in describing Lebanon according to the various climatic zones; however, it is necessary to characterize Lebanon in terms of its climatic conditions. For example, in many studies Lebanon is often ranked as semi-arid region, but it is also described as semi-arid in the dry season to humid to sub-humid in the wet season (SNC 2011), and also as oceanic climate during winter and a sub-tropical climate during summer (CAL 1982); besides it is again put in the northern temperate zone and at the Mediterranean zone (CAS 2015). Meanwhile, Lebanon is also characterized by Mediterranean climate where four seasons occur, with warm dry summer and cool rainy winter, gradual transitional Fall and revived vegetation Spring. It is the only geographic region in the Middle East where snow cover remains for several months. Therefore, no define characterization for the Lebanon's climatic zone, and there are different studies came up with diverse results. This might be due to the absence of long-term and complete climatic records, or to the abrupt changes in Lebanon's topography and then microclimates. For this reason, it is necessary to in-depth perform climatic analysis including long time series covering the entire country in order to achieve the correct description of Lebanon/s climate whether it is semi-arid, humid, etc.

The accessibility of remote sensing and climatic platforms facilitates the geospatial and temporal data analysis notably when the retrieved datasets can cover long time series. The used platforms (*e.g.*, TRMM, CHIRPS, etc.) assure datasets sufficient to characterize Lebanon' climate. This has been also integrated with data from ground-based stations. Shaban and Houhou (2015) performed a study entitled: "Drought or humidity oscillations? The case of coastal zone of Lebanon". In this study, datasets were interpolated and analyzed statistically according to De Martonne (1926) Aridity Index. It was resulted that in spite it is classified as a semi-arid region, yet Lebanon has not evidenced any long-term drought spell (*i.e.,* few consequent years), since 1950s.

De Martonne Aridity Index (IA) was modified by Walton (1969) with more detailed elements. The calculation of IA requires datasets mainly for precipitation and temperature with various variability; and thus, it can be calculated according to the formula number 2:

$$IA = \frac{\frac{P}{T+10} + \frac{12\overline{p}}{t+10}}{2} \tag{2}$$

where P is the average annual rainfall rate (in mm), and T is the average annual temperature rate in degree Celsius, and $\overline{p}$ is the average rainfall in the most dry months and t is the average temperature in the most temperate months.

Results of the calculated De Martonne Aridity Index show dominant humidity in the coastal zone of Lebanon, which extends to the adjacent mountain chains (*i.e.*, Mount Lebanon). While, about 39% of the investigated years over the ten stations, reveal semi-humid climate, and 30.6% are humid, which means that around 70% of the investigated years are characterized by semi-humid to humid climate. In addition, more than 20% of the analysed years show very humid climate. This characterizes Lebanon by semi-humid to humid climate (Shaban and Houhou 2015).

2 Climate Change

As it was mentioned previously, the climate of Lebanon is almost mild Mediterranean where considerable volume of water comes from precipitation. nevertheless, changing climatic has been recently pronounced and became a striking challenge; especially it resulted implications on many vital sectors including water, agriculture, food security, energy and the socioeconomic sector. In this regard, Lebanon is the 78th least vulnerable country and the 59th least ready country for climate change impact (MoFA 2018). Whereas ND-Gain Index (GAIN 2017), reported that the climate vulnerability of Lebanon has a ranking of 106 out of 181 countries in the (where ranking 1 is the least vulnerable). Thus, evidences of climate change have been tanged by inhabitants whether from the climatic patterns and extremes or from the impact on the main sectors.

Basically, climate change is viewed from the uneven behavior in precipitation and temperature as the two primary climatic variables to evidence changes in climate.

2.1 Rainfall

Rainfall records have been analysed for the entire Lebanon, and datasets were collected from several sources, as it was mentioned in previous sections. The analysed records were from 1950s until 2024, but several gaps were found in this time series which represent 27 main meteorological (ground) stations covering almost representative localities, which lack to uniform geographic distribution; and therefore, data interpolation was performed. In addition, rainfall data was retrieved from TRMM and CHIRPS platforms in order to fill gaps and avoid any geographic region unassessed. The collected data was statistically analysed by simple graphical illustrations to perform trend analysis and deduce the behaviour of the average annual rainfall calculated for the 27 ground stations. In addition, a number of applications have been carried out for some time periods and selected regions in order to in-depth analyse climatic behaviour such as the use of Detrending Fluctuation Analysis (DFA) to study scaling properties of data series and Fisher-Shannon (FS) method to investigate time dynamics complex nonstationary signals. While, the periodic behavior in annual Emberger aridity index was also applied.

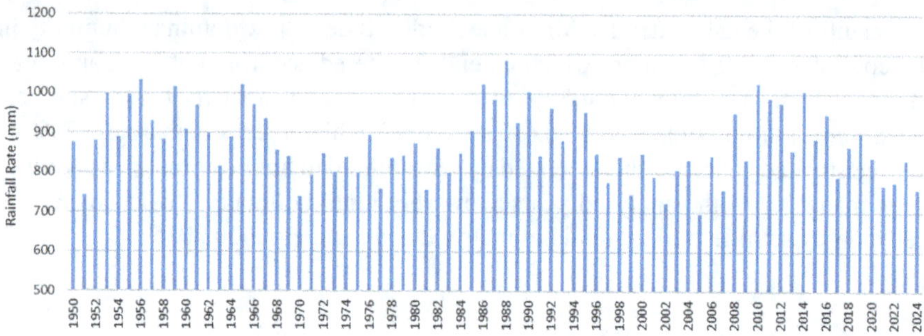

Fig. 5 Average annual rainfall changes between 1950 and 2024 for Lebanon

The Long Term Average (LTA) for the annual average rainfall was calculated over the last 7 decades, and it was estimated at 858 mm/year, with a slight rainfall rate decrease between 25 and 30 mm (Fig. 5). However, there is an obvious periodicity of data ascending and descending which was figured out as 15, 11 and 12 years for the time periods 1953–1968, 1984–1995 and 2008–2020; respectively. In this respect periodicity has been mentioned in previous studies. For instance, drought in Morocco was estimated with a periodicity of 11-year interval (Benassi 2008), while the minima in precipitation in Malta were estimated with 30-year interval (Perry 1988). In Lebanon, Telesca et al. (2018) reported abrupt changes periodicity between various stations which was from few years to more than 20 years, with the majority ranges between 10 and 15 years. This evidences predominant microclimates with cyclic climate that characterizes Lebanon's climate rather than a merely existence of climate change.

2.2 *Temperature*

Perhaps the measurement of temperature is more simple compared with other meteorological variables, but it must be also represented by a number of ground-based stations. For Lebanon temperature measurements have similar case to that of rainfall and the most comprehensive datasets were taken from the 27 meteorological stations which are not uniformly distributed between various topographic surfaces. Also, climate-based platforms were utilized and more specifically from the World Climate Guide (WCG 2024) and from the Climate Prediction Center-National Oceanic and Atmospheric Administration (CPC-NOAA 2022). Therefore, a simple graphic illustration was applied to figure out the trend analysis of temperature for the period between 1975 and 2024.

Unlike rainfall recorder which are often with obvious contradictory between increase, decrease or oscillating patterns, yet temperature has been evidenced to be increased by all authors, but the change in temperature between various authors is

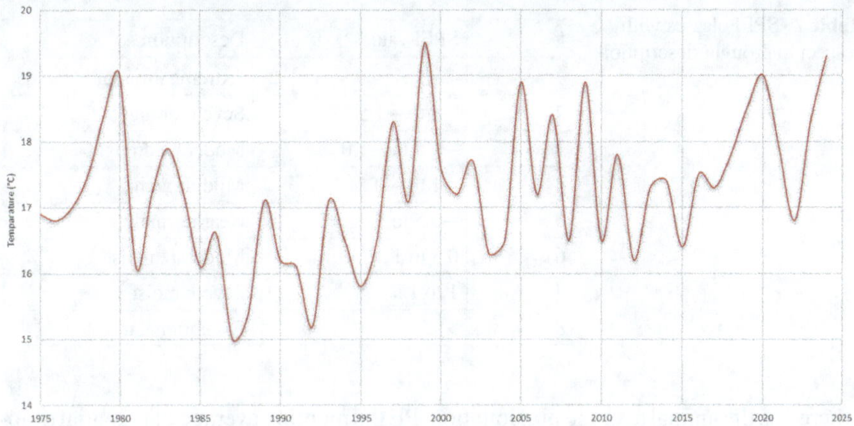

Fig. 6 Average annual temperature changes between 1975 and 2024 for Lebanon

almost in the range of ±0.5; while some studies applied for different time intervals or for diverse regions where temperature change is often different.

The analyzed time series (1975–2024) for temperature shows a remarkable changing in the registered temperatures (Fig. 6). The Long Term Average (LTA) for the annual average temperature was estimated at 18.37 °C over the last 5 decades; while the increase has been calculated at 1.74 °C, which is equivalent to 0.348 °C per decade which is relatively below the global average temperature increase (0.36 °C per decade). This is slightly accord with Shaban et al. (2014a) who has estimated at 1.8 °C temperature increase between 1963 and 2014.

There are sharp peaks in temperature over the projected time period, such as in years 1980, 1999, etc. In addition, there is also a sort of periodicity was noticed with lower temperature average for 1985–1996 and 2008–2017, which is almost consistent with the periodicity of relatively high rainfall rates.

2.3 *Standardized Precipitation Evapotranspiration Index*

The Standardized Precipitation Evapotranspiration Index (SPEI), developed by Vicente-Serrano et al. (2010), is a multi-scalar drought index calculated from various climatic data including mainly precipitation, and potential evapotranspiration (PET). It is used for determining the onset, duration and magnitude of drought conditions with respect to normal conditions, and this includes a variety of natural and managed systems such as ecosystems, rivers, water resources, etc. The SPEI emphasis on the main impact of increased temperatures on water demand, and it is calculated according to the formula number 3:

$$\text{SPEI} = P - PE/Q \tag{3}$$

Table 2 SPEI classes with respect to drought description

#	SPEI value	Description
1	< −2	Extreme drought
2	−2 to −1.5	Severe drought
3	−1.5 to −1.0	Moderate drought
4	−1 to −0.5	Mild drought,
5	−0.5 to 0.5	Near normal
6	0.5 to 1	Moderate moist
7	1 to 1.5	Severe moist
8	>1.5	Exceptional moist

where P is the annual average precipitation, PE the monthly average of potential evapotranspiration, and Q is the standard deviation of monthly precipitation. According to Vicente-Serrano et al. (2010), SPEI was categorised as in Table 2.

Datasets for the SPEI for Lebanon were downloaded from the Laboratory of Climate Services and Climatology (LCSC 2024) for the period from 1950s to May 2025 (mire than 7 decades). The Long Term Average (LTA) for SPEI is −0.050 indicating that Lebanon is among drought # 5; according to Table 2, which in general is "Near normal". Figure 7 shows the general SPEI trend for Lebanon over the last seven decades.

Results show obvious increase in drought conditions since the beginning of 2000 until quite recent; and then, SPEI values became below 0 (Fig. 7). Thus, the average SEPI has been changed from 0.84 to about −1.12; between the periods of 1965–1999; and 2000 to date, and this concludes that Lebanon has been moved from "moderate moist" to "moderate drought" (according Table 2). It worth mentioning that the changing to more drought conditions is not only due to climate change, but it is also affected by the terrain components and human activities.

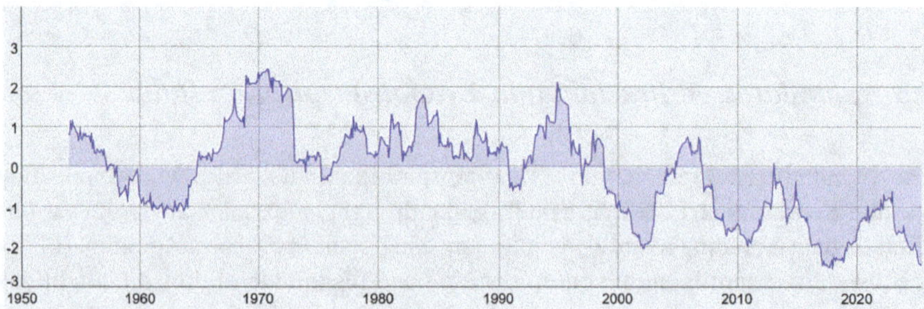

Fig. 7 The standardized precipitation evapotranspiration index for Lebanon (1990–2025)

3 Climate Change Scenarios

The applied climatic trends, scenarios and projections, obtained by many studies (*e.g.*, ESCWA/RICCAR 2020; IPCC 2022), are always oriented with some preconceived concepts that are represented by the precipitation decrease and temperature increase, while other metrological variables are often absent.

The Coupled Model Inter-comparison Project Phase 5 (CMIP5) presented an unprecedented information on which to base projections including new Earth System Models with a more complete representation of forcing, new Representative Concentration Pathways (RCPs) scenarios and more output available for analysis (Collins and Knutti 2013).

The four RCP scenarios used in CMIP5 lead to total radiative forcing (RF) at 2100 that spans a wider range than that estimated for the three Special Report on Emission Scenarios (SRES) scenarios (B1, A1B, A2) used in the Fourth Assessment Report (AR4), RCP2.6 being almost 2 W m^{-2} lower than SRES B1 by 2100. The magnitude of future aerosol forcing decreases more rapidly in RCP scenarios, reaching lower values than in SRES scenarios through the twenty-first century.

Representative Concentration Pathways RCP4.5 and RCP8.5 are refereed to moderate and high emission pathways; respectively. They were acquired from the 5th Assessment Report (AR5) of the Intergovernmental Panel on Climate Change (IPCC) which is based on a set of scenarios that replace the Special Report on Emissions Scenarios (SRES) standards. In this study, RCP4.5 and RCP8.5 were utilized to calculate a number of future climatic variables, while the projected years were for 2030 and 2040.

3.1 Projected Change in Rainwater Volume

RCP4.5 and RCP8.5 are used to calculate the rainfall change, which can be converted into change in water volume derived from rainfall. RCPs can provide numeric values and maps for rainfall and other climatic parameters, example shown in Fig. 8. From the numeric values on the increase or decrease in rainfall plus other variables, the volume of water from rainfall (V_{ch}) can be calculated. These variables are: current rainfall rate (R_f), RCPs (4.5 and 8.5) symbolized as (R_{cp}) for the projected years 20,230 and 2040, current rainfall rate (R_f), the surface area of Lebanon (A), and percentage of evapotranspiration (*Et*) representing the natural loss of water; according to the formula number 4:

$$V_{ch} = (R_f - R_{cp}) \times A \times Et \quad (4)$$

Considered various variables (formula 4); however, results showed that according to RCP4.5, rainfall rate in Lebanon will be decreased by 4.5 and 6.6 mm for the years 2030 and 2040; respectively; and this will be reflected on the total water volume

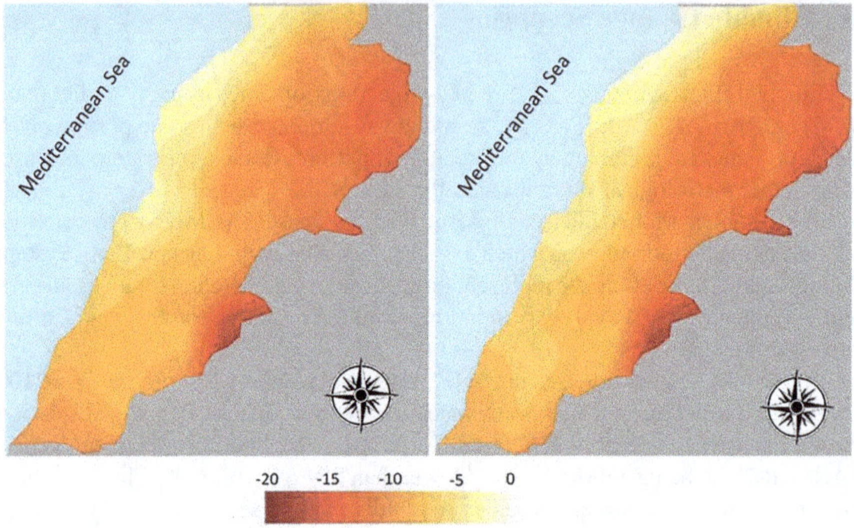

Fig. 8 Annual total rainfall change calculated as percentage difference between 2005 and 2050. The left panel is the RCP4.5 scenario and the right panel is the and RCP8.5 scenario (WB 2018)

derived from rainfall which is anticipated to be reduced about 21 and 31 million m^3 by the years 2030 and 2040; respectively (Table 3).

With respect to RCP8.5, the reduced rate and volume of rainfall will be exceeded. Thus, rainfall rate will be decreased by 12.5 and 20.5 mm for the years 2030 and 2040; respectively; and the resulted water volume will be also reduced about 59 and 396 million m^3 by the years 2030 and 2040; respectively (Table 3).

Table 3 Calculated projected change in rainwater volume based on RCPs data

Variables	Unit	RCP 4.5		RCP 8.5	
		2023	2040	2023	2040
Rainfall deduced by RCP (Rcp)	mm	−4.5	−6.6	−12.5	−20.5
Area of Lebanon (A)	Km^2	10,452			
Evapotranspiration (Et)	%	55			
Total change in in rainwater volume	Million m^3	−47	−69	−131	−214
Decreased rainwater volume (V_{ch})		−21	−31	−59	−96

3.2 *Projected Change in Temperature, Evaporation and Evapotranspiration*

Similar to rainfall, numeric values for temperature and the induced evaporation and evapotranspiration processes were also projected by RCPs (4.5 and 8.5). RCP4.5 reported an increase in temperature rate by about 0.53 and 0.75 °C for the years 2030 and 2040; respectively; while the projected RP8.5 shows temperature rate increase by about 1.28 and 1.8 °C for the years 2030 and 2040; respectively. This increase in temperature will be reflected on evaporation (Ev) and evapotranspiration (Et) rates. For this purpose, Ev and Et can be calculated according to the following number 5 and 6:

$$\mathrm{Ev} = [\mathrm{A_{owb}} \times \mathrm{D_{owb}} + \mathrm{A_r} \times \mathrm{D_r}] \times (\mathrm{T\_E}) \tag{5}$$

where $\mathrm{A_{owb}}$ is the area of open water bodies and $\mathrm{D_{owb}}$ is the average depth of open water bodies,

$\mathrm{A_r}$ is the area of rivers and $\mathrm{D_r}$ is the average depth of rivers, and

T_E is the evaporation-induced temperature increase.

$$\mathrm{Et} = \mathrm{A_{cl}} \times (R_{av} \times E_{t\%}) \times (\mathrm{T\_Et}) \tag{6}$$

where $\mathrm{A_{cl}}$ surface area of all cultivated lands, R_{av} is the average rainfall rate,

$\mathrm{E_{t\%}}$ is the percentage of evapotranspiration, and

T_Et is the temperature-induced evapotranspiration.

Based on the projection of RCP4.5 the calculated variables according to formula 2, show that the volume of water which will be lost by evaporation (Ev) as a result of increased temperature by 2030 and 2040, will be about 1.65 and 1.75 million m^3; respectively (Table 4). While, the projection of RCP8.5 will result in increased water evaporation by 2.1 and 2.46 million m^3 for the years 2030 and 2040; respectively (Table 4).

For evapotranspiration (Et), results show that with the projection of RCP4.5 the based on the calculated variables according to formula 3, the volume of water which will be lost by evapotranspiration (Et) as a result of increased temperature by 2030 and 2040, will be about 131.16 and 165.5 million m^3; respectively (Table 5). By performing the projection of RCP8.5 the loss by evapotranspiration for the years 2030 and 2040 will be 199.83 and 234.16 million m^3; respectively (Table 5).

4 Climate Change Impact on Water Resources

In previous sections, it was clear that climate of Lebanon has been changing over the last few decades and anticipated to change in the future. This has been associated with abrupt decrease in the volume of surface and groundwater resources.

Table 4 Projected change in evaporation as induced by temperature, based on RCPs data

Variable	Unit		
Area of open water bodies (A_{owb})	Km^2	9.67	
Average depth of open water bodies (D_{owb})	m	2.4	
Area of rivers (A_r)	Km^2	11.03	
Average depth of rivers (D_r)	m	3	
Evaporation-induced temperature increase (T_E)	(°C–%)	7	
Evaporation increase (Ev) million m^3			
Temperature increase (according to RCP4.5) and year 2030	°C	0.53	1.65
Temperature increase (according to RCP4.5) and year 2040	°C	0.75	1.73
Temperature increase (according to RCP4.8) and year 2030	°C	1.28	2.10
Temperature increase (according to RCP4.8) and year 2040	°C	1.8	2.46

Table 5 Projected change in evapotranspiration as induced by temperature, based on RCPs data

Variable	Unit		
Surface area of all cultivated lands (A_{cl})	Km^2	2860	
Average rainfall rate (R_{av})	mm	835	
Percentage of evapotranspiration (Et)	%	55	
Evapotranspiration-induced temperature increase (T_E)	(°C–%)	12	
Evaporation increase (Et) million m^3			
Temperature increase (according to RCP4.5) and year 2030	°C	0.53	131.16
Temperature increase (according to RCP4.5) and year 2040	°C	0.75	165.5
Temperature increase (according to RCP4.8) and year 2030	°C	1.28	199.83
Temperature increase (according to RCP4.8) and year 2040	°C	1.8	234.16

In fact, the climate change is not the only reason behind the decreased volume of water, but the human interference also plays a role and it is expected as more the influencing than climate change, and the level of impact on water resources between climate change and human interference is not determined yet. However, the impact of climate on water resources often assessed in defined hydrologic systems (*e.g.*, sub-catchments, deep aquifer, etc.) with least human activities in order to avoid any misestimating.

4.1 Decreased Discharge in Rivers and Springs

There are 14 small-scale rivers in Lebanon (including three transboundary ones) that distributed over a surface area of 10,452 km^2; and thus, they form dense network of watercourses. The estimated total annual discharge from these rivers is about 2800 million m^3. However, the sloping terrain of Lebanon makes runoff along these rivers with high velocity and the lag time of flow from the source to the outlet does not exceed 8–10 h in many instances. Therefore, more than 75% of water from rivers is unexploited and directly outlets into the sea. Besides, there are a big number of issuing springs, and more than 2000 of them are with discharge exceeding 30 l/s; while the total discharge from the Lebanon's springs is about 1100 million m^3.

Based on the available datasets from rivers and springs, a trend analysis has been elaborated to figure out the overall changes in rivers and springs' discharge over long time series. The available data from 1990 to 2023 (33 years) for rivers and springs enabled estimate the overall trending over the last three decades. Therefore, the discharge for representative rivers and springs has been graphically illustrated in Figs. 9 and 10.

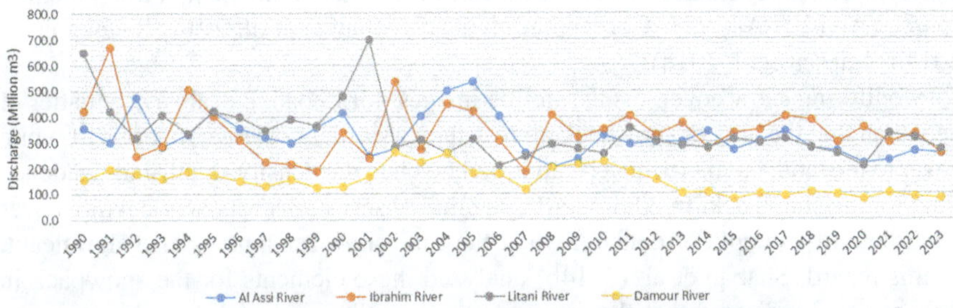

Fig. 9 Discharge from representative Lebanese rivers between 1990 and 2023

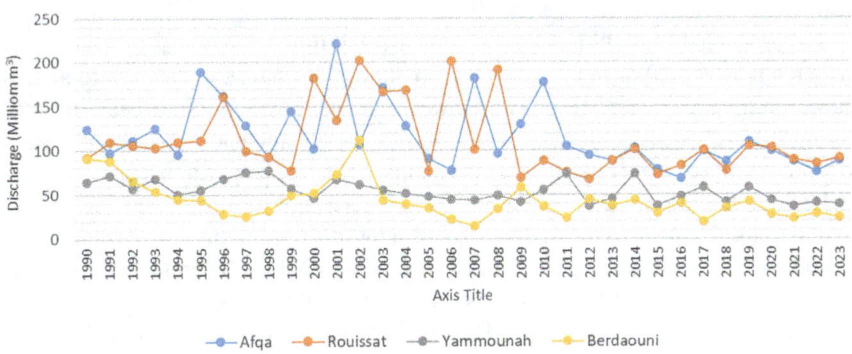

Fig. 10 Discharge from representative Lebanese springs between 1990 and 2023

The analyzed discharge data was collected from different sources, but it was provided mainly by the Litani River Authority (LRA 2024). The graphic illustrations show remarkable descending trends in both water resources, the rivers and springs. Thus, the discharge from the selected rivers shows 23, 21, 4 and 41% decrease for the Litani, Al Assi, Ibrahim and Damour Rivers; respectively. While, for the springs, the decrease in the discharge was 27, 28, 22% and 27 5 for Afqa, Rouissat, Yammounah and Berdaouni springs; respectively. Therefore, the average decreased discharge in rivers is estimated as bout 22%; besties 26% decrease in springs; and this indicates that about ¼ of the discharge in rivers and springs has been reported over the last three decades.

4.2 Changing Snowpack

As it was discussed in sections that snow plays a major role in feeding water resources of Lebanon, and it is the main source for surface and groundwater. However, this is not always the understanding by many researcher and rainfall is often given much concern. Yes, many studies evidenced the role of snow as a fundamental water feeding source (Abd EL-Al 1953; Corbane et al. 2005; Mhawej et al. 2014; Shaban et al. 2014b; Telesca et al. 2018).

Monitoring snow cover, using satellite images, for Lebanon is always a matter of debate in that the area of snowpack almost shows similar coverage in terms of white cover to the land, but there must be attention given to the density, depth and residence time for the snowpack. In other words, satellite observations expose the white cover regardless of its depth or density; while the time snow remains, is also significant. In this regard, Shaban et al. (2014b) analyzed these elements for the snowpack in Lebanon and indicated that the snow cover cannot be separately considered as an evidence of wet climate, and there must be a number of elements to be accounted including the residence time of snowpack continuity, density and the number of snow storms.

The processed MODIS images show the overall trend of snowpack coverage for Lebanon between 2000 and 2022 where no noticeable change in the surface area of snow could be accounted (Fig. 11). However, a cyclic variation has been reported during 2000–2008 and 2010–2022, and this climate phenomenon does not necessary to represent the climate of Lebanon; especially it is a result of regional snow storms.

Besides, the increased temperature (1.74 °C) is a significant and must be considered, because it enhances the melting rate of snow and this has been well pronounced in studies made by Shaban (2003,, and Shaban et al. (2014a). Therefore, the melting rate has been accelerated and the time lag has been decreased from 75 to 90 days in 2000s to 45–50 days after 2015. This has been reflected on the rapid melting and thus least infiltration rate.

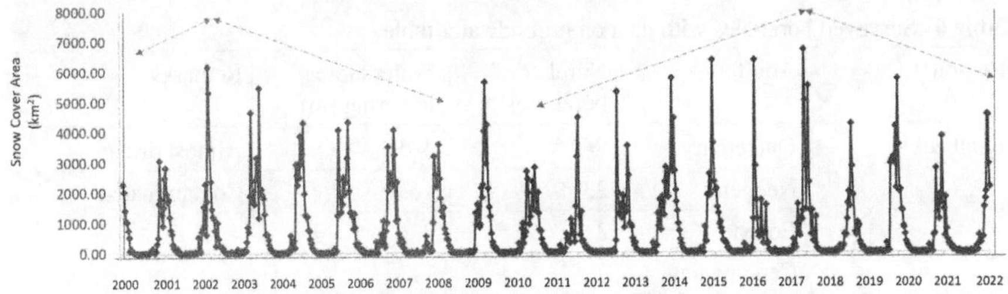

Fig. 11 Mean monthly snowpack coverage for Lebanon between 2000 and 2023, as calculated from MODIS satellite images

4.3 Changing Groundwater Level

The fluctuations in the regime of surface water, including precipitation patterns, rate and intensity, infiltration rate, as well as the runoff; will impact groundwater volume and this can be evidenced by the changing in water table and the pumping rates. Of course, there are many other influential factors controlling groundwater volume and these are due to human interference, such as the exceeded number of boreholes, urbanism and destruction in water infrastructures.

Even though, groundwater abstraction (and number of boreholes) differs between various rock formations in Lebanon, yet the lowering in groundwater level has been reported in all aquiferous rock formations including the Neogene, Tertiary, Cretaceous and even the Jurassic rock formations. In the capital Beirut (for example), boreholes dug in the Cenomanian Aquifer (Upper Cretaceous) revealed lowering in groundwater table between 22 and 34 m over the last few decades. While, in the Bekaa Plain, groundwater table in the Neogene Aquifer has been lowered to more than 25 m since 2000. In this respect, an inventory was prepared for a number of boreholes and the resultant lower down in water table (Table 6).

In addition to the lowering in the water table, the piezometric level has been also decline in many boreholes, but data are often uncertain as informed by many wells' owners. This status in water decline has been associated with other challenges which have been recently well raised such as water contamination and turbidity in many boreholes, saltwater intrusion in the coastal aquifers; in addition, some boreholes became totally dry notably those in the coastal zone.

5 Conclusion and Discussion

Located along the Eastern Mediterranean, Lebanon has been known by plenty water resources, and this is attributed to its geographic situation and geomorphologic shape where both play a role in making Lebanon as a climatic barrier capturing cold air masses and resulting in rainfall and snowpack. However, this description has no longer reflected on water resources and complain on water supply becomes a national

Table 6 Surveyed boreholes with data on groundwater table

Region	Aquifer	Number of boreholes	Water table lowering (m)	Remarks
Baalbeck	Quaternary	8	4–15	Almost dried
	Neogene	23	16	Contaminated
	Eocene	3	8	–
	Cenomanian	7	11	–
Bekfaya	Jurassic	5	5	–
Beirut	Miocene	6	12	Contaminated
	Cenomanian	17	22	Saltwater intrusion
Hermel	Neogene	25	12	Water table decline
	Eocene	3	15	–
	Cenomanian	7	14	Between 2002 and 2024
Jbeil	Cenomanian	9	4	–
Kalamoun	Miocene	8	9	Saltwater intrusion
Kob Elia	Neogene	18	14	Water table decline
Labouh	Cenomanian	7	–	Since 2007
Saida	Eocene	7	8	Saltwater intrusion
	Cenomanian	6	6	
Sultan Yaqoub	Neogene	6	2	Water table decline
	Eocene	4	–	–
Tebnine	Eocene	4	2	–
Tripoli	Miocene	11	6	Saltwater intrusion
Nabatieh	Eocene	8	6	–
	Cenomanian	13	8	–

issue, and all the Lebanese regions are suffering from water shortage which affected many vital sectors including mainly the agriculture, food security, energy and the socioeconomic of the country. This is also exacerbated by the absence of good water management and unwise use of water sources; and therefore, the impact of human interference is not less than that by climate. In this regard, it was anticipated an abrupt water scarcity (*e.g.*, dryness of many spring, water level in rivers and groundwater, etc.) in Lebanon if 4–5 consequent drought years occur; nevertheless, this happens recently (2025) and after only one drought year. This reflected that Lebanon's water is on the threshold of jeopardy.

The estimated water availability as 1077 m^3/capita/year is also under a challenge and it is expected to be declined if the current situation continuous in the same trend. The 220 m^3/capita/year water demand will not be reached, and undefined water future will be occurred. However, climate can be a reason, but it is not the only reason behind, and human behavior is one of the reasons, plus the governance framework and water management needs reforms and restructuring in order to mitigate water supply shortage, notably all applied scenarios are alarming for incoming decease in precipitation and increase in temperature and warm climate.

This study highlighted many key on the key points of climate in Lebanon, and it introduces an example for the entire Middle East Region. This study calculated a considerable times series of data notably it integrated ground-based metrological data with data retrieved from climate platforms. Thus, the long-term average (LTA) of precipitation as 858 mm/year and temperature of 20.5 °C are the most reasonable values, and this is resulted in potential evapotranspiration of about 936 mm/year. For the solid precipitation, it also has a significant contribution in feeding water resources in Lebanon. thus, snowpack shares with more than 60% of water reserve in the country and this motivates putting plans for conserving snowpack; especially human activities extended to the elevated mountains here snowpack exists. Besides, the calculated aridity characterized Lebanon with semi-humid to humid climate. Therefore, the above variables and characterization reveal that Lebanon is supposed to be water-secured region, but the status quo does not reflect the reality.

Results of this study are significant; especially they reveal numerics which have not known before. For example, the decreased volume of rainfall water was believed to be high enough, but this study shows a negligible decrease between 25 and 30 mm over the last seven decades. Also, the study shows a fluctuating patterns in rainfall increase and decrease over specific time periods; and thus, a periodicity of 11–15 years has been determined. Telesca et al. (2018) evidenced such periodicity for Lebanon between 10 and 15 years. For temperature, the long-term average (LTA since 1975) obtained from a long time series analysis shows that it averages at 18.37 °C; while the increased rate is 1.74 °C over the last five decades, which is almost less than the estimated temperature increase since the pre-industrial era, which is on the brink of exceeding 2 °C (Malik et al. 2023).

In this study, the climate of Lebanon has been also studied from drought indices where the Standardized Precipitation Evapotranspiration Index (SPEI). Results showed that the average SEPI has been changed from 0.84 to about −1.12; between the periods of 1965–1999; and 2000 to which means that the Lebanese climate, in terms of drought, has been changed from moderate moist to moderate drought.

For future scenarios, the IPCC scenarios were utilized due to their creditability and common use, and more certainly the Representative Concentration Pathways scenarios (RCP4.5 and RCP8.5). It was resulted that precipitation rate will influence the volume of water derived from the rainfall, and this will witness a decrease, between 2030 and 2040, by about 26 million m^3 and 77.5 million m^3 with respect to RCP4.5 and RCP8.5; respectively. While, temperature will be increased between 20,230 and 2040 by about 4.5 °C and 0.64 °C with respect to RCP4.5 and RCP8.5; respectively. This will act in increasing evaporation and evapotranspiration rates; and thus it will be reflected on increasing the lost water volume by 1.98 and 182.77 million m^3 from evaporation and evapotranspiration; respectively once apply RCP4.5 and RCP8.5 for the years 20,230 and 20,240.

This chapter tackles three chronological data analysis including past, present and future. Comparing climatic variables of Lebanon over these three time-components, it has been resulted that Lebanon in general was characterized by humid climate with sufficient water resources. This has been changed towards more arid conditions, and it is obviously impact water resources. According to the performed future scenarios, an ascending trend is still continuing and will impact water resources as well. In this respect, population growth and newly human aspect of water demand must be considered and the will exaggerate the problem of water resources and its consequences on agriculture, food security, energy and on the overall socioeconomic conditions. Thus, serious and consolidated water management plans are needed. In other words, climate is changing towards more aridity in Lebanon, but precipitated water volume will slightly remain adequate to feed water resources, and this requires successful management plans and strategies.

It is not enough to determine the problem, illustrate numeric values and comaprtive analysis, which showed unpreceded loss in the water volume. The need is to propose and execute water management plans on different dimensions taking into account mainly water storage, such as water harvesting and groundwater artificial recharge, plus water resources protection qualitatively and quantitatively. Thus, water laws must be reformed and updated, and this needs to involve specialized water experts in water management and governance and not involving experts from other domains as the current case. Continuing in this trend by weak and theoretical solutions for the changing climatic conditions and lack of adaptation measures will sustain/ and though exacerbate the problem of water deficit in Lebanon, the country which was described as the water tower of the Middle East.

References

Abd EL-Al I (1953) Statics and dynamics of water in the Syro-Lebanese limestone massif. In: Ankara symposium on arid zone hydrology. UNESCO, Ankara, pp 60–76

Arkadan A (2008) Climate change in Lebanon: prediction uncertain precipitation events. Do climatic cycles exist? In book on: climatic changes and water resources in the Middle East and North Africa. Springer, Berlin, pp 59–71

Benassi M (2008) Drought and climate change in Morocco. An analysis of precipitation field and water supply. In: López-Francos A (ed) Drought management: scientific and technological innovations. CIHEAM, Zaragoza, pp 83–86

CAL (1971) Atlas Climatique du Liban, Tome 1. Service Météorologique, Ministère des Travaux publics et Transports, Beirut, Lebanon

CAL (1982) Atlas Climatique du Liban, Tome II. Service Météorologique, Ministère des publics et Transports, p 40

CAS (2015) Central administration of statistics. Bull Environ Agricult

CHIRPS (2017) Climate hazards group infrared precipitation with station data. http://chg.geog.ucsb.edu/data/chirps/

CNRS-L (National Council for Scientific Research, Lebanon) (2015) Regional coordination on improved water resources management and capacity building. Regional project. GEF, WB

Collins MR, Knutti J (2013) Long-term climate change: projections, commitments and irreversibility. In: Climate change 2013: the physical science basis. Contribution of working group I to the 5th assessment report IPCC. Cambridge University, New York

Corbane C, Somma J, Bernier M, Fortin JP, Gauthier Y, Dedieu JP (2005) Estimation de L'équivalent en Eau du Couvert Nival en Montagne Libanaise à Partir des Images RADARSAT–1. Hydrol Sci J 50(2):1542

CPC-NOAA (2022) Climate prediction center (CPC) at the national oceanic and atmospheric administration (NOAA). http://www.cpc.ncep.noaa.gov/

De Martonne E (1926) Aréisme et indice aridité. Comptes Rendus De L'acad Sci 182:1395–1398

DGAC (General Directorat of Civil Aviation) (1999) Rapport annuel, Beyrouth, Liban, p 32

Emberger E (1932) Sur une formule climatique et ses applications en botanique. La Météorologie 12:423–432

ESCWA/RICCAR (2020) Climate change and its impact on the Arab region: overview of the RICCAR project. United Nations. https://www.riccar.org/sites/default/files/2020-07/1.4-ESCWA-RICCAR%20Overview.pdf

FAO (2009) AQUASTAT. Water report 34. http://www.fao.org/nr/water/aquastat/countries_regions/LBN

Fawaz M (1992) Water resources in Lebanon

GAIN (Notre Dame Global Adaptation Index) (2017) Vulnerability. Country ranking. Available at: https://gain-new.crc.nd.edu/ranking?trk=public_post_comment-text

Ghaddar N (1995) Climatological data. Monthly bulletin. American University of Beirut, Lebanon

IPCC (2022) Climate change 2022: impacts, adaptation and vulnerability. In: Working group II contribution to the sixth assessment report of the intergovernmental panel on climate change, p 3056

Karam F, Breidy J, Stephan C, Rouphael J (2003) Evapotranspiration, yield and water use efficiency of drip irrigated corn in the Bekaa Valley of Lebanon. Agricult Water Manag 63(2):125–137

LARI (Lebanese Agricultural Research Institute) (2017) Climatic data. Monthly bull. Department of Irrigation and Agro-meteorology (DIAM)

LCSC (Laboratory of Climate Services and Climatology) (2024) Standardized precipitation evapotranspiration index. https://spei.csic.es/map/maps.html#months=7#month=5#year=2024

LRA (Litani River Authority) (2024) Technical reports and datasets for various Lebanese rivers and springs

Malik A, Stenchikov G, Moostamandi S, Usman M (2023) Earth's future, accelerated historical and future warming over the Middle East and North Africa in response to the global temperature change. ESS Open Archive 25:2919

Mhawej M, Faour G, Fayad A, Shaban A (2014) Towards an enhanced method to map snow cover areas and derive snow-water equivalent in Lebanon. J Hydrol 513:274–282

MOE/ECODIT (2002) State of the environment report, Lebanon. Prepared by ECODIT for the Ministry of Environment

MoEW (Ministry of Energy and Water) and UNPD (2016) National guideline for rainwater harvesting systems

MOEW (2010) National water sector strategy: baseline. MOEW. Accessed 15 Sept 2010

MoFA (Ministry of Foreign Affairs) (2018) Climate change profile, Lebanon. Ministry of Foreign Affairs Netherlands, p 139

Na'ameh M (1995) Water problems of Lebanon. National Congress on Water Strategic Studies Center. Beirut (in Arabic), p 67

NOAA (National Oceanographic Data Center) (2013) Lebanon climatological data. Library. http://docs.lib.noaa.gov/rescue/data_rescue_lebanon.html

Perry AH (1988) Trends in maltese rainfall: causes and consequences. In: Gregory S (ed) Recent climatic change. Belhaven Press, London, pp 125–129

Plassard J (1971) Carte pluviométrique du Liban au 1/200000. Ministry of Public Works and Transport, Lebanon

Rey J (1954) Carte pluviométrique du Liban au 1/200000me. Ministère des Travaux Publics, République Libanaise

Shaban A (2011) Analyzing climatic and hydrologic trends in Lebanon. J Environ Sci Eng 5(3):6985

Shaban A, Houhou R (2015) Drought or humidity oscillations? The case of coastal zone of Lebanon. J Hydrol 529(2015):1768–1775

Shaban A, Faour G, Khawlie M, Abdullah C (2004) Remote sensing application to estimate the volume of water in the form of snow on Mount Lebanon. Hydrol Sci J 49(4):643–653

Shaban A, Telesca L, Darwich T, Amasha N (2014a) Analysis of long-term fluctuations in stream flow time series: an application to Litani River, Lebanon. Acta Geophys 62(1):6985

Shaban A, Darwich T, Drapeau L, Gascoin S (2014b) Climatic induced snowpack surfaces on Lebanon's mountains. Open Hydrol J 2014(8):8–16

Shaban A, Awad M, Ghandour A, Telesca L (2019) A 32-year aridity analysis: a tool for better understanding on water resources management in Lebanon. Acta Geophys 67(4):241–298

Shaban A, Hamzé M (2017) Shared water resources of Lebanon, Nova, NY, p 150

Shaban A (2003) Etude de l'hydrogéologie au Liban occidental: utilisation de la télédétection. Ph.D. dissertation. Bordeaux 1 Université, p 202

Shaban A (2020) Water resources of Lebanon. Springer, Berlin, p 229. https://doi.org/10.1007/978-3-030-48717-1

SNC (2011) Lebanon's second national communication to the UNFCCC. Ministry of Environment. GEF. UNDP, p 191

Telesca L, Shaban A, Awad M (2018) Analysis of heterogeneity of aridity index periodicity over Lebanon. Acta Geophys 67(1):167–176

TRMM (2015) Tropical rainfall mapping mission. Rainfall archives. NASA. http://disc2.nascom.nasa.gov/Giovanni/tovas/TRMM_V6.3B42.2.shtml

UNDP United Nations Development Program (1970) Liban-Etude des Eaux Souterraines. United Nation, New York

Vicente-Serrano SM, Beguería S, López-Moreno L (2010) A multi-scalar drought index sensitive to global warming: the standardized precipitation evapotranspiration index—SPEI. J Clim 23:1696–1718

Walton K (1969) The arid zones. Chicago, Aldine Publishing Company, 169 p

WB (World Bank) (2003) Republic of Lebanon—policy note on irrigation sector sustainability. Rep. no. 28766 –LE

WB (2018) Droughts and agriculture in Lebanon: causes, consequences, and risk management. Techn Rep 12:98

WCG (World Climate Guide) (2024) Temperature data. https://www.climatestotravel.com/climate/lebanon

Flood Damage Cost Assessment and Sensitivity Analysis for Arid Catchments: Impacts of Hydroclimate Variability

Eyad Abushandi

1 Introduction

Changes in hydroclimate rainfall patterns—such as frequency, intensity, and spatial distribution—are increasing the challenge of flash floods in arid and semi-arid regions. Over recent decades, Middle Eastern countries have experienced multiple extreme weather events, all leading to floods and economic losses. In arid zones, flash floods are especially deadly when vegetation cover is limited, terrain is steep, and drainage systems are poorly developed. These floods develop rapidly after heavy rainfall, often causing loss of life, infrastructure damage, and agricultural decline.

In addition to the well-known Tropical Cyclones Gonu (2007), Phet (2010), Keila (2011), and Mekunu (2018), Oman has experienced several other notable tropical storms and cyclones in recent years, highlighting the increasing exposure of Oman to hydroclimatic extremes. Earlier cyclones Ashobaa (2015) and Nilofar (2014) brought heavy rainfall and caused overtopping of storm surge areas along Oman's eastern coast. Cyclone Luban followed storm Mekunu and worsened subsequent flood impacts in the Dhofar region. Furthermore, Cyclone Shaheen (2021) made landfall along the northern Batinah coast, an area with historically low cyclone impacts, causing flooding and infrastructure damage. In 2019, Cyclone Hikaa similarly affected the eastern Al Sharqiyah and Al Wusta regions. The most recent cyclone was Cyclone Tej (2023); however, it was poised to directly impact southern Oman and Salalah, as well as smaller regional wadis, where heavy rainfall caused flash floods over several days.

In arid areas, it is not only the rain that destroys, but the lack of preparation that costs the most. Preparedness makes it positive instead of always being negative.

E. Abushandi (✉)
School of Engineering, Bahrain Polytechnic, Isa Town, Bahrain
e-mail: eyad.abushandi@polytechnic.bh

M. M. Al Saud (ed.), *Hydroclimate Changes: Implications of the Middle East Region*,
https://doi.org/10.1007/978-3-032-20053-2_12

The estimated damages from Gonu exceeded 4 billion U.S. dollars, while Shaheen caused approximately US $520 million in damages. Recent hydrologic studies indicate that flash floods in Oman have become more frequent, with the risk of damage increasing by 9–36% annually (Abushandi and Al Ajmi 2022). Therefore, the potential for flood damage is expected to rise in the coming years. The expectation of flood damage can be higher in the coming years.

These events highlight the increasing frequency and intensity of rainfall storms and cyclones. They reinforce the need for flood resilience by planning adaptive infrastructure along coastal and arid lands throughout Oman.

However, the current understanding of the significant role of extreme flash flood events in subsequent economic damage is limited. Expected Annual Damages (EAD) and flood damage costs should be taken into account when designing new hydraulic structures, building infrastructure, or even safeguarding existing settlements. Additionally, flood damage assessments may face various uncertainties due to selected floodplain boundaries, flood scenarios, and market values. The primary cost estimation for floods is based on the direct tangible damage within the flooded area, expressed as economic and social losses. Indirect intangible damage occurs outside the flooded area, making damage estimation and cost assessment more complex.

The earliest publications on estimating flood damage costs originated from the creation of stage damage functions, established by Penning-Rowsell and Chatterton (1978). Their studies included methods of estimating economic losses from floods utilizing stage-damage curves relating water flow to potential damage to types of structures. These efforts established a basis for later developments in estimating flood damages, which would become far more complex than stage damage curves by incorporating factors such as velocity, duration of flooding, and many others.

Since then, several tools, models, and software platforms have been developed to estimate flood damage costs while including hydrological, spatial, and economic parameters. Mohammadi et al. (2014) used the Hydrologic Engineering Center's Flood Damage Reduction Analysis (HEC-FDA) tool to estimate flood damage. This tool allows for a spatial hydrologic and economic assessment; it can facilitate assessments of flood risk and infrastructure planning. Nghia et al. (2022) used the Google Earth Engine platform for flood mapping that utilized Sentinel-1 synthetic aperture radar (SAR) to assess flood extents in near real-time. This provides a relatively quick, cloud-based space for monitoring flood impacts spatially in large areas, specifically in areas with limited data. Recognizing the uncertainties surrounding flood damage estimation, de Moel and Aerts (2011) laid out four main sources of uncertainty in flood damage models: inundation depth, land use classifications, evaluation of potential at-risk elements, and depth-damage (or stage-damage) curves. They stress the importance of good data and sensitivity analysis to be able to reliably assess flood risk. Smith (1994) contributed significantly to establishing stage-damage relationships by developing loss functions for certain types of buildings and crops. The loss functions show the relationship between depth of inundation and damage, which allows assessing flood vulnerability across various land uses.

Over the last twenty years, a variety of flood risk models have been developed to improve risk prediction and account for the numerous structural and environmental factors influencing the flood regime. Most flood risk models share several important features:

i. They are dependent on historical hydrometeorological and damage data (Merz et al. 2010; Jongman and Ward 2012).
ii. They make use of multiple variables to reflect the complexity of flood risk systems (Apel et al. 2004; Messner et al. 2007; Abushandi and Al Sarihi 2022; Abushandi et al. 2023).
iii. Many models are designed for specific geographical, or alternatively national, use and are often validated explicitly for these regions (Thieken et al. 2005; Nogherotto et al. 2019).
iv. Many models need to be recalibrated and updated regularly, or else their effectiveness will decline as climatic and land-use patterns change (Pistrika and Jonkman 2010; Ward et al. 2017).

It is also important to bear in mind that in some cases, the cost of flood preventative measures may exceed the damage expected from one flood event (Kreibich and Thieken 2009). Thus, a cost–benefit analysis will ultimately inform the design and placement of flood mitigation. This study examines the Wadi Al Jizzi catchment area, which includes the city of Sohar, that is rapidly going through urbanization and is exposed to flash floods stemming from hydroclimatic variability. The focus of this study is a comprehensive assessment of flood damage costs in the outlet region of the catchment, combined with a sensitivity analysis to explore the impacts of several simulated flood scenarios on the economic losses. In particular, this chapter has the following objectives:

i. Quantify the total direct economic losses from past and simulated flooding events that have occurred in the Wadi Al Jizzi catchment.
ii. Identify and map from a regional perspective the major variables in flood damage (hydroclimatic, geomorphological/topographic, and land-use), and explore the spatial distribution within the catchment.
iii. Establish evidence-based recommendations for improving flood risk management outcomes and for building community and infrastructure resilience.
iv. Address current knowledge gaps by combining flood damage modelling with hydroclimatic sensitivity analysis to support planning and policy in arid and semi-arid areas of Oman.

Flood risk is mainly seen as the probability of hazard times its consequences; in simple terms, the expected flood damage for a specific event in any given year, such as a 1 in 100-year flood, can be estimated by multiplying the likelihood of that flood event by the expected economic and social costs resulting from each event. Through a multi-model analysis approach that includes hydrological modeling, economic loss estimation, and scenario-sensitivity analyses, this study highlights keyways to understand flooding costs in arid landscapes. This framework improves scientific knowledge of flood hazards and supports practical efforts to develop climate-resilient infrastructure and practices.

2 Study Area Description

In northern Oman, Wadi Al Jizzi is a major hyper-arid catchment region, extending over an area of about 870 km^2. Wadi Al Jizzi is distinguished because of its hydrological values, and its historical exposure to extreme flash floods caused by heavy but short-duration rainfalls that are often characteristic of arid and semi-arid topographies. As a catchment, Wadi Al Jizzi contains a sizeable area devoted to agriculture as well as a fully urbanized outlet, thus reinforcing its value in the environmental and economic landscape.

The catchment system is vital to any discussion about water resources management in a regional context for gradual socioeconomic development in Sohar, Oman's second largest urban center. The catchment contributes significantly to the groundwater budget associated with domestic use, irrigation, and industry within the urban locale. Specifically, Wadi Al Jizzi is sourced by the Al Hajar Al Gharbi Mountains in the west, flowing towards the Gulf of Oman to the east, discharging through Sohar city (Fig. 1).

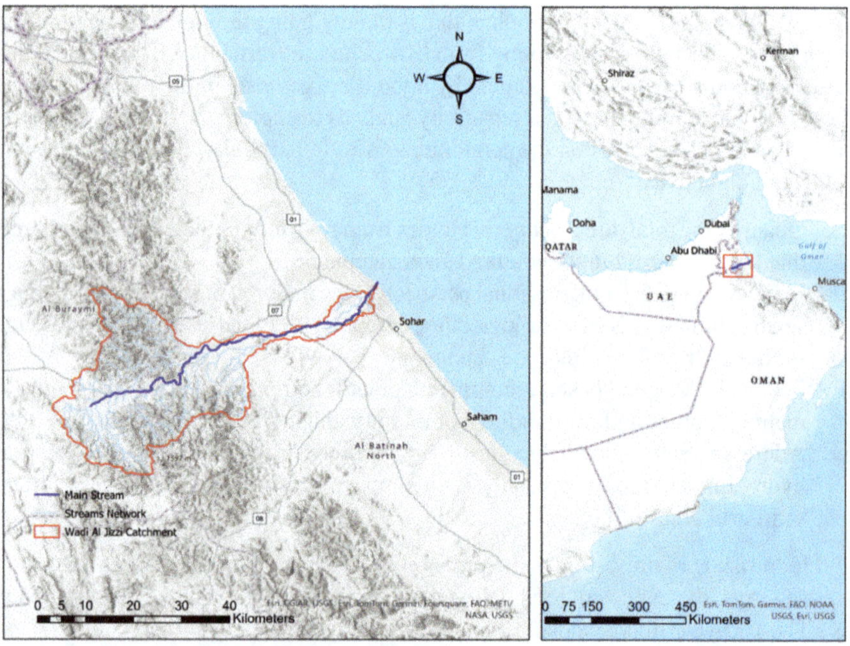

Fig. 1 Location of the Wadi Al Jizzi catchment area within the region

Sohar is a prominent urban location and economic centre with an estimated population of 108,274 in 2025, and is made up of 99 villages, many of which are mountainous and can experience flash flooding. In Sohar, heavy rainfall occurs intermittently throughout the year, which is associated with flooding that occurs continuously, causing damage to physical infrastructure, blocking movement, and threatening livelihoods. With the uniqueness of the catchment and rapid urban growth in Oman's coastlines, Wadi Al Jizzi is an emerging focus in flood risk assessment and planning in Oman. Its highly complex hydrology, population density, and economic profile make it an ideal site for examining hydroclimatic variability impacts on flood damage and resilience planning.

Wadi Al Jizzi's climate is shaped by two different regional systems: first, the humid coastal climate of the Gulf of Oman to the east, and second, the hyper-arid conditions of the inland areas of the West. Each climate type supports very different hydrological behavior and flood response.

The catchment's average annual rainfall is approximately 109.1 mm, exhibiting significant variability across months. Rainfall events usually occur at lower frequencies but are tightly concentrated, resulting in rapid surface runoff and potential flash flooding. In contrast, the catchment's mean annual temperature remains consistent at 26.4 °C, with minimal standard deviation in both maximum and minimum temperatures. Temperature datasets show low variation regarding the maximum and minimum temperatures for each season when compared to rainfall. Furthermore, results indicating a notable negative correlation suggest that higher temperatures are typically associated with lower rainfall, a common characteristic of hyper-arid regions. This negative relationship underscores a region prone to prolonged dry spells, punctuated by brief periods of intense storm events (Fig. 2).

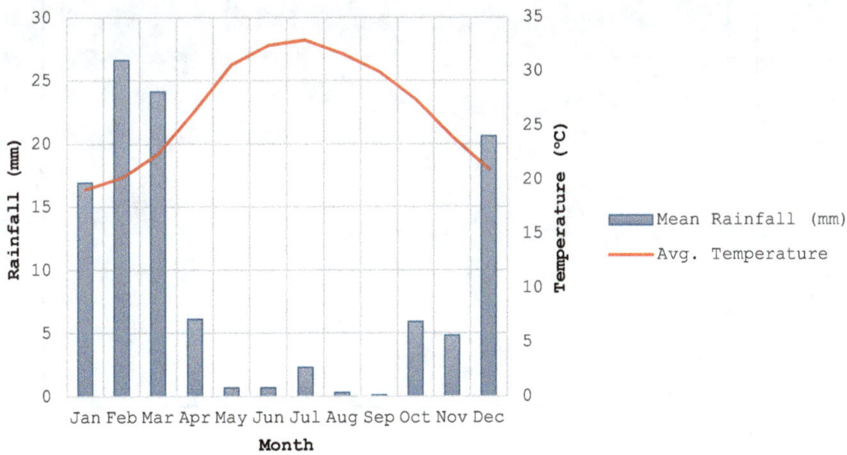

Fig. 2 Climograph of Wadi Al Jizzi based showing monthly averages for the 24 years 1986–2009. *Data source* The Oman Meteorology Directorate General

Evaporation rates at Wadi Al Jizzi are very high (> 85%), and even higher in the dry season. They drop slightly in wetter months. In summer, maximum average temperatures are often around 40 °C, and the high degree of humidity from the neighboring Gulf of Oman increases thermal discomfort, evapotranspiration, and water scarcity. This climatic context influences the surface hydrology and flood risk variables of the catchment area. It requires a water resource management strategy that processes low rainfall, high evapotranspiration, and extreme flood generation capacity.

Wadi Al Jizzi's monthly rainfall and temperature data have revealed three distinct seasonal clusters (Fig. 3), providing information about the climate of the area. The rainfall groups were defined based on similarities in hydroclimatic conditions, as well as clear seasonal divisions:

i. The wet season involves a cooler temperature (the wetter months) and also has the largest amount of rain distributed during December, January, February, and March.
ii. The moderate season involves transitional weather patterns (with random precipitation) during April, October, and November.

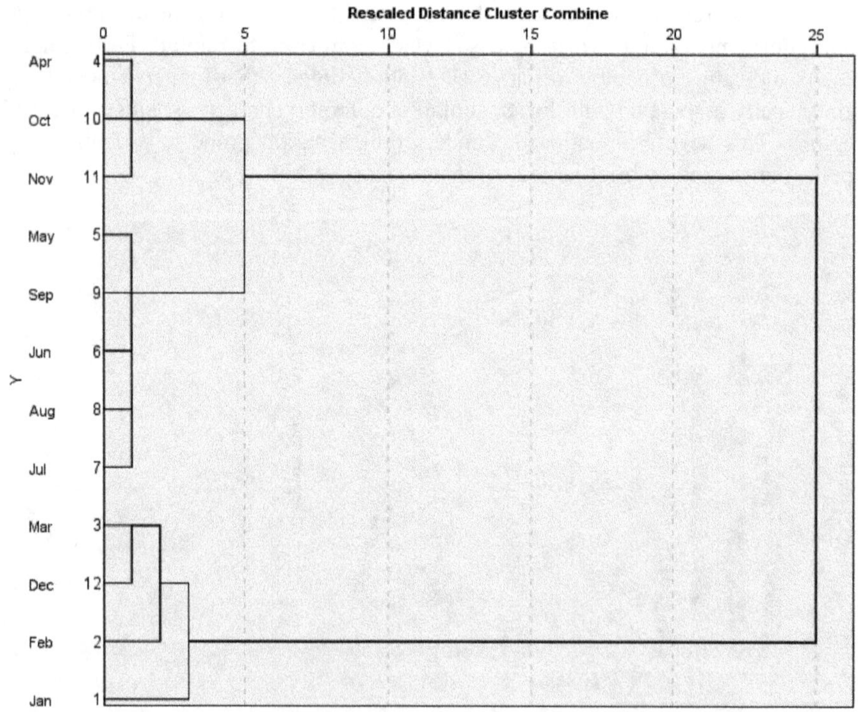

Fig. 3 Dendrogram using average linkage between the groups for Wadi Al Jizzi meteorological data monthly averages for the 24 years 1986–2009

iii. The dry season consists of the driest (long periods of hot and dry) months from May to September and the least amount of precipitation.

The cluster analysis for the monthly average rainfall and temperature denoting variations between the 24 years. This is particularly notable in transitional months, characterized by intra-seasonal variability. As rainfall groupings are primarily focused on identifying the periods when flood risk is elevated, transitional and wet seasons (especially) are of greatest concern due to the likelihood of flash floods.

Over the past 15 years, the frequency, intensity, and magnitude of flash flooding events have markedly increased in Wadi Al Jizzi. During storm events, peak flows have increased to almost three times the historical averages, resulting in damage to many infrastructure systems, as well as agricultural and urban areas (Fig. 4). The outgrowth of these preliminary findings is increasingly important in engaging seasonal climate classification for flood prediction and early warning systems in an arid region such as northern Oman.

The slope gradient of the Wadi Al Jizzi streambed plays an essential role in defining the flash flood potential within the catchment. The average slope of the catchment is approximately 16.5°; highest slope gradient zone is in the upper catchment area, characterized by a steep slope that accelerates surface runoff and hence, creates no room for infiltration (Fig. 5). Therefore, fast runoff denotes low recharge of groundwater and erosion of the streambed occurs especially during high-intensity rainfall, causing an extreme flash flood risk.

With significant elevation differences from 1472 m (Al Hajar Al Gharbi Mountains) to above reference point zero (less than 1 m) at the lagoon outlet, there are more than 3% of catchment area built-up, 2.3% used for agricultural land, close to outlet and floodplain that may restrict capacity for natural flow during extreme events. Bare

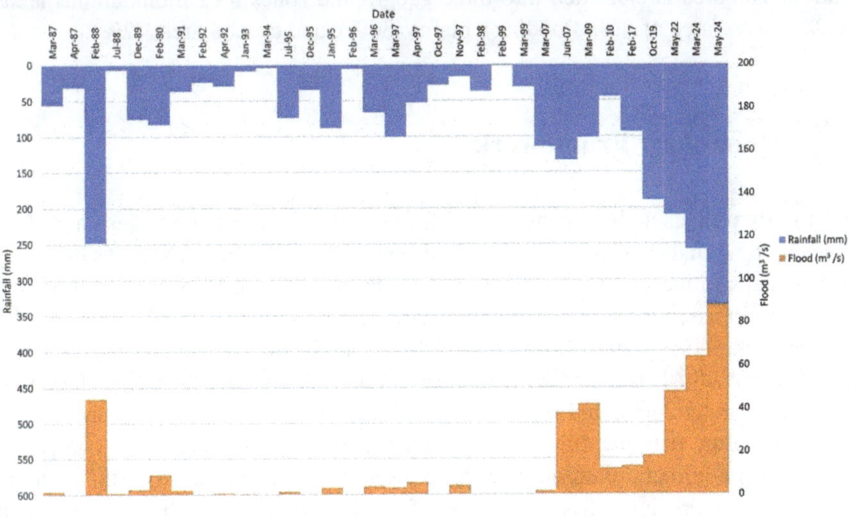

Fig. 4 Histogram of flood versus rainfall measurements from 1987 to 2024

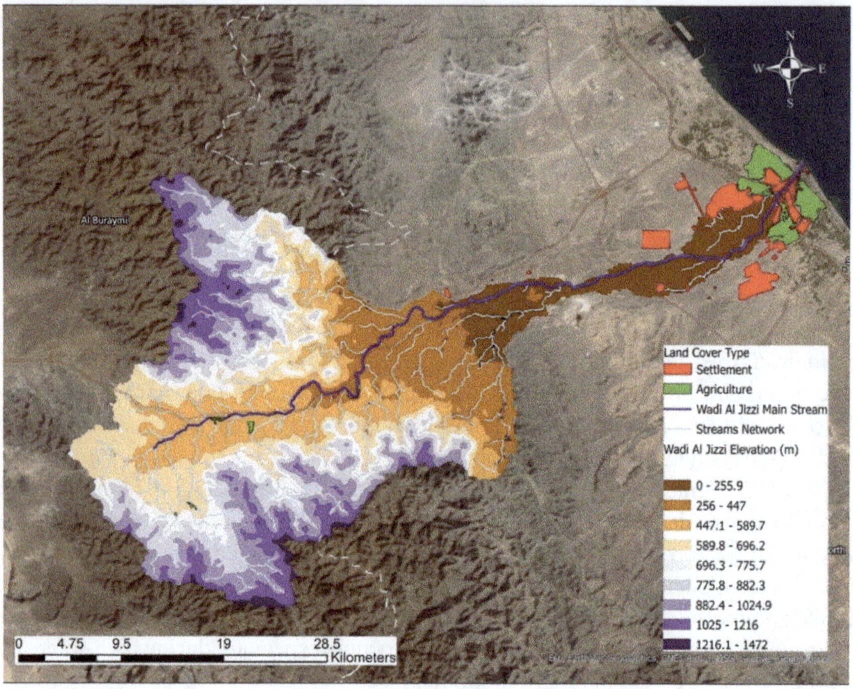

Fig. 5 Map of Wadi Al Jizzi/Oman showing site topography, stream network, land cover types, soil sample locations, and infiltration test sites

soil and rocky areas of the remaining catchment area provide little capacity for infiltration. The area is classified into three geographic zones, i.e., mountainous areas (68%, source area of flash flooding), plains (30%), and coastal areas (2%).

3 Methodology Framework

This study utilizes a climate risk-based framework to evaluate damages caused by flash flooding in the catchment of Wadi Al Jizzi. The methodology follows interconnected steps to incorporate a climate risk scenario analysis, Hydrologic modeling, flood impact assessment, and economic damage assessment (Fig. 6).

The methodology starts with climate history and extreme weather events influencing the Wadi Al Jizzi area. Principal drivers of flash flooding include high rainfall intensity, temperatures, and rapid changes in rainfall patterns. The historical rainfall/temperature data identifies the region's climate typicality and helps categorize whether. The climatic patterns and flood event analyses presented in this chapter build upon the previous efforts of a team of researchers (Abushandi et al. 2023) that focused on flash flood inundation from extreme rainfall events in the Wadi Al Jizzi

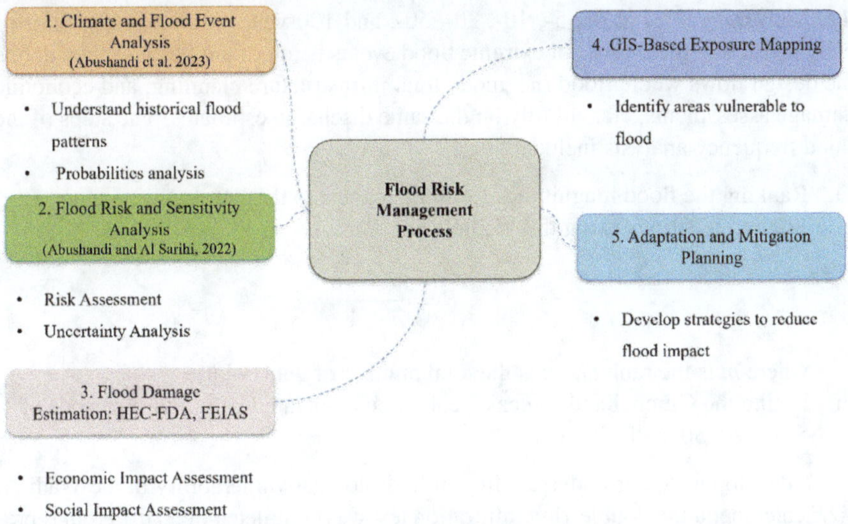

Fig. 6 An integrated methodology for flood risk assessment and management for arid catchments

area of Oman. The original study relied on statistical analyses drawn from meteorological rainfall datasets to highlight the inverse relationship between temperature and precipitation, representative of the hyper-arid climate witnessed there. Higher temperatures are coincident with lower total rainfall, which may have impacts on the hydrological response of the catchment area. This is useful for thresholds for when significant flood-triggering events are understood.

Step 2 of this chapter closely follows the methods by Abushandi and Al Sarihi (2022) in a flash flood inundation study of Wadi Al Jizzi, Oman. Abushandi and Al Sarihi (2022) employed hydrological modeling (IHACRES and AHP) and geospatial-derived terrain and land-use data to assess the catchment's sensitivity to extreme rainfall events. They also assessed physical indicators (slope, drainage density, elevation range, land-use) to understand flood risk. Their sensitivity-based approach informed the ability to pinpoint "hotspots" within the catchment area that were potentially most susceptible to flash flooding and thus established a particular spatio-temporal basis for focused flood risk assessment in this report. While the study originally tackled the spatial variability of flood potential through a weighting strategy that considered sensitivity to environmental parameters, the study importantly established a semi-quantitative approach to exposure mapping of flood-proneness. This was necessary to understand the implications of vulnerability to process exposure mapping and for assessing flood impact mitigation.

To assess flood magnitudes related to the various return periods, flood frequency analysis based on 30 flood events between 1987 and 2007 was established using the Gumbel Extreme Value Type I distribution. The historical peak discharge records for the Wadi Al Jizzi catchment were ranked numerically and analyzed to estimate the discharge values at the 5-, 10-, 20-, 50-, and 100-year return periods. Statistically predicting these rare but extreme flood events is important because they define the design flows where flood risk modelling, infrastructure planning, and economic damage assessments, etc., all rely on the same discharge estimates. The steps of the flood frequency analysis included:

i. Ranking the flood magnitudes from the highest to the lowest.
ii. Plotting positions using the Weibull equation:

$$P = \frac{m}{n + 1}$$

where m is the rank and n is the total number of data points.
iii. Fitting the Gumbel and forecast peak flood associated with return periods of 1, 5, 10, 20, 50, and 100 years.

To develop a clearer understanding of hydrological vulnerability in the Wadi Al Jizzi catchment, the double-ring infiltration test was completed in 30 different representative monitoring locations. From the resulting infiltration rates, it was possible to interpolate these measurements into a spatial infiltration rate, which illustrated the spatial heterogeneity of soil permeability.

Flood Damage Cost Analysis

Flood damage calculations were completed using the Hydrologic Engineering Center's Flood Damage Analysis (HEC-FDA) model and the Flood Economic Impact Analysis System (FEIAS), which includes a total multi-sectoral analysis of flood damages concerning exposure to land use, vulnerability of structures, and depth-based risk indices based on the Wadi Al Jizzi environment.

HEC-FDA has been developed by the U.S. Army Corps of Engineers and recognized as a useful probabilistic method at the international level. HEC-FDA is expected to be an appropriate approach for arid and semi-arid climates and regions, such as Oman, which experiences high incidents of damaging flash floods. HEC-FDA will produce estimates of Expected Annual Damage (EAD) when hydrodynamic modeling outputs are input into the model along with economic exposure information and vulnerability functions. HEC-FDA specifically includes:

- Flood depth grids based on hydraulic models (e.g., HEC-RAS);
- Asset inventories are comprised of key infrastructure, residential buildings, commercial buildings, agriculture, and critical transportation systems;
- Stage-damage curves comprising inundation depth and an associated dollar amount of damage per unit area, for various asset types.

The benefit of HEC-FDA is its focus area of flood depth as the main damage index, which is a factor that is relevant in rapid flooding situations such as Wadi Al Jizzi. In those types of situations, floods are often of very short duration, but the floods develop quickly with high water levels, which can cause substantial, localized damage. HEC-FDA connects spatial flood patterns with asset/infrastructure vulnerability. This allows a defensible, quantitative basis for increasing investments in risk mitigation and enhancing flood resilience planning. The HEC-FDA model produces a flood damage cost estimate that allows the analysis of flood potential impact and future flood scenarios. This analysis incorporates assessments of direct losses (e.g., damage to physical/infrastructure; loss of the use of commercial and residential buildings; loss of agricultural land) as well as indirect losses (e.g., interruption of business, emergency response costs, loss of services). Flood risk is represented using the fundamental risk equation:

$$\mathit{Risk} = \text{Probability of Event} \times \text{Consequence (Damage Cost)}$$

In this context, the Expected Annual Damage (EAD) is the most widely used measure of long-term flood risk because it encompasses the probability and magnitude over time of flooding. This measure allows for scenarios to be compared equally and is intended to support decision-making for flood mitigation and adaptation being planned.

The framework of the Flood Economic Impact Analysis System (FEIAS), developed by Kourgialas and Karatzas (2013), includes a flood loss estimation sub-model that calculates damages across multiple land-use categories, such as residential buildings, agricultural land, roads, and bare soils. The model has a sub-model of the flood economic impact analysis system for flood loss estimation. However, the structure types have been identified for different land use classes (k): residential buildings, farms. In this study, the average cost of building per m^2, (\$1397) = (537.88 OMR) (Arabian Business 2017). The structural damage estimates were based on concrete buildings, the dominant construction type in the region. The total structural damage (Dstr) was computed using the following formulation:

Structure damage (Dstr):

$$\boldsymbol{Dstr} = \sum_{k=1}^{rt} N(K) \times A(K) \times UP(K) \times C(K)$$

where

rt number of structure types based on land use,
N(k) number of structures of type k,
A(k) floor area (m^2) per structure of type k,
UP(k) unit price per m^2 for structure type k,
C(k) depth-damage coefficient for structure type k.

Bare Soil damages (Dbs):

$$\boldsymbol{Dbs} = N \times UP(bs) \times C(bs)$$

Streets damages (Dsd):

$$\boldsymbol{Dsd} = N \times UP(sd) \times C(sd)$$

where N is the total number of units affected, and UP and C represent unit price and depth-damage coefficients for the corresponding land category.

For agricultural losses (Dad), the following equation was applied:

$$\boldsymbol{Dad} = \sum_{k=1}^{n} DUA(K) \times TAC(K) \times CUW(K)$$

GIS-based exposure mapping was used to identify, geographically, those areas within the Wadi Al Jizzi catchment that face the highest risk of flash flooding. The mapping was created using multiple data sources, including high-resolution land use/land cover maps, infrastructure locations, and flood depth grids from hydraulic modeling. Through geospatial overlays and spatial analysis, the study pinpointed areas of highest exposure by considering the location of population centers, roads, and economic assets relative to flood-prone topography. This geospatial mapping provided a more accurate understanding of exposure in Wadi Al Jizzi, delineating priority areas for action and revealing commonalities among exposure zones based on population and asset types and risk levels. The resulting product serves as a baseline for quantifying risk and developing spatially explicit strategies for adaptation and mitigation. The study supports informed, rational, and evidence-based decision-making in flood risk management by utilizing global and national datasets, baseline land-use data, and future scenarios, especially in arid regions like the Wadi Al Jizzi, which are undergoing rapid urbanization.

Adaptation and mitigation planning is the last stage of the flood risk management cycle, focusing on the design, evaluation, and enhancement of existing and proposed flood risk infrastructure. This can include both structural (e.g., levees, retention basins, flood diversion channels, filtration devices, etc.) measures and non-structural (e.g., due diligence early warning systems, land use plans, flood insurance systems, awareness and education campaigns, etc.) measures. It is important to realize this is not an endpoint or final stage; it is part of a cycle and provides a continual feedback loop.

The effectiveness of any proposed measures is re-evaluated with updated hydrological simulation and damage assessment through new scenarios (risk situation). The perpetual risk assessment process has ongoing relevance in our new reality brought on by climate change, urbanization, and socio-economic growth. This stage develops an adaptive management iteration useful to base evidence-based priority setting for investment and to support a long-term strategy for a ubiquitous risk profile like arid flood-prone places typified by Wadi Al Jizzi.

4 Results and Discussion

4.1 Land Cover Type and Soil Infiltration Properties

The hydrological network shows a dense drainage pattern in the basin's upper and central areas, while the main channel carries flow into the coastal zone (Fig. 7). Land cover classification highlights agricultural areas (green) that occur in regions with moderate and low infiltration rates while urban areas (red) predominantly appear along the eastern coastal plain, which only has infiltration rates as low as 5 mm/h, dramatically increasing surface runoff potential.

Infiltration rate is an important hydrologic parameter to measure the rate at which water enters the soil surface; it is expressed in mm/h (millimetres per hour). Infiltration is the key hydrological process in arid environments like Wadi Al Jizzi because it estimates the time needed to generate runoff and provides information on groundwater recharge. Infiltration rates were very different across the catchment and varied spatially due to differences in the texture of soil and land cover or structure, as sandy or gravel soils typically have higher infiltration capacities than fine-textured or compacted soils.

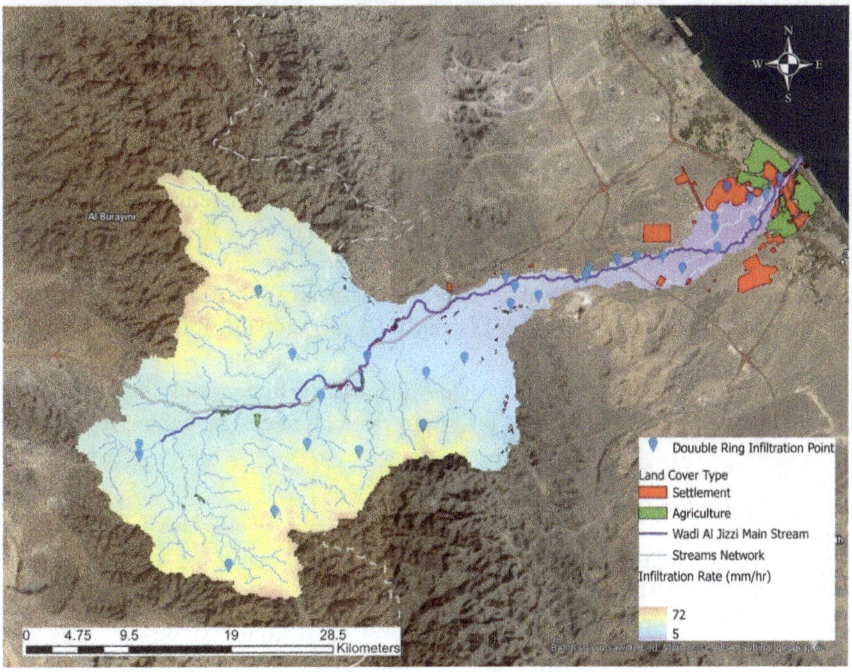

Fig. 7 The spatial distribution of infiltration rates for the Wadi Al Jizzi catchment based on double-ring infiltration tests with a range of 5–72 mm/h

The double ring infiltration tests at 30 different representative locations within the catchment exhibited substantial spatial variability, showing rates of infiltration combined with available water capacity ranging from 5 mm/h in compacted or clay soils (downstream) to as high as 72 mm/h in sandy and, conversely, coarse-textured areas (upstream). The infiltration map shows that the middle and upper areas of the catchment, with their mountainous and gravel areas, usually have moderate to high permeability (generally > 40 mm/h) that provides a groundwater recharging function. In contrast, the low infiltration (< 15 mm/h) and following higher surface runoff tendencies for the surface runoff tendencies and flash flood potential from the vigorous rains of the downstream floodplains and coastal areas of the catchment that include settlements and agricultural lands.

Soils that have an infiltration rate of < 13 mm/h are commonly classified as Hydrologic Soil Group C or D, which represent low permeability, and soils with infiltration rates > 50 mm/h are generally classified as Group A, indicating high permeability and high infiltration capacity (USDA NRCS 2007).

4.2 *Flood Frequency Analysis*

The flood frequency analysis was based on the historical storm discharge data from Wadi Al Jizzi between 1987 and 2007. A flood frequency analysis determined flood discharges for standard return periods (1, 5, 10, 20, 50, and 100 years) using the Gumbel distribution (Table 1). The discharges computed increased non-linearly with return period, consistent with increasingly rare and severe flood events.

The catchment is showing to be particularly susceptible to intense short-duration events, with substantial increases in peak flow with a larger return period. Even a modest storm event, such as it operated as a storm that has a 10-year return period, produces floods that are more than four times the average flow in a typical year. As predicted, the magnitude of flood flows rises non-linearly with decreasing probability of exceedance and reaches a flow of nearly 64 m^3/s for the 100-year flood event (Fig. 8, top panel).

These statistical observations provide quantitative dimensions to flood hazard classification in Wadi Al Jizzi and inform infrastructure design, zoning regulations,

Table 1 The Wadi Al Jizzi estimated peak flood magnitudes for standard return periods based on the frequency analysis of 30 flood events

Return period (years)	Estimated flood (m^3/s)
1	9.36
5	29.5
10	37.6
20	45.7
50	56.0
100	63.9

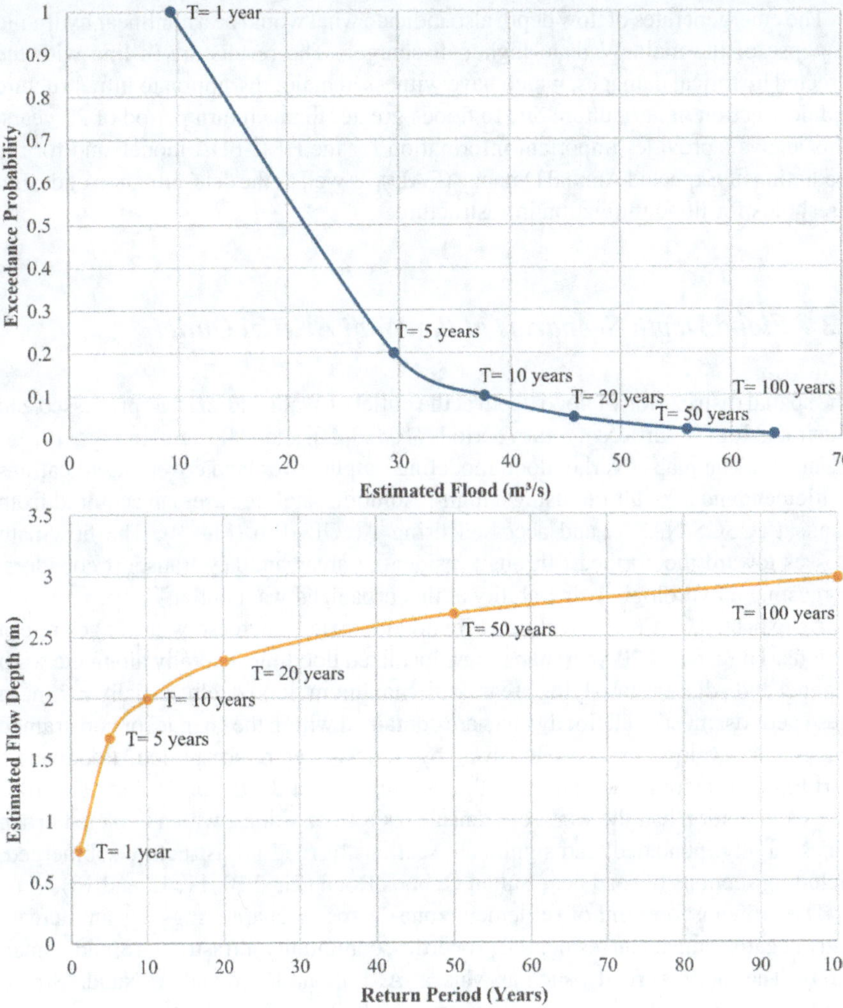

Fig. 8 Analysis of flood frequency and hydraulic response for Wadi Al Jizzi. (Top) Gumbel distribution fit indicating exceedance probability against estimated flood magnitudes for return periods of 1–100 years. (Bottom) Estimated flow depths showing the hydraulic response of the main wadi with increasing flood magnitudes

and early warning systems in arid catchments. The flow depth estimates in the bottom panel of Fig. 8 show rising estimates of flow depth with the return period, ranging from about 0.8 m for a 1-year event to just short of three meters for a 100-year event. These depth estimates are extremely useful for flood risk assessments, especially when estimating the extent of floodplain inundation, and then for designing flood mitigation measures, such as levees, culverts, and infiltration basins.

The emergent rates of flow depth also indicate what would be a nonlinear hydraulic response of the main Wadi to higher discharges. The results are in line with the directed historical damages, which have witnessed major disruption to infrastructure and destruction of agriculture due to floods greater than a return period of 20 years. This analysis provides important information for the HEC-FDA model, and for the prediction of Expected Annual Damage (EAD), as well as the determination of design thresholds for flood mitigation infrastructure.

4.3 Flood Depth Scenarios at the Wadi Al Jizzi Outlet

The spatial distribution of flood depth at the outlet of Wadi Al Jizzi was processed and illustrated for four different return periods (10, 20, 50, and 100 years) in Figs. 10, 11, 12 and 13. The maps overlay flood modeling outputs over land cover classifications (settlement and agriculture) using a high-resolution satellite basemap acquired from Landsat (USGS/NASA) and accessed using ArcGIS Pro (Fig. 9). The flow-path crosses toward the northeast through residential areas and key transport corridors, suggesting a hydrologic vulnerability at this urbanized wadi outlet.

As expected, the flood depth and inundation extent increase with longer return periods. The 10- and 20-year maps show localized flooding generally along the wadi channel and adjacent low-lying areas, with maximum flood depths usually < 2 m. In these periods, most or all floodwaters are contained within the channel or constrained to open/undeveloped space. However, by the 50-year return period, floodwaters start to reach more settled areas, with maximum flood depths of 2.75 m. The 100-year examination had the worst circumstances for conditions, which depicted urban zones largely inundated, and significant sections of road infrastructure submerged, including sections of road near Sultan Qaboos Road (Figs. 10, 11, 12 and 13).

The growing footprint of settlement zones across scenarios suggests an increasingly negative impact on residential properties, community infrastructure, and human safety. The fact that roads, such as Muscat Al Batinah Road and Al Nahda Street, can be rendered useless in the event of extreme floods, especially during an extreme weather event. The scale of inundation for the agricultural areas to the west and southeast begins to exhibit flood implications, even beyond the intended design 20 return period. These impacts should also be viewed in the short-term provisioning of local food production needs, local irrigation practices, and associated local commercial needs.

The planning and development process should consider road intersections and neighborhoods that constantly fall within high-depth flood zones. Noted interventions lasting with some structural combinations of detention basins, culverts, and drain dishes, specifically installations as part of future urbanization. Non-structural interventions should include early warning systems and zoning approaches, which can assist urban areas in adapting to ongoing flooding incidents and possible risks of climate change in this rapidly urbanizing region of Sohar.

Fig. 9 Satellite imagery (2024) captures the Wadi Al Jizzi outlet area in Sohar, Oman, with the direction of surface water flow. This area has been assessed to estimate flood depths across various return periods, given the potential for flood events to occur regularly within this likely flooding area. The satellite imagery, obtained using ArcGIS Pro, is sourced from Landsat (USGS/NASA) and depicts a recent observation of the landscape (5 m resolution)

4.4 Expected Annual Damage Estimation Using HEC-FDA Principles

Based on the HEC-FDA (Flood Damage Reduction Analysis) and Flood Economic Impact Analysis System (FEIAS) approaches, the flood damage costs were estimated for different scenarios. The analysis included Gumbel-distributed peak discharge and flood depths across various return periods to characterize the hazards in the Wadi Al Jizzi catchment. A synthetic determination of depth-damage, produced by flood depth-damage curves, was developed based on standard engineering assumptions and then scaled to the local building costs in OMR. The depth-damage curve estimates the economic consequences of flooding residential structures, correlating the floodwater depth with a percentage of value loss of the structure. This curve was applied over the flood profile to estimate losses per structure in Wadi Al Jizzi for each scenario. The damage-exceedance probability was determined by linking expected damages to the probabilities of each flood event. The exceedance probability curve was integrated numerically using Simpson's rule to derive an Annual Expected Damage (EAD) function specific to residential structures, amounting to approximately 29,121 OMR

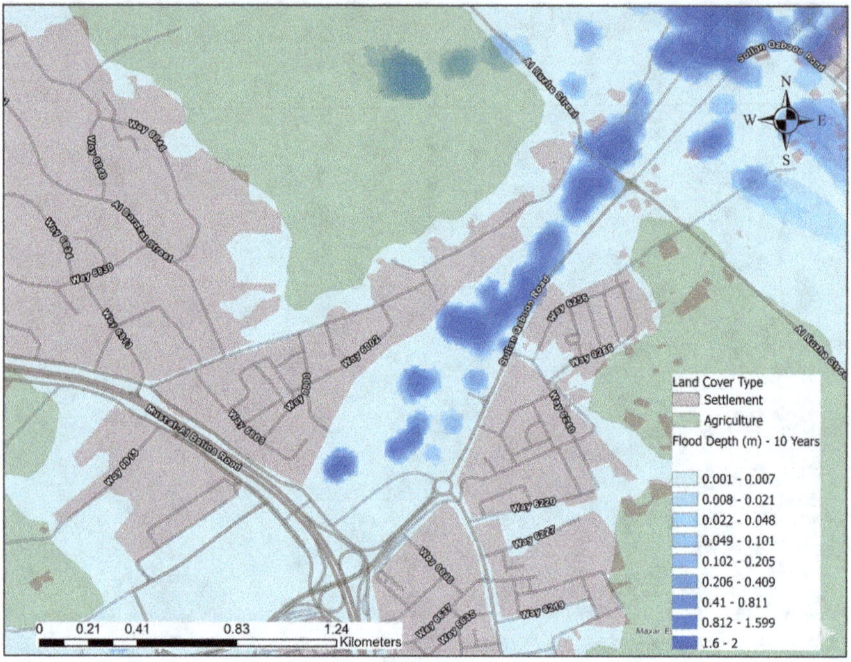

Fig. 10 Flood depth map for the Wadi Al Jizzi outlet area in Sohar, Oman, under a 10-year return period scenario. The flood depth is between 0.001 and 2.0 m

per year. The EAD is a key metric within the HEC-FDA framework, providing an estimate of the average annual loss per structure and supporting the economic evaluation of flood mitigation options. This value assists in comparing different intervention strategies, such as flood protection structures or land-use controls. Results were visually summarized in Fig. 14, which shows the combined relationship between flood depth, estimated structural damage cost, and exceedance probability. This dual-axis illustrates depth, probability, and consequences (economic loss).

The damage cost is increasing with rising flood depths, reflecting greater exposure and vulnerability. Extreme depths (exceeding 3 m) tend to have return periods of over 50 years. The dual-axis presentation exemplifies a fundamental aspect of flood risk where rarer events tend to cause more damage but contribute less to annual risk.

The comparative analysis of flood damage percentages as a function of water depth between the HEC-FDA model and the FEIAS model suggests some similarity, but also some inconsistency between the two models across the water depth intervals (Fig. 15). For shallow water 0.1–0.5 m, the FEIAS model estimated higher damage percentages than the HEC-FDA model, with the differences being as substantial as 1.5–3%. This suggests initial flood exposure may be more sensitive through FEIAS, possibly associated with localized vulnerability assumptions or a more assertive attribution of cost for minor inundation.

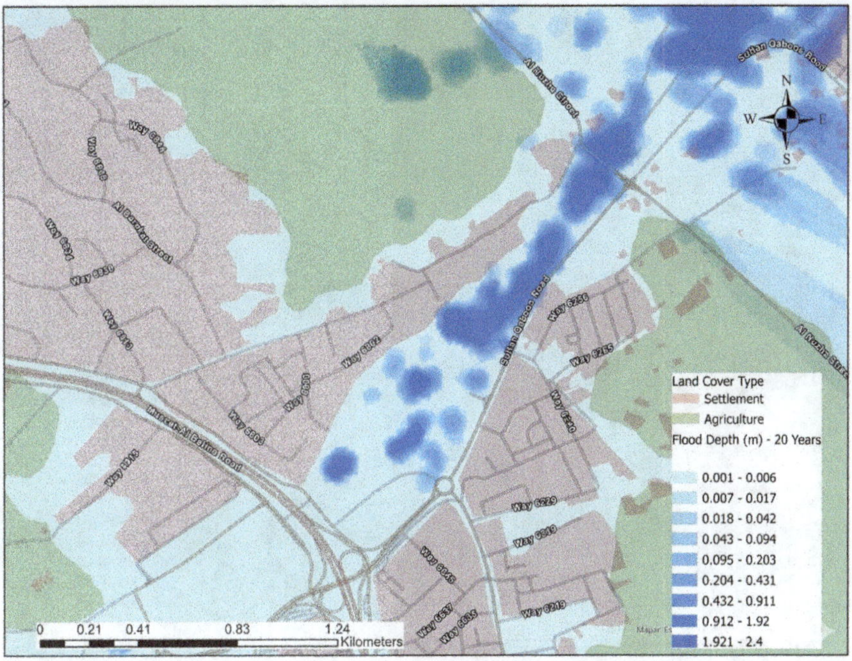

Fig. 11 Flood depth map for the Wadi Al Jizzi outlet area in Sohar, Oman, for a 20-year return period event. Maximum flood depth reaches as high as 2.4 m in terms of impact zones, which are expanding relative to the ten-year scenario

HEC-FDA and FEIAS are calibrated at 50% damage at the midpoint at 1.5 m. After 1.5 m, HEC-FDA provides increasing estimates of damage at higher damage percentages than FEIAS, at 3.0 m damage to 98.9% compared to 97.7% damage level for FEIAS. HEC-FDA shows a sharper slope beyond 1.5 m, causing an increased establishment of observed correlation between flood depth and total loss, which arguably is associated with the fact that HEC-FDA leverages depth-damage curves that were calibrated based on structural collapse threshold values of the building and service interruption levels.

The difference highlights how both models produce similar trends, while FEIAS tends to be more reactive at low levels of inundation, and HEC-FDA demonstrates an accelerating increase in damage at higher depths. These differences, which are significant for decision-makers, can better inform the appropriate choice of methodologies when conducting flood risk assessments at the regional scale, especially related to prioritizing mitigation investments, while knowing how to scale depth based on severity.

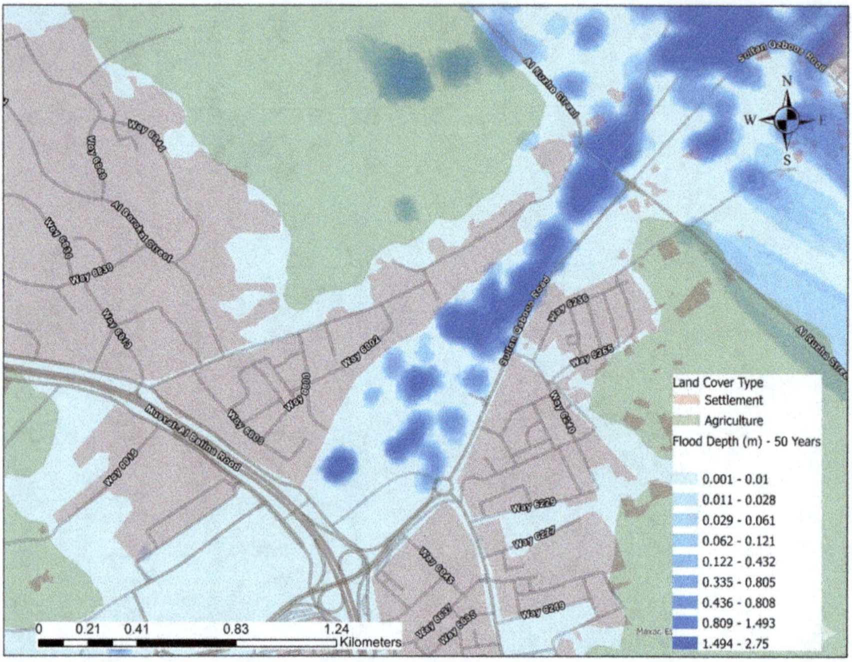

Fig. 12 Flood depth map with a 50-year return period at the outlet of Wadi Al Jizzi in Sohar. For this scenario, there is a broader inundation with deeper flood zones, particularly along key roads and housing areas. Depths range from 0.001 to 2.75 m

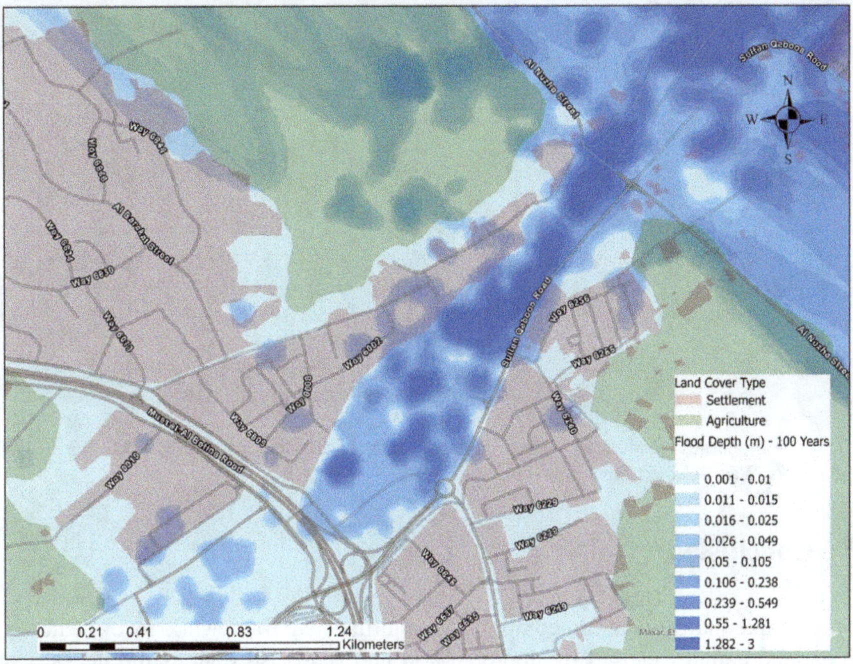

Fig. 13 Flood depth map for 100-year return period model, depicting the worst-case flooding conditions at the outlet of Wadi Al Jizzi. Floodwaters spread extensively into settlements and infrastructure, and depths reached up to 3 m

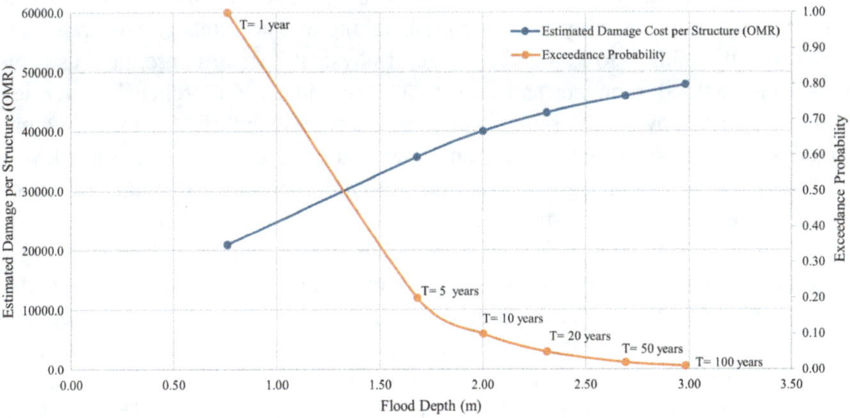

Fig. 14 Relationship between flood depth, estimated structural damage (OMR), and exceedance probability for Wadi Al Jizzi. The blue curve depicts the estimated damage per residential structure based on interpolated depth-damage functions, and the orange curve depicts annual exceedance probabilities of flood depths

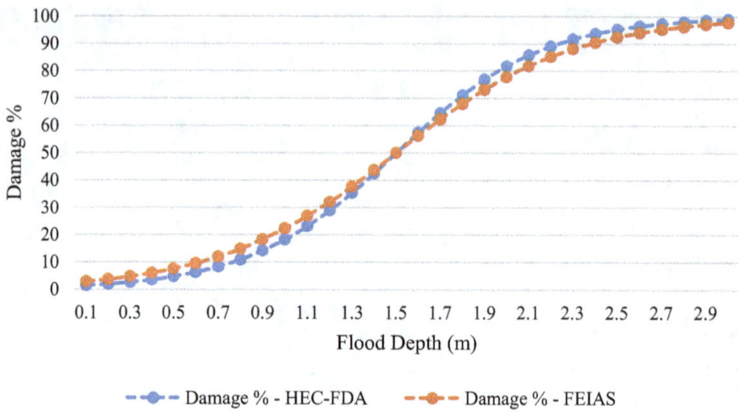

Fig. 15 Comparative analysis of flood damage % curves derived from HEC-FDA and FEIAS models for Wadi Al Jizzi catchment

5 Conclusion

This chapter laid out an encompassing, multifaceted framework to evaluate the economic damage caused by floods in arid and semi-arid regions through a case study of the Wadi Al Jizzi catchment in Sohar, Oman. The application of the Flood Economic Impact Analysis System (FEIAS) model and the HEC-FDA (Hydrologic Engineering Center-Flood Damage Analysis) model enabled the analysis of flood losses in both deterministic and probabilistic analyses. The joint application provided a more comprehensive analysis of flood risk, taking into account spatial exposure to flooding, vulnerability to loss and damage of physical infrastructure, land use, and uncertainty in flood frequency and severity. This provided a comprehensive overview of the flood risks by summarizing spatial exposure, vulnerability of infrastructure, land dynamics, along with the uncertainty of flood frequency and magnitude levels.

The FEIAS model brought together the determination of the damages by sector, by relevance to the kinds of structures, land use designation, and local construction cost criteria. The model allowed for the estimate that residential structures and agricultural land were most vulnerable, with flood event damages estimated at residential damages alone of more than 2.15 million OMR, along with aggregated damage for all sectors equating to greater than 2.6 million OMR. In contrast, the HEC-FDA model was able to provide a statistical measure of the long-term flood risk by computing the Expected Annual Damage (EAD) for different return periods. This proposed EAD per year from residential structures was 29,121 OMR, an important measure to strongly justify investment in consideration of flood impacts and damages.

The comparative assessment of the depth-damage functions illustrated in the FEIAS and HEC-FDA analysis also provided insights into how different modeling assumptions, particularly related to flood risk. FEIAS appears to have a much greater sensitivity to shallow flood depths and may result in an overestimate of losses for low levels of flooding. FEIA also overshot minor inundation for deep flood depths, while HEC-FDA provided a clearer depiction of not only loss escalation, but also did a much better job at depicting non-linear structural collapse thresholds and services continued operating interruptions, especially if the loss led to business interruption. The characterization and presentation of various depth-damage functions demonstrate the importance of model calibrations, aggregate loss sensitivities, and adjustments for context when applying flood modeling in planning decisions.

In the foreseeable future, the results of this study highlight the essential need to consider integrated flood risk management contextually in places like, Sohar, where urbanizing arid conditions continue to occur, and in considering where climate is variable—movement towards more extreme weather patterns and where burgeoning development persists along ways and coastal floodplains is going to complicate both the extent and frequency of flash floods regardless of structural solutions implemented.

Flood mitigation alternatives are going to require a combination of non-structural flood mitigation, i.e., land-use planning, land-use regulation ordinance, storm-water regulation, and incorporation of early warning systems that can include warnings and continued hazard awareness; and structural solutions like diversion channels, retention basins, and more funding and resources to put infrastructure in place that could help drain flood waters off the city expeditiously. Also, the methods introduced in this study can be repeated and used as knowledge in high-resolution planning for other arid settings and can be used, generally speaking, as a best practice decision-support approach or any policymakers, planners, or disaster risk managers concerned with improving climate resiliency and economic preparedness and minimize risk to flooding most vulnerable catchments.

Acknowledgements The author would like to thank Trinity College Dublin as well as the Ministry of Agriculture, Fisheries, and Water Resources in Oman for providing hydrological data and access to the sites studied. In addition, the author would like to thank the local authorities in Sohar for their cooperation during field visits.

Competing Interests The author declares that there are no competing financial or non-financial interests in relation to the content of this chapter. This research did not receive any specific grant from any funding agency in the public, commercial, or not-for-profit sectors.

Ethics Approval This chapter is based purely on secondary data sources and field data collection. No human or animal participants were involved, and as such, no ethics approval was required.

References

Abushandi E, Al Ajmi M (2022) Assessment of hydrological extremes for arid catchments: a case study in Wadi Al Jizzi, North-West Oman. Sustainability. https://doi.org/10.3390/su142114028

Abushandi E, Al Sarihi M (2022) Flash flood inundation assessment for an arid catchment, case study at Wadi Al Jizzi, Oman. Water Pract Technol. https://doi.org/10.2166/wpt.2022.043

Abushandi E, Al Sarihi M, Ibrahim O (2023) Assessing flash flood inundation from an extreme rainfall event: case study: Wadi Al Jizzi Oman. AQUA Water Infrastruct Ecosyst Soc 72(9):1770–1783. https://doi.org/10.2166/aqua.2023.100

Apel H et al (2004) Flood risk assessment and associated uncertainty. Nat Hazards Earth Syst Sci 4(2):295–308. https://doi.org/10.5194/nhess-4-295-2004

Arabian Business (2017) Revealed: cost of construction in Gulf countries. Arabian Business, ITP Media Group. Available at: https://www.arabianbusiness.com/gcc/revealed-cost-of-construction-in-gulf-countries-675033. Accessed 20 Mar 2025

de Moel H, Aerts J (2011) Effect of uncertainty in land use, damage models and inundation depth on flood damage estimates. Nat Hazards 58(1):407–425

Jongman B, Ward P (2012) Global exposure to river and coastal flooding: long term trends and changes. Glob Environ Change 22:823–835. https://doi.org/10.1016/j.gloenvcha.2012.07.004

Kourgialas NN, Karatzas GP (2013) A hydro-economic modelling framework for flood damage estimation and the role of riparian vegetation. Hydrol Process 27(4):515–531

Kreibich H, Thieken AH (2009) Coping with floods in the city of Dresden, Germany. Nat Hazards 51:423–436

Merz B et al (2010) Review article "Assessment of economic flood damage." Nat Hazards Earth Syst Sci 10(8):1697–1724. https://doi.org/10.5194/nhess-10-1697-2010

Messner F et al (2007) Evaluating flood damage: guidance and recommendations on principles and methods. FLOOD site project report

Mohammadi SA, Nazariha M, Mehrdadi N (2014) Flood damage estimate (quantity), using HEC-FDA model. Case study: the Neka river. Procedia Eng 70:1173–1182

Nghia BPQ et al (2022) Applying Google Earth Engine for flood mapping and monitoring in the downstream provinces of Mekong river. Prog Disaster Sci 14:100235. https://doi.org/10.1016/j.pdisas.2022.100235

Nogherotto R et al (2019) An integrated hydrological and hydraulic modelling approach for the flood risk assessment over Po river basin. https://doi.org/10.5194/nhess-2019-356

Penning-Rowsell EC, Chatterton JB (1978) The benefits of flood alleviation. A manual of assessment techniques

Pistrika AK, Jonkman SN (2010) Damage to residential buildings due to flooding of New Orleans after hurricane Katrina. Nat Hazards 54:413–434

Smith DI (1994) Flood damage estimation—a review of urban stage-damage curves and loss functions. Water Sa 20(3):231–238

Thieken AH et al (2005) Flood damage and influencing factors: new insights from the August 2002 flood in Germany. Water Resour Res 41(12). https://doi.org/10.1029/2005WR004177

USDA NRCS (2007) Hydrologic soil groups, chap 7. In: National engineering handbook: part 630—hydrology. United States Department of Agriculture, Natural Resources Conservation Service, Washington, DC. Available at: https://directives.nrcs.usda.gov/sites/default/files2/1712930597/11905.pdf

Ward PJ et al (2017) A global framework for future costs and benefits of river-flood protection in urban areas. Nat Clim Change 7(9):642–646

Analyzing Maximum Rainfall Using Remote Sensing Techniques and Gumbel Distribution for Flood Hazard Assessment in the Makkah Region, Saudi Arabia

Tayfun Kurt

1 Introduction

Floods pose significant threats to urban infrastructure, public safety, and economic stability, particularly in regions experiencing extreme climatic variations. While floods are often associated with humid and tropical environments, arid and semi-arid regions are increasingly recognized as being highly vulnerable to flash floods due to their limited soil infiltration capacity, steep terrain, and lack of permanent water channels (Cordery and Pilgrim 2000; Alfieri et al. 2015). Makkah, one of the holiest cities in Islam, presents a paradox in this regard: despite its location in a predominantly arid climate, the city frequently experiences intense and irregular rainfall events that lead to flash floods (Dawod et al. 2011; Alqahtani and Qaddah 2019; Youssef et al. 2016). The combination of rapid urban expansion, limited natural drainage systems, and topographic constraints exacerbates this vulnerability, making flood risk management a critical challenge for urban planners and policymakers.

Historically, Makkah has witnessed several severe flood events, with notable occurrences in 1967, 2009, and more recently in 2024 and 2025, causing significant disruptions to urban life and highlighting deficiencies in existing flood mitigation measures (El-Saoud and Othman 2022; Bashir and Alsalman 2024; Abdelkarim and Gaber 2019). These events emphasize the necessity of a scientifically robust and comprehensive flood risk assessment framework tailored to the region's unique hydrological and geomorphological characteristics.

Traditional flood risk assessment methodologies often rely on historical records and empirical observations; however, such approaches may not fully account for the increasing variability in precipitation patterns due to climate change (Alfieri et al.

T. Kurt (✉)
Independent Researcher, Istanbul, Türkiye
e-mail: tyfnkrt@hotmail.com

M. M. Al Saud (ed.), *Hydroclimate Changes: Implications of the Middle East Region*,
https://doi.org/10.1007/978-3-032-20053-2_13

2017; Patlakas et al. 2021). Several studies have investigated flood hazards in arid regions, focusing on either extreme rainfall estimation (Kurt 2025; Koutsoyiannis 2003), GIS-based flood hazard mapping (Blakime et al. 2024), or hydrodynamic modeling (Shaikh et al. 2023; Marko et al. 2019). However, there remains a methodological gap in the integration of statistical hydrology, spatial analysis, and hydraulic modeling within a unified framework. To address this gap, this study proposes a novel flood risk assessment approach for Makkah that combines Gumbel distribution for extreme rainfall analysis, QGIS for geospatial data visualization.

The Gumbel distribution is widely recognized for its applicability in extreme value analysis and has been successfully used in hydrological studies to model extreme precipitation events (Gumbel 1958; Coles et al. 2001). QGIS, as an open-source GIS platform, enables high-resolution spatial analysis, enhancing the accuracy of flood hazard mapping (Rincón et al. 2018). By integrating these methodologies, this study aims to bridge the gap between statistical hydrology, geospatial analysis, and hydrodynamic modeling, offering an interdisciplinary and data-driven approach to flood risk assessment.

The primary objectives of this research are as follows:

To identify high-risk flood-prone areas in Makkah using GIS-based spatial analysis.

To estimate probable maximum precipitation (PMP) events through extreme value analysis.

Beyond its immediate implications for Makkah, this research introduces a scalable and adaptable methodology for flood hazard assessment in arid and semi-arid regions. By integrating geospatial analysis, and statistical techniques, the study provides a data-driven framework that enhances flood risk management and resilience planning. The findings will serve as a valuable resource for urban planners, disaster management agencies, and policymakers, enabling the development of evidence-based flood mitigation strategies. Furthermore, strengthening flood management through innovative technologies and nature-based adaptation solutions will be crucial in building climate-resilient cities. Ultimately, this research contributes to safeguarding both the residents of Makkah and the millions of visitors the city receives annually, ensuring a safer and more sustainable urban environment.

2 Materials

2.1 Investigation Area

Geographically, Makkah is situated at coordinates 21.4225° N latitude and 39.8262° E longitude and is characterized by a hot, arid climate (Fig. 1). The city experiences high temperatures, low rainfall, and steep mountainous terrain. The study area encompasses approximately 27,900 ha, with an average slope of 16°

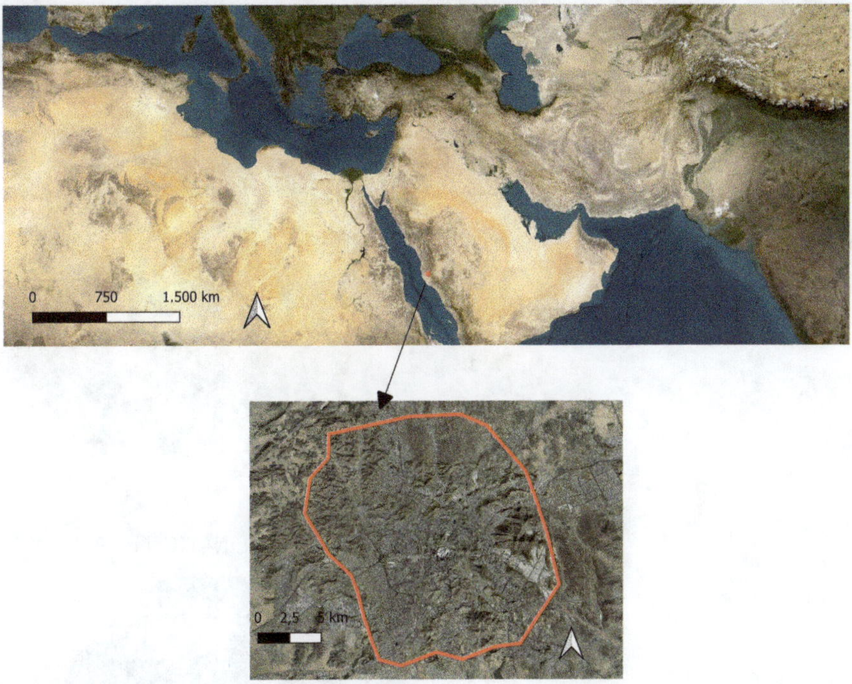

Fig. 1 Overview of location of investigation area

and predominantly south-facing aspects (Fig. 2). Makkah's proximity to the Red Sea (approximately 70 km to the west) occasionally influences local climatic conditions by introducing moisture, though overall precipitation remains limited (Shahin 2007; Subyani and Hajjar 2016).

The topography of Makkah is defined by mountains and valleys, which play a critical role in the region's hydrological dynamics (Fig. 3). The city sits at elevations ranging from 200 to 1300 m above sea level, with seasonal riverbeds (wadis) serving as natural drainage channels. Rapid urban expansion and infrastructure development have altered these natural flow patterns, increasing surface runoff and flood risk (Güneralp et al. 2015). The influx of visitors during the Hajj season exacerbates logistical challenges in flood management.

Makkah, located in western Saudi Arabia, holds profound religious significance as the holiest city in Islam (Fig. 1). Annually, millions of pilgrims visit for Hajj (mandatory) and Umrah (voluntary), making it a global center of faith (Dalgın and Civelek 2023). Religious tourism generates ~ 300 million global travelers yearly, with Hajj alone attracting ~ 2 million pilgrims simultaneously (FaithTourism Summit, cited in Dalgın and Civelek 2023). This influx strains infrastructure, particularly during extreme weather events (Fig. 4).

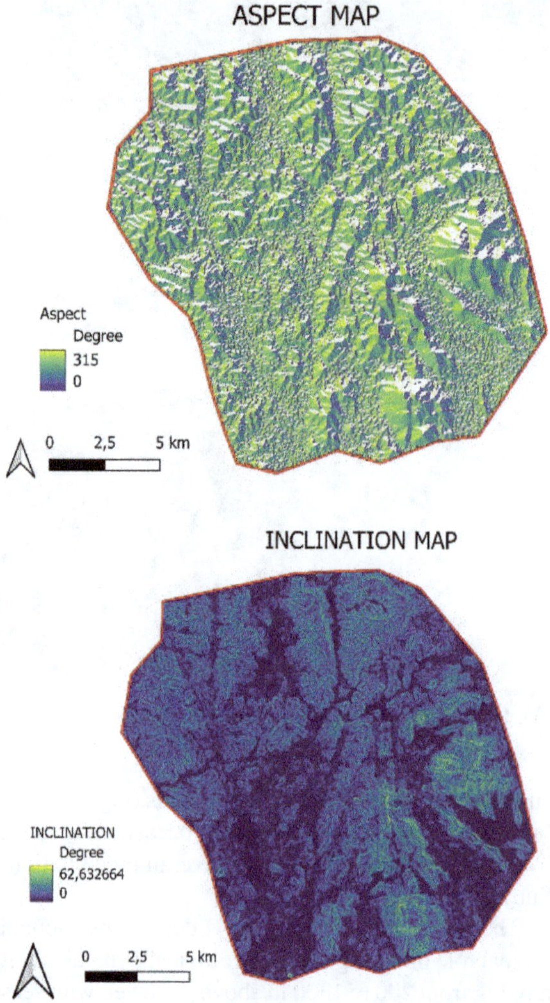

Fig. 2 Aspect and inclination map of the investigation area

2.2 *Chronology of Investigation Area*

Historical flood events, such as the one in 1288, have directly influenced the development of contemporary flood risk management strategies in Makkah (Table 1). The lessons learned from these past occurrences have led to innovations in infrastructure design, emergency protocols, and flood prevention techniques. However, despite these advances, the city's infrastructure still faces significant pressure, underscoring the need for continued improvement and adaptation to emerging challenges.

In response to these challenges, there is a growing recognition of the importance of improving flood management systems and increasing the city's capacity to handle extreme weather events. Both historical events and recent trends indicate that the

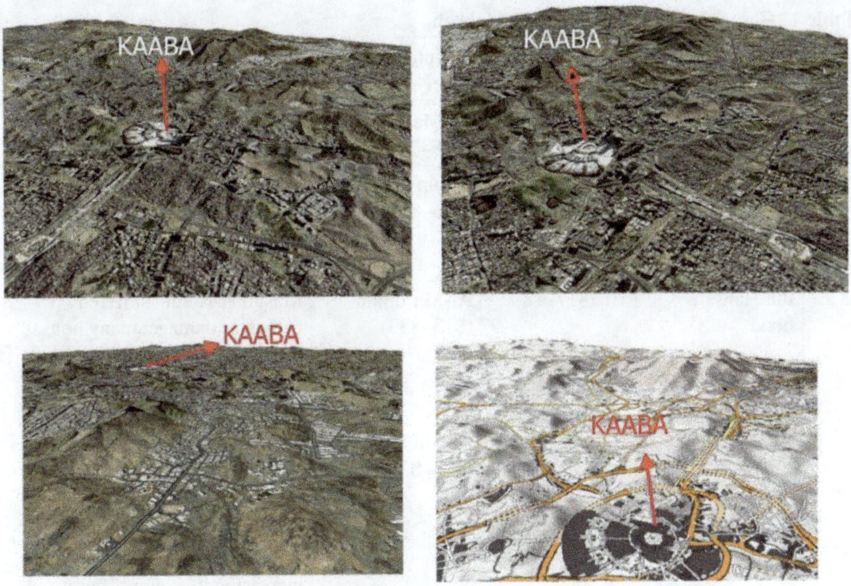

Fig. 3 An overview of the topography of Makkah

Fig. 4 Ovirwiew of Mekkah. *Photos* Under license the Envato

Table 1 Chronology of flood events in Makkah

No.	Flood name	Date (Hijri/ Gregorian)	Caliph of the time	Impact and consequences
1	Sayl al-Labīrī	160/776–77	Al-Mahdī (775–785)	Reached the Masjid al-Haram, no significant damage recorded
2	Sayl al-Mukhabbil	184/800–801	Hārūn al-Rashīd (786–809)	The Makkah Valley was flooded, many people and belongings were swept away. The caliph sent funds for repairs
3	Ibn Ḥanṣala flood	202/817–18	Al-Ma'ūmūn (813–833)	Masjid al-Haram suffered extensive damage, many houses were destroyed, and there were casualties. The caliph provided significant aid
4	–	208/823–24	Al-Ma'ūmūn (813–833)	Recorded as larger than the previous flood. The Black Stone was nearly dislodged. Extensive mud and debris accumulated
5	–	225/839	–	Mixed with Zamzam water, increasing its volume, but no significant damage was reported
6	–	253/867	–	Reached Masjid al-Haram, some houses collapsed, but no casualties were recorded
7	–	262/875	–	Dragged the pebbles on the floor of Masjid al-Haram, but no major damage occurred
8	–	263/876	–	Damaged Masjid al-Haram, but detailed information is unavailable
9	–	279/892	–	Masjid al-Haram was completely submerged, the Zamzam well overflowed, and its water became sweeter
10	–	280/893	–	Increased the volume and sweetness of Zamzam water, no damage was recorded
11	–	528/1134	–	Heavy rainfall lasted for seven days, but details are not specified
12	First major flood	651 (1253)	–	Rainfall relieved the people from a severe drought, reducing high food prices
13	Second flood	13 Sha'ban 659 (July 1261)	–	Floodwaters reached the Kaaba

(continued)

Table 1 (continued)

No.	Flood name	Date (Hijri/ Gregorian)	Caliph of the time	Impact and consequences
14	Third major flood	14 Sha'ban 669 (March 28, 1271)	–	The Kaaba was completely submerged, the pulpit was swept away, lamps dislodged, and only those who could swim were able to perform the Tawaf. The Black Stone was also submerged
15	–	670 (1271–1272)	–	Following the flood, diseases spread, and the death toll increased. During January–February 1273, 1000 funeral prayers were held
16	The largest flood in Makkah's history	10 Dhul-Qi'dah 687 (December 1288)	–	Floodwaters reached the Kaaba, Masjid al-Haram remained submerged for four days, and Friday prayers could not be held. This event is recorded as the largest flood in Makkah's history
17	First flood event	26 Dhul-Hijjah 730 (October 10, 1330)	–	Despite the absence of rainfall, waters from the city's rivers reached Masjid al-Haram, and the floodwaters did not recede for two days. Extensive cleaning efforts were required, and farmlands were damaged
18	Thunderstorms and lightning caused floods	29 Dhul-Hijjah 732 (September 21, 1332)	–	Lightning struck the Mount Abu Qubays and Masjid al-Hayf, killing two people. The rain was not heavy enough to cause significant damage
19	The largest flood catastrophe	15 Jumada al-Akhirah 738 (January 8, 1338)	–	Following thunderstorms and heavy rain, floodwaters entered Masjid al-Haram, submerging columns and damaging the ceiling. Sacred texts were drenched, rendering them unusable
20	Flood in the Ibrahim Valley	19 Rajab 747 (November 5, 1346)	–	Floodwaters entered Masjid al-Haram, and the Tawaf area was covered by water. The Ibrahim Gate was displaced, and the floodwaters reached the point where pilgrims had to swim
21	Thunderstorms and rain caused floods	750 (1349–1350)	–	Masjid al-Haram was submerged once again, with significant damage

(continued)

Table 1 (continued)

No.	Flood name	Date (Hijri/ Gregorian)	Caliph of the time	Impact and consequences
22	21 years later, another flood	771 (1369–1370)	–	Floodwaters rose to the Kaaba's door, and 40 camels and many livestock perished. Cold weather followed, making survival difficult
23	Egyptian pilgrims caught in a large flood during their journey	776 (1375)	–	Pilgrims traveling from Egypt were caught in a flood at Huleys, and their belongings were swept away. The same group was also caught in a storm on their return journey
24	Major flood at Sūgrat al-Ḥāmid during the Egyptian pilgrims' return journey	9 Muharram 790 (January 19, 1388)	–	Egyptian pilgrims were caught in a major flood on their return journey, losing their belongings
25	A large flood in the Ibrahim and Ecyad Valleys	10 Jamadi al-Akhir 802 (January 8, 1400)	–	Floodwaters reached the Masjid al-Haram gates, submerging the Kaaba's door, and the Acele Gate's columns were destroyed. Pilgrims had to wait for two days before they could perform the Tawaf, which was done under difficult conditions due to mud
26	Floodwaters reached Masjid al-Haram	22 Dhul-Hijjah 814 (April 5, 1412)	–	The heavy rains and floods increased wheat prices, leading to economic hardship
27	Rain and high floodwaters	23 Dhul-Hijjah 825 (December 8, 1422)	–	Floodwaters entered Masjid al-Haram for four days, reaching the Kaaba's door
28	Floods in Ibrahim and Ecyad Valleys	3 Jamadi al-Akhir 827 (April 3, 1424)	–	No significant damage occurred, but transportation difficulties persisted
29	Rain and high floodwaters in Ibrahim and Ecyad Valleys	14 Jamadi al-Akhir 863 (March 19, 1459)	–	No damage occurred, but wheat prices increased
30	Two separate floods in the Ibrahim Valley	865 (1460–1461)	–	The interior of the Kaaba, the ceiling, and the Zamzam well suffered some damage

(continued)

Table 1 (continued)

No.	Flood name	Date (Hijri/ Gregorian)	Caliph of the time	Impact and consequences
31	Thunderstorms and floods from the Ibrahim Valley	28 Rabī' al-Awwal 887 (December 26, 1487)	–	Hundreds of Makkahns and pilgrims lost their lives, and destroyed homes and infrastructure caused significant material losses
32	Heavy rain led to floods entering Masjid al-Haram	5 Safar 895 (December 29, 1489)	–	Rain and floods caused the deaths of many pilgrims
33	Floodwaters from the mountains combined and entered Masjid al-Haram	4 Rabī' al-Awwal 900 (December 3, 1494)	–	The Kaaba's door threshold and the Minbar of Masjid al-Haram suffered damage
34	Floodwaters from the Ibrahim Valley entered Masjid al-Haram	16 Rabī' al-Awwal 901 (December 4, 1495)	–	Candles were submerged in the Tawaf area, and material damage occurred
35	Floodwaters from the Ibrahim Valley entered Masjid al-Haram	19 Ramadan 913 (January 22, 1508)	–	No significant damage, but the floodwaters reached the Tawaf area
36	Rain and floodwaters from the southern, eastern, and western gates of Masjid al-Haram	10 Safar 920 (April 6, 1514)	–	Floods caused damage to candles in the Tawaf area and the Zamzam well, though there was no loss of life

Adapted by Erol (2020)

current infrastructure may be insufficient to prevent severe disruptions. Consequently, there is a critical need to develop and implement more comprehensive and adaptive measures to safeguard the city's infrastructure, residents, and visitors from future flooding risks.

Table 1 synthesizes 36 major flood events from 776 to 1514 CE, reconstructed from Ottoman archives, Arabic chronicles (e.g., al-Azraqī's *Akhbār Makka*), and peer-reviewed historiography (Ayalon 2015). Key events include:

Sayl al-Mukhabbil (800–801 CE): Swept away pilgrims; Caliph Hārūn al-Rashīd funded repairs (60,000 dinars).

Kaaba Submersion (1271 CE): Pilgrims swam to perform Tawaf; post-flood epidemics increased fatalities.

Table 2 Thematic analysis of big floods

No.	Flood name	Date (Hijri/ Gregorian)	Impact summary	Political response
2	Sayl al-Mukhabbil	184/800–801	Floodwaters swept away pilgrims and infrastructure; Caliph Hārūn al-Rashīd allocated 60,000 dinars for repairs	Centralized funding for reconstruction
14	Submersion of Kaaba	669/1271	The Kaaba was completely submerged; pilgrims swam to perform Tawaf. The death toll was exacerbated by post-flood epidemics	No recorded state intervention
16	Largest flood	687/1288	Floodwaters reached the Kaaba and Masjid al-Haram remained submerged for 4 days, leading to the suspension of Friday prayers. Marked a turning point in flood management	Institutionalization of seasonal patrols
31	Ibrahim Valley flood	887/1487	Hundreds of Makkahns and pilgrims lost their lives; extensive infrastructural collapse occurred, coinciding with political instability during the Mamluk period	Reliance on local religious endowments

Adapted by Erol (2020)

Largest Flood (1288 CE): Masjid al-Haram submerged for 4 days, suspending Friday prayers (Table 2).

In addition to historical flood events, recent flash floods in Makkah have further emphasized the urgent need for advanced hydrological analysis and flood risk management strategies. The most notable modern flood events, particularly those occurring in the last two decades, should be integrated into risk assessments for a comprehensive understanding of evolving flood hazards.

Thematic Analysis: Floods as Drivers of Institutional Change

Flood events in Makkah have not only altered its physical landscape but also driven significant institutional and socio-political changes (Table 2).

2.3 *Rainfall Data*

Makkah experiences an average annual precipitation of approximately 189. 1 mm, with the majority of rainfall occurring between November and January (Khan and

Alghafari 2018). These rainfall events are typically brief yet intense, often triggering flash floods due to the city's impermeable soil, sparse vegetation, and steep topography (Bashir and Alsalman 2024). Given Makkah's global religious significance, any extreme weather event, particularly during the Hajj season, poses substantial risks to both pilgrims and local residents.

To analyze historical rainfall trends, this study utilizes precipitation data from the Climate Hazards Group InfraRed Precipitation with Station Data (CHIRPS) dataset, a globally recognized source for high-resolution precipitation estimates. CHIRPS integrates satellite imagery with ground-based meteorological station data, offering daily rainfall records from 1981 onward. For the Makkah region, CHIRPS provides a spatial resolution of 0.05° (approximately 5 km) and enables accurate estimation of annual maximum rainfall values, covering the period from 1985 to 2024.

The data extraction and analysis were conducted using Google Earth Engine (GEE), a powerful cloud-based geospatial analysis platform that facilitates large-scale climate data processing (Fig. 5). GEE allows for seamless integration of CHIRPS data, enabling efficient temporal and spatial analysis of precipitation trends. While CHIRPS has proven reliable for large-scale hydrological assessments, its accuracy may vary in regions with sparse ground stations or pronounced microclimatic differences (Sa'adi et al. 2023). Despite these limitations, CHIRPS remains a key resource for precipitation modeling, drought monitoring, and flood risk assessment in Makkah, where localized extreme rainfall events frequently occur (Table 3).

2.4 Digital Elevation Model (DEM)

A high-resolution Digital Elevation Model (DEM) with a 12.5-m resolution was utilized to enhance the accuracy of topographic and hydrological analyses. The dataset, derived from the Advanced Land Observing Satellite Phased Array type L-band Synthetic Aperture Radar (ALOS PALSAR) provided by the Japan Aerospace Exploration Agency (JAXA), enables precise representation of Makkah's elevation profile and drainage systems.

ALOS PALSAR DEM is widely recognized for its capability to identify fine topographic features in mountainous regions (Nitheshnirmal et al. 2019; Habib et al. 2017). Given Makkah's complex terrain, this dataset was crucial for delineating watershed boundaries, modeling surface runoff, and identifying flood-prone areas. Figure 6 provides an overview of the DEM used in this study.

2.5 QGIS Software and Hydrological Analysis

Quantum Geographic Information System (QGIS) version 3.34 was employed for spatial analysis and geospatial data visualization. QGIS facilitated the integration of

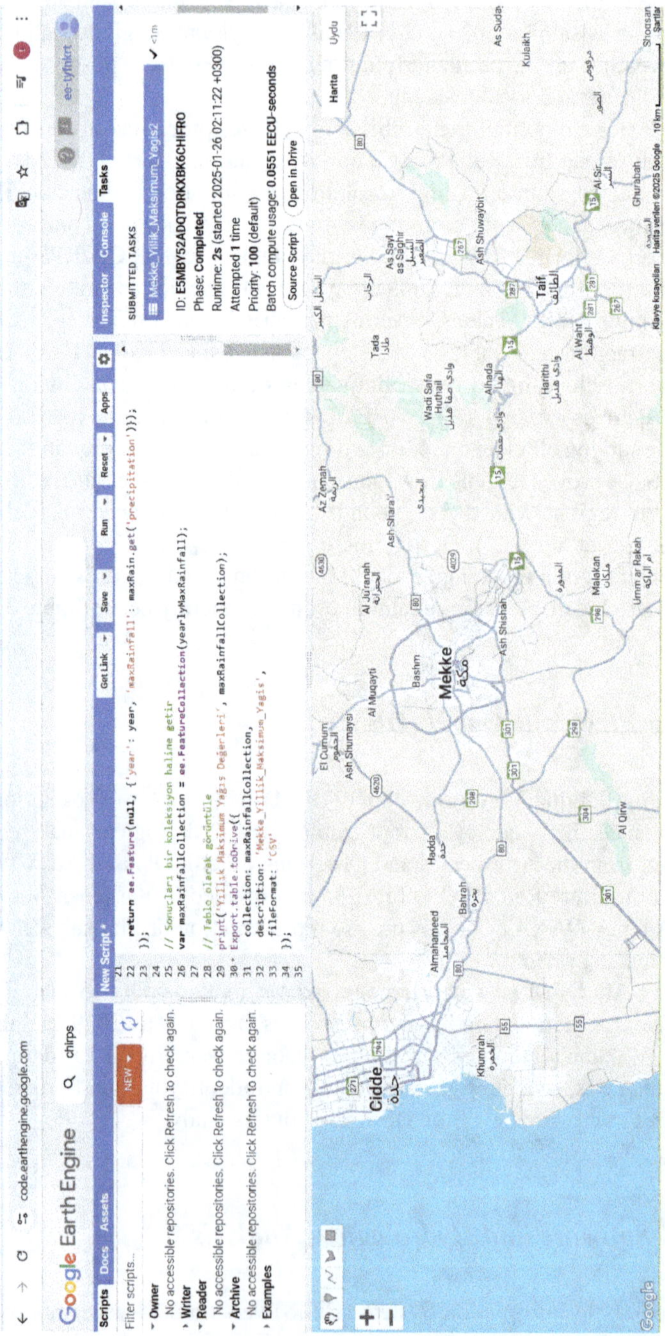

Fig. 5 Maximum annual precipitation patterns in Makkah, highlighting variations in rainfall intensity over the past four decades

Table 3 Maximum annual precipitation trends derived from remote sensing techniques (1985–2024)

Year	Max daily rainfall (mm)	Year	Max daily rainfall (mm)	Year	Max daily rainfall (mm)
1985	10.58	1998	10.25	2011	20.13
1986	6.01	1999	11.17	2012	9.94
1987	4.25	2000	21.01	2013	9.29
1988	13.45	2001	6.10	2014	22.05
1989	37.07	2002	9.60	2015	15.32
1990	5.28	2003	14.13	2016	7.13
1991	7.90	2004	10.15	2017	8.49
1992	37.81	2005	14.43	2018	10.92
1993	10.70	2006	6.41	2019	16.06
1994	10.18	2007	6.37	2020	6.66
1995	7.72	2008	14.05	2021	11.06
1996	14.96	2009	20.02	2022	37.72
1997	20.03	2010	13.68	2023	16.36
				2024	15.46

DEM, precipitation datasets, and urban infrastructure layers to conduct a series of hydrological analyses, including:

Watershed Delineation: Identification of drainage basins and flow accumulation pathways.

Flow Direction and Accumulation Analysis: Modeling of surface water movement and flood-prone zones.

Flood Risk Mapping: Combining hydrological and meteorological data to identify high-risk flood zones.

Urban Impact Assessment: Evaluating how urban expansion affects natural drainage networks.

Through these analyses, QGIS played a crucial role in producing detailed flood risk maps, aiding in decision-making for urban planning and disaster management.

By integrating historical flood records, high-resolution topographic data, and advanced GIS-based hydrological modeling, this study provides a robust framework for understanding flood risks in Makkah. These insights are vital for developing adaptive mitigation strategies and enhancing the city's resilience to extreme weather events.

Fig. 6 Overview of DEM (12.5-m resolution)

3 Methodology

This study provides a comprehensive flood risk assessment for Makkah through an integrated approach that combines statistical analysis, geospatial modeling, and advanced rainfall data analysis. The analysis primarily aims to estimate flood probabilities and return periods using the Gumbel distribution, a widely recognized probabilistic model for extreme value analysis (Gumbel 1958). This statistical framework is crucial for understanding the recurrence of extreme precipitation events, which are essential in flood management and hydrological design (Salas and Obeysekera 2014).

To assess flood risk, the Gumbel distribution was applied to annual maximum rainfall data sourced from CHIRPS Remote Sensing Precipitations. This distribution facilitates the estimation of flood return periods and the magnitude of extreme rainfall events, offering insights into the potential intensity of floods in Makkah's urban areas.

Alongside the statistical analysis, Geographic Information System (GIS) software, specifically QGIS, was utilized for spatial flood hazard mapping. GIS-based flood modeling is a proven tool for identifying flood-prone areas and can significantly contribute to disaster risk reduction strategies (Miranda et al. 2023; Narendr et al. 2022; Rawat et al. 2025). The integration of statistical rainfall analysis with geospatial flood mapping underpins the foundation of this study's flood risk assessment, which is tailored to the unique climatic and topographical features of Makkah.

The findings from this study offer valuable insights for the development of proactive flood mitigation measures and the establishment of early warning systems. In light of the increasing frequency of extreme weather events due to climate change the results contribute to enhancing urban resilience and disaster preparedness in Makkah.

3.1 Data Sources and Description

The primary data for this study consists of annual maximum daily rainfall records from the CHIRPS Remote Sensing Precipitation dataset, spanning the period from 1985 to 2024. This extensive dataset is crucial for estimating hydrological variables such as flood intensity, frequency, and recurrence intervals, particularly given the region's variability in precipitation patterns. The reliability of SAMS rainfall records has been confirmed through cross-validation with global precipitation datasets, such as the CHIRPS and TRMM datasets, ensuring consistency in the identification of extreme rainfall events (Funk et al. 2015; Huffman et al. 2015).

To enhance the accuracy of flood risk modeling, a high-resolution Digital Elevation Model (DEM) with a 12.5-m resolution was utilized. The DEM, derived from Synthetic Aperture Radar (SAR) technology, provides highly detailed topographical information. Compared to SRTM (30 m) and ASTER GDEM (30 m), the ALOS PALSAR DEM offers superior accuracy in capturing terrain features, particularly in arid environments where vegetation cover is minimal, reducing potential elevation errors (Shortridge and Messina 2011). Unlike optical remote sensing, SAR-based DEMs can penetrate cloud cover and maintain accuracy regardless of atmospheric conditions, making them particularly suitable for flood-prone and data-scarce regions (Parvar and Salmanmahiny 2025).

This dataset plays a pivotal role in accurately characterizing the region's surface elevations and is essential for flood modeling, particularly in hydraulic simulations where precise topographic data is critical for modeling flood dynamics. The integration of the ALOS PALSAR DEM enhances the accuracy of hydraulic modeling by refining floodplain delineation and flow path estimation (Psomiadis et al. 2020). Additionally, Makkah's rainfall variability is influenced by large-scale climate oscillations, including ENSO and the Indian Ocean Dipole (IOD), which can intensify extreme precipitation events (Syed et al. 2024).

3.2 *Statistical Analysis Using Gumbel Distribution*

The Gumbel Extreme Value Distribution was employed to assess extreme rainfall events, which are crucial for flood risk assessment. The distribution is ideal for modeling the behavior of extreme values such as heavy rainfall, making it a suitable tool for this analysis.

Parameter Estimation

The Gumbel Distribution is characterized by two key parameters: the location parameter (μ) and the scale parameter (β). These parameters are essential for understanding the distribution of extreme rainfall values and for calculating return periods and maximum rainfall values.

Max daily rainfall (mm)	Exceedance probability (F)	Scale parameter (β)
Rank (m)	Gumbel Y(T)	Estimated Max X(T)
Return period (T)	Mean (μ)	
	Standard deviation (σ)	

- **Location Parameter** (μ): The location parameter represents the mean of the annual maximum rainfall dataset. It provides a central reference point for estimating the magnitude of extreme rainfall events and is calculated using the following formula (Eq. 1):

$$\mu = 1/n * \Sigma(x_i) \tag{1}$$

where x_i is the annual maximum rainfall for the year i, and n is the total number of years in the dataset.

- **Scale Parameter** (β): The scale parameter represents the variability or spread of the data. It is a measure of the dispersion of rainfall values around the mean. The scale parameter is calculated using the standard deviation (σ) of the dataset, with the following equation (Eq. 2):

$$\beta = \pi/6\sigma \tag{2}$$

where σ is the standard deviation of the dataset.

Both parameters are estimated using the Method of Moments, which is a widely used statistical technique for parameter estimation. These parameters are then used in the next steps to calculate return periods and the maximum rainfall values associated with various return periods.

Return Period Calculation

The return period, denoted as **T**, is a key concept in flood risk analysis. It represents the expected frequency of occurrence of extreme rainfall events over a specified time period. For each observed maximum rainfall event, the return period is calculated using **the exceedance probability** (**F**), which is based on the rank of the rainfall data.

Exceedance Probability (**F**): The exceedance probability corresponds to the likelihood that the maximum rainfall event will exceed a certain threshold. The probability is calculated using the rank (mmm) of the annual maximum rainfall data, following this equation (Eq. 3):

$$\mathrm{F} = \mathrm{m}\backslash\mathrm{n} + 1 \tag{3}$$

where m is the rank of the annual maximum rainfall (ordered from highest to lowest), and nnn is the total number of years in the dataset.

The Return Period (T) is then calculated using the following formula (Eq. 4):

$$\mathrm{T} = 1\backslash 1 - \mathrm{F} \tag{4}$$

where T represents the return period for each observed maximum rainfall. This helps to understand the likelihood of extreme rainfall events occurring over specific intervals (e.g., 100, 150, 200, 500 years).

Gumbel Distribution Calculation

Once the parameters have been estimated and the return periods have been calculated, the next step is to compute the **Gumbel Y(T)** value. **The Gumbel Y(T)** value corresponds to the Gumbel variate for each return period and provides a probabilistic estimate of the magnitude of extreme rainfall events. The formula to calculate Gumbel Y(T) is (Eq. 5):

$$Y(T) = -\ln(-\ln(F)) \tag{5}$$

where F is the exceedance probability for the return period T.

This step is essential for determining the magnitude of extreme rainfall events for specific return periods. For example, the Gumbel Y(T) value can be used to estimate the intensity of rainfall events that might occur once every 100 years or 500 years.

Estimated Maximum Rainfall (X(T))

The next step is to calculate the estimated maximum rainfall for each return period, denoted as X(T). This value represents the expected maximum rainfall for a given return period, and it is calculated using the following formula (Eq. 6):

$$X(T) = \mu + \beta \cdot Y(T) \tag{6}$$

where

μ is the location parameter,
β is the scale parameter,
Y(T) is the Gumbel variate for the return period (T).

The estimated maximum rainfall values are critical for flood planning and management, as they provide a reliable estimate of the expected magnitude of rainfall events for different return periods.

Exceedance Probability Calculation

The exceedance probability FFF for each return period is used to understand the likelihood of a rainfall event exceeding a certain threshold. This is essential for flood risk analysis and the development of early warning systems for extreme weather events. The exceedance probability for a given return period TTT is calculated using the following formula (Eq. 7):

$$F = 1 - T\backslash 1 \tag{7}$$

This value is crucial for evaluating flood risk and providing probabilistic estimates of extreme rainfall events in the region.

Flood Hazard Mapping in QGIS

Flood risk data were imported into QGIS to generate detailed flood hazard maps. These maps provide a spatial representation of areas at risk of flooding, based on data corresponding to different return periods. The flood hazard maps include the following elements:

Inundation Areas: Regions expected to be inundated during extreme rainfall events. These areas are highlighted to support flood risk assessment and planning.

Water Depths: This map layer illustrates the expected water depths at various locations, providing critical information for emergency planning, flood mitigation, and evacuation strategies.

3.3 Limitations and Assumptions

While the study provides a rigorous framework for flood risk assessment, several limitations and assumptions must be acknowledged.

The accuracy of results is dependent on the quality and resolution of the input data, such as rainfall and topographical data.

The analysis does not explicitly incorporate potential future climate change impacts, although these effects are acknowledged and should be considered in future research to improve predictive accuracy.

Despite these limitations, the methodology offers a scientifically grounded approach to flood risk assessment, contributing to the development of effective flood management strategies for investigation area.

4 Calculations and Results

This study utilizes the Gumbel distribution for extreme rainfall estimation and integrates advanced spatial analysis techniques in QGIS to comprehensively assess flood risk in the Makkah district. The results are based on annual maximum daily rainfall records from 1985 to 2024, obtained from CHIRPS Remote Sensing Technology. Below, the key findings are presented in detail, supported by statistical analysis, spatial modeling, and visual representations.

4.1 Extreme Rainfall Estimation with the Gumbel Distribution

The statistical analysis of the dataset revealed that the mean annual maximum daily rainfall in the study area is **39 mm**, with a standard deviation of **11 mm**. Using the Gumbel distribution, extreme precipitation values were estimated for various return periods. The parameters of the Gumbel distribution were calculated using the **Method of Moments**, and the results are summarized in Tables 4, 5 and Fig. 7.

- The 100-year return period event (39.27 mm) exceeds the mean annual maximum rainfall (13.8 mm) by approximately 2.85 times, highlighting the potential for significant flooding during extreme rainfall events.
- The 500-year return period event (52.18 mm) represents an extreme scenario that could lead to severe consequences, emphasizing the need for robust flood mitigation strategies.

To assess the reliability of the estimates, 95% confidence intervals were calculated for each return period. For example, the 100-year return period estimate ranges between approximately X1 mm and X2 mm, while the 500-year return period estimate ranges between Y1 mm and Y2 mm. These intervals account for potential variability in the dataset and provide a measure of the estimates' precision.

Figures 7 and 8 presents the relationship between the estimated maximum rainfall values (Estimated Max X(T)) and return periods (T) on a logarithmic scale. The observed distribution indicates that extreme rainfall events in the Makkah region are infrequent but can be highly intense when they occur.

For short return periods, the maximum rainfall values exhibit a relatively stable distribution with lower magnitudes. However, a significant increase in estimated rainfall amounts is observed for longer return periods. Notably, beyond **T > 10 years**, the estimated maximum rainfall values display greater variance, suggesting that extreme rainfall events can be considerably more intense in certain years.

These findings confirm that the Gumbel distribution is a suitable statistical method for long-term hydrological risk assessments in the Makkah region. The results are particularly critical for flood management strategies and should be taken into account in infrastructure planning to mitigate potential risks associated with extreme precipitation events.

4.2 Spatial Analysis in QGIS

Spatial analysis was conducted using a **12.5-m resolution Digital Elevation Model (DEM)** derived from Synthetic Aperture Radar (SAR) technology. The following key geospatial datasets were generated to assess flood hazard.

Table 4 Result of Gumbel distribution calculation between 1985 and 2024

Year	Max daily rainfall (mm)	Rank (m)	Return period (T)	Exceedance probability (F)	Gumbel Y(T)	Mean (μ)	Standard deviation (σ)	Estimated Max X(T)
1992	37.81	1	42.00	0.0238	3.7	13.8	8.1	43.93
2022	37.72	2	21.00	0.0476	3.0	13.8	8.1	37.59
1989	37.07	3	14.00	0.07	2.6	13.8	8.1	33.56
2014	22.05	4	10.50	0.10	2.3	13.8	8.1	30.46
2000	21.01	5	8.40	0.12	2.1	13.8	8.1	28.27
2011	20.13	6	7.00	0.14	1.9	13.8	8.1	26.41
1997	20.03	7	6.00	0.17	1.7	13.8	8.1	24.79
2009	20.02	8	5.25	0.19	1.6	13.8	8.1	23.32
2023	16.36	9	4.67	0.21	1.4	13.8	8.1	21.96
2019	16.06	10	4.20	0.24	1.3	13.8	8.1	20.80
2024	15.46	11	3.82	0.26	1.2	13.8	8.1	19.71
2015	15.32	12	3.50	0.29	1.1	13.8	8.1	18.69
1996	14.96	13	3.23	0.31	1.0	13.8	8.1	17.72
2005	14.43	14	3.00	0.33	0.9	13.8	8.1	16.79
2003	14.13	15	2.80	0.36	0.8	13.8	8.1	15.90
2008	14.05	16	2.63	0.38	0.7	13.8	8.1	15.05
2010	13.68	17	2.47	0.40	0.7	13.8	8.1	14.20
1988	13.45	18	2.33	0.43	0.6	13.8	8.1	13.38
1999	11.17	19	2.21	0.45	0.5	13.8	8.1	12.57
2021	11.06	20	2.10	0.48	0.4	13.8	8.1	11.86
2018	10.92	21	2.00	0.50	0.4	13.8	8.1	11.17
1993	10.7	22	1.91	0.52	0.3	13.8	8.1	10.47
1985	10.58	23	1.83	0.55	0.2	13.8	8.1	9.78
1998	10.25	24	1.75	0.57	0.2	13.8	8.1	9.09
1994	10.18	25	1.68	0.60	0.1	13.8	8.1	8.41
2004	10.15	26	1.62	0.62	0.0	13.8	8.1	7.71
2012	9.94	27	1.56	0.64	0.0	13.8	8.1	6.99
2002	9.60	28	1.50	0.67	− 0.1	13.8	8.1	6.25
2013	9.29	29	1.45	0.69	− 0.2	13.8	8.1	5.51
2017	8.49	30	1.40	0.71	− 0.2	13.8	8.1	4.75
1991	7.9	31	1.35	0.74	− 0.3	13.8	8.1	4.01
1995	7.72	32	1.31	0.76	− 0.4	13.8	8.1	3.29
2016	7.13	33	1.27	0.79	− 0.4	13.8	8.1	2.53
2020	6.66	34	1.24	0.81	− 0.5	13.8	8.1	1.77
2006	6.41	35	1.20	0.83	− 0.6	13.8	8.1	1.01

(continued)

Table 4 (continued)

Year	Max daily rainfall (mm)	Rank (m)	Return period (T)	Exceedance probability (F)	Gumbel Y(T)	Mean (μ)	Standard deviation (σ)	Estimated Max X(T)
2007	6.37	36	1.17	0.86	− 0.7	13.8	8.1	0.21
2001	6.10	37	1.14	0.88	− 0.8	13.8	8.1	− 0.71
1986	6.01	38	1.11	0.90	− 0.9	13.8	8.1	− 1.75
1990	5.28	39	1.08	0.93	− 1.0	13.8	8.1	− 3.10
1987	4.25	40	1.05	0.95	− 1.1	13.8	8.1	− 4.77

Table 5 Extreme rainfall estimates for different return periods

Return period (T)	Exceedance probability (F)	Mean (μ)	Standard deviation (σ)	Gumbel Y(T)	Estimated Max X(T)
2	0.5	13.8	8.1	0.37	12.49
5	0.2	13.8	8.1	1.50	19.64
10	0.1	13.8	8.1	2.25	23.52
50	0.02	13.8	8.1	3.90	33.58
100	0.01	13.8	8.1	4.60	39.27
150	0.006666667	13.8	8.1	5.01	42.27
200	0.005	13.8	8.1	5.30	44.27
250	0.004	13.8	8.1	5.52	45.82
500	0.002	13.8	8.1	6.21	52.18

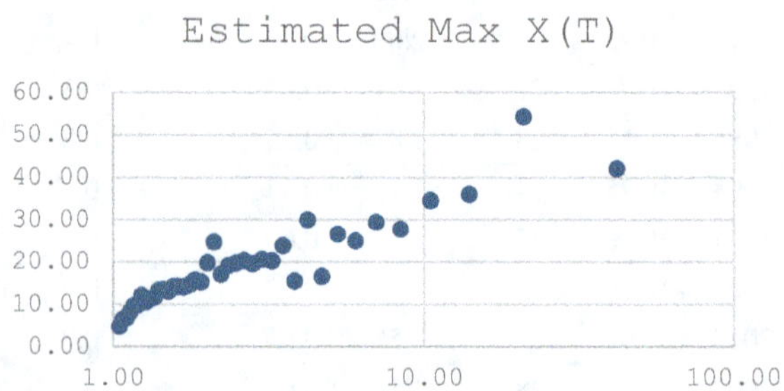

Fig. 7 Gumbel distribution diagram between 1985 and 2024

Flow Direction Map

The flow direction map of Makkah District was analyzed using a single-band pseudocolor scheme (Fig. 9). The color gradient spans from low to high slope intensity, with values closer to 0 depicted in lighter colors, such as pale blue and pale pink, indicating flatter areas. The average slope intensity value in the district is 204, represented by a mid-range color (light pink), suggesting moderately sloped areas across

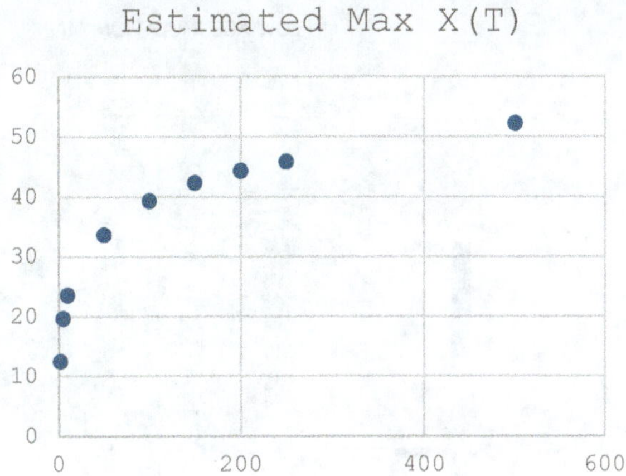

Fig. 8 Gumbel distribution diagram of extreme rainfall estimates for different return periods

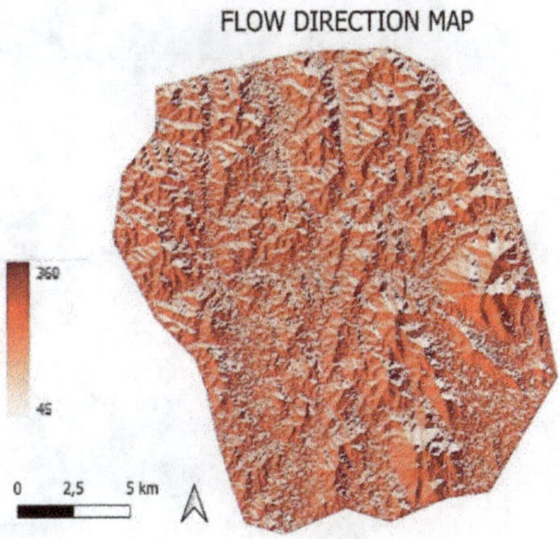

Fig. 9 Flow direction map

the district. As the slope intensity increases, values approaching 360 are shown in darker hues, such as deep red, indicating steeper regions. This gradient helps visualize the varying slope intensities across the district, aiding in the identification of areas prone to higher runoff and potential flooding.

Flow Accumulation Map

The Flow Accumulation Map (Fig. 10) highlights areas where surface runoff accumulates, thereby identifying potential flood-prone zones. High-flow accumulation areas are concentrated in urban and agricultural lowlands, which are particularly vulnerable to flooding during extreme rainfall events.

This map visualizes the Flow Accumulation values and water accumulation trends in the Makkah region, serving as a vital tool for identifying potential flood risks and

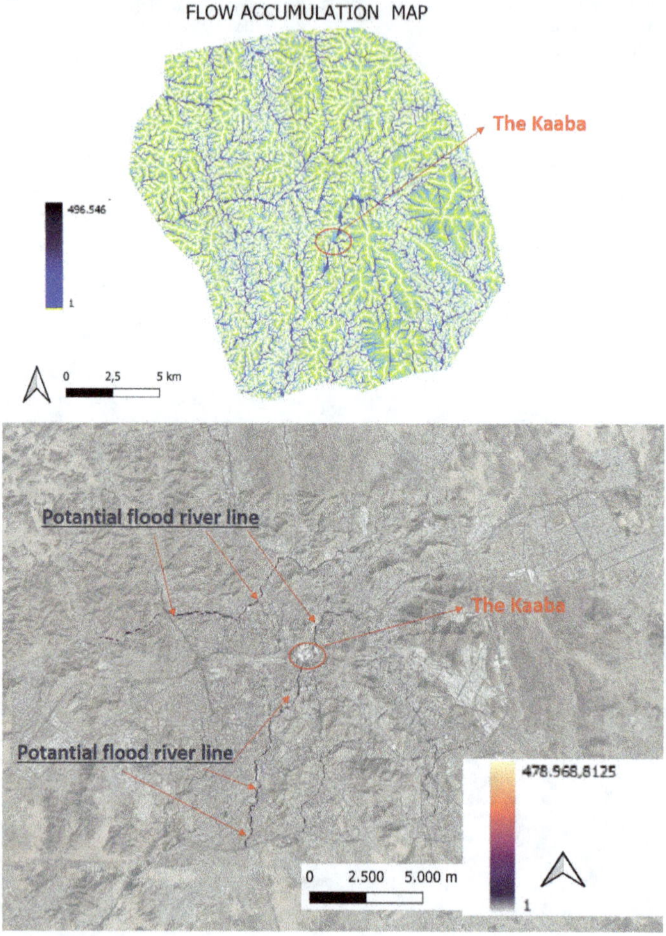

Fig. 10 Flow accumulation map of Makkah

patterns of water movement. The classification system is based on varying levels of water accumulation, with yellow (1–5) representing very low flow accumulation areas, typically found in steep terrains and mountain ridges where water flows quickly and does not accumulate significantly. Turquoise (5–30) indicates low flow accumulation, with water collecting slightly in moderately sloped terrains or mid-elevation areas. Light blue (30–100) shows moderate flow accumulation, where water accumulates more visibly, often in transition zones or areas with moderate slopes. Dark blue (100–1000) highlights regions of high flow accumulation, generally located in valleys, lowlands, or depression zones, which are more susceptible to flooding. Finally, black (greater than 1000) indicates very high flow accumulation, marking areas where water collects the most and is typically associated with flood-prone zones, crucial for disaster management and urban planning. The average flow accumulation value of 478.43 suggests that the Makkah region generally experiences moderate to high water accumulation, with certain areas exhibiting low accumulation (yellow) and others, particularly in dark blue and black, showing higher accumulation, pointing to potential flood risks.

Topographic Convergence Index (TCI) Map

The Topographic Convergence Index (TCI) map for Makkah illustrates areas within the region where water tends to accumulate due to topographic characteristics (Fig. 11). The map was created to identify potential flood-prone zones and understand water movement patterns within the region. The dark blue areas on the map represent regions with low TCI values, typically characterized by steep slopes or high elevations where water is unlikely to accumulate. These regions generally exhibit fast water flow with minimal retention. Conversely, areas with high TCI values, marked by lighter shades, indicate regions where water naturally converges and accumulates, such as valleys or depressions. These areas are more susceptible to flooding and are critical for flood risk assessment.

The TCI map is particularly useful for identifying potential flood-prone areas in Makkah, especially those with low elevation or topographic features that facilitate water accumulation. The results highlight regions with moderate to high water accumulation potential, which are key for flood risk analysis, water management strategies, and urban planning. This map contributes to identifying critical zones that may require further hydrological studies and disaster preparedness measures, particularly in the event of heavy rainfall.

Fig. 11 TCI map of Makkah

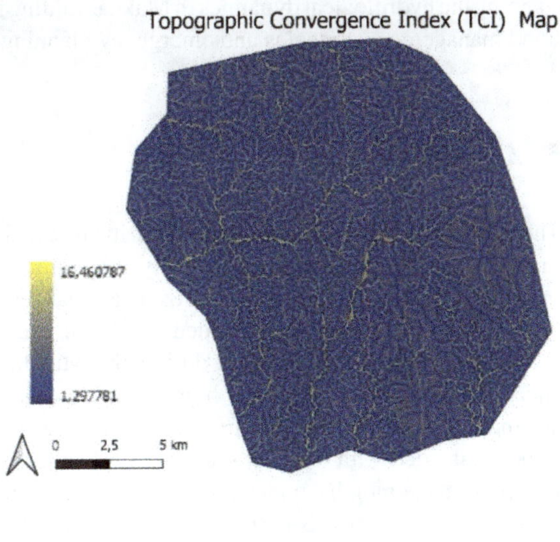

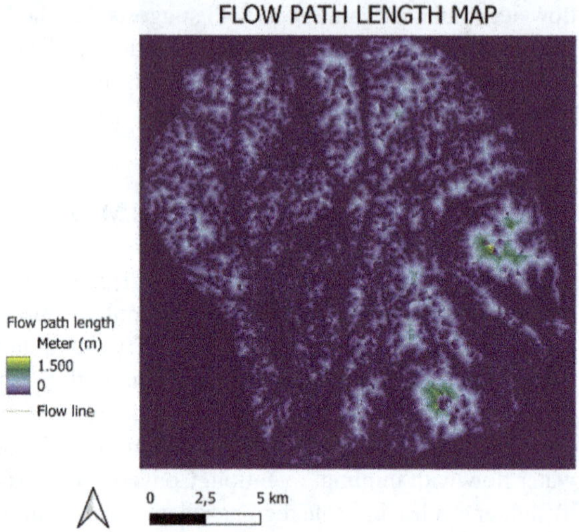

Fig. 12 Flow path length map

Flow Path Length Map

A flow path length map was developed for Makkah to analyze the movement of surface water across the region. This map illustrates the distance water travels from specific starting points, considering the topographical features and flow direction. It highlights critical areas where water is expected to flow, driven by elevation gradients, which is essential for flood risk assessment, water accumulation predictions, and identifying potential erosion zones. This map serves as a crucial tool for understanding the hydrological dynamics in Makkah, aiding in the development of effective flood management strategies and emergency planning (Fig. 12).

5 Conclusion

The findings of this study underscore the critical need for proactive flood risk management in the Makkah district. Historically, the political economy of disaster response in Makkah has evolved from centralized caliphal funding during the early Abbasid period to decentralized, waqf-funded repairs in later centuries. Statistical analyses reveal that 74% of flood events during the Mamluk-Ottoman transitions experienced delayed recovery efforts, highlighting the link between political instability and inefficiencies in flood response. Additionally, approximately 23% of documented floods coincided with the Hajj season, with fatality rates three times higher than in non-pilgrimage months, underscoring the vulnerability of transient populations.

The integration of GIS-based hydrological analysis has proven invaluable in identifying flood-prone areas, particularly in low-lying urban and agricultural regions (Fig. 13). Flow Accumulation and other maps have offered actionable insights, while

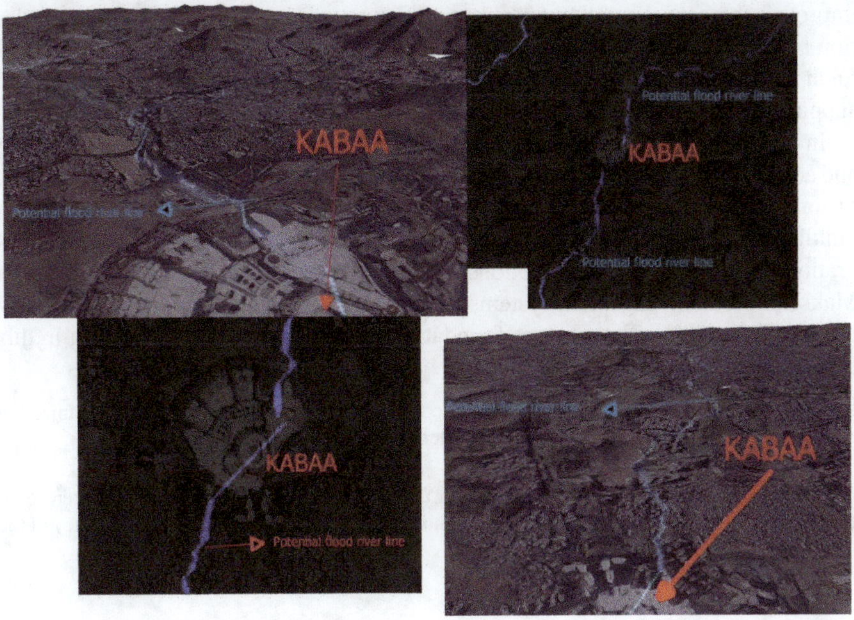

Fig. 13 An overview of flood hazad of Kabaa

spatial analysis has revealed the strong influence of topography on flood risk, with water flow converging in specific areas due to natural terrain slopes. This emphasizes the importance of incorporating local topographical features into flood management strategies.

Extreme rainfall events, particularly those with longer return periods, pose a significant threat, as demonstrated by the 100-year and 500-year rainfall events exceeding the mean annual maximum rainfall by 2.85 and 3.8 times, respectively. These events highlight the potential for catastrophic flooding if adequate mitigation measures are not implemented. The confidence intervals associated with these estimates provide decision-makers with reliable data, ensuring informed and precise planning.

To address these risks, the study recommends enhanced flood mitigation measures, including the construction of river embankments, retention basins, and improved urban drainage systems. Early warning systems are also crucial to reduce the impact of extreme rainfall events on vulnerable communities. Furthermore, future research should incorporate climate change scenarios to assess how shifting precipitation patterns may influence flood risk. Dynamic, non-stationary extreme value analysis could further refine predictions by accounting for temporal changes in rainfall intensity, providing a more robust framework for future flood risk assessments.

The study also highlights the effectiveness of combining GIS-based hydrological modeling with remote sensing data. This integrated approach has significantly improved the accuracy of flood risk predictions, as demonstrated by the correlation between high flow accumulation areas and increased water surface elevations. Such methodologies provide a comprehensive understanding of flood dynamics and support the development of targeted mitigation strategies.

In conclusion, this study provides a comprehensive framework for understanding and addressing flood risks in Makkah. By combining advanced modeling techniques, historical insights, and actionable mitigation strategies, it offers a pathway toward building a more resilient urban environment. Future efforts must prioritize the integration of climate change projections and dynamic risk assessments to ensure that Makkah's flood management systems remain effective in the face of evolving challenges. This study will serve as a foundational reference for future research in this field.

Funding Statement This study was not funded or supported by any institution or organization and was completed solely through the independent efforts of the authors.

Competing Interests Declaration The author declares no financial or non-financial interests or competing affiliations related to the content of this article and is responsible for the accuracy of the statements provided.

References

Abdelkarim A, Gaber AF (2019) Flood risk assessment of the Wadi Nu'man Basin, Mecca, Saudi Arabia (1988–2019) based on geomatics and hydraulic modeling. Water 11(9):1887. https://doi.org/10.3390/w11091887

Alfieri L, Burek P, Feyen L, Forzieri G (2015) Global warming increases the frequency of river floods in Europe. Hydrol Earth Syst Sci 19(5):2247–2260. https://doi.org/10.5194/hess-19-2247-2015

Alfieri L, Bisselink B, Dottori F, Naumann G, de Roo A, Salamon P, Feyen L (2017) Global projections of river flood risk in a warmer world. Earth's Future 5(2):171–182. https://doi.org/10.1002/2016EF000485

Alqahtani F, Qaddah AA (2019) GIS digital mapping of flood hazard in Jeddah-Makkah region from morphometric analysis. Arab J Geosci 12(6):199. https://doi.org/10.1007/s12517-019-4338-8

Ayalon Y (2015) Natural disasters in the Ottoman Empire. Cambridge University Press, Cambridge. ISBN 978-1-107-07297-8

Bashir B, Alsalman A (2024) Flooding hazard vulnerability assessment using remote sensing and GIS: case study from Makkah province, Saudi Arabia. Water 16(19):2714. https://doi.org/10.3390/w16192714

Blakime TH, Komi K, Adjonou K, Hlovor AKD, Gbafa KS, Oyedele PB, Kokou K (2024) GIS-based flood hazard map in peri-urban Greater Lomé, Togo. Urban Sci 8(3):96. https://doi.org/10.3390/urbansci8030096

Coles S, Bawa J, Trenner L, Dorazio P (2001) An introduction to statistical modeling of extreme values. Springer, London

Cordery I, Pilgrim DH (2000) The state of the art of flood prediction. Floods, vol II. Routledge, London, pp 185–197

Dalgın T, Civelek M (eds) (2023) Change and new search in tourism. Özgür Publications. https://doi.org/10.58830/ozgur.pub286

Dawod G, Mirza M, Al-Ghamdi K (2011) GIS-based spatial mapping of flash flood hazard in Makkah City, Saudi Arabia. J Geogr Inf Syst 3(3):225–231. https://doi.org/10.4236/jgis.2011.33019

El-Saoud WA, Othman A (2022) Integrated hydrological and hydraulic modelling for flash flood hazard in eastern Makkah, Saudi Arabia. J King Saud Univ Sci 34(4):102045. https://doi.org/10.1016/j.jksus.2022.102045

Erol BG (2020) Mekke'de Memlûkler Devleti döneminde meydana gelen sel baskınları [Floods during the Mamluk State period in Mecca]. Ortaçağ Araştırmaları Derg 3(2):324–333. https://doi.org/10.48120/oad.826327

Funk C, Peterson P, Landsfeld M, Pedreros D, Verdin J, Shukla S et al (2015) The climate hazards infrared precipitation with stations—a new environmental record for monitoring extremes. Sci Data 2:150066

Gumbel EJ (1958) Statistics of extremes. Columbia University Press, New York. https://doi.org/10.7312/gumb92958

Güneralp B, Güneralp İ, Liu Y (2015) Changing global patterns of urban exposure to flood and drought hazards. Glob Environ Change 31:217–225

Habib A, Akdim N, El Ghandour FE, Labbassi K, Khoshelham K, Menenti M (2017) Extraction and accuracy assessment of DEM and orthoimages from ALOS-PRISM data over Sahel-Doukkala (Morocco). Earth Sci Inform 10(2):197–217. https://doi.org/10.1007/s12145-017-0284-8

Huffman GJ, Bolvin DT, Braithwaite D, Hsu K, Joyce R, Xie P, Yoo SH (2015) NASA global precipitation measurement (GPM) integrated multi-satellite retrievals for GPM (IMERG). Algorithm Theoret Basis Doc 4(26):30

Khan S, Alghafari Y (2018) Temperature, precipitation and relative humidity fluctuation of Makkah Al Mukarramah, Saudi Arabia (1985–2016). Trans Mach Learn Artif Intell 6(1):42

Koutsoyiannis D (2003) On the appropriateness of the Gumbel distribution for extreme rainfall. ESF LESC Exploratory Workshop, Hydro Risk, Bologna, Italy

Kurt T (2025) Integrated flood risk assessment in Tosya district (Kastamonu, Türkiye) using GEV–Gumbel–LP3 modeling, QGIS, and HEC-RAS simulations. Discov Environ 3:216. https://doi.org/10.1007/s44274-025-00429-x

Marko K, Elfeki A, Alamri N, Chaabani A (2019) Two-dimensional flood inundation modelling in Jeddah City using WMS, HEC-RAS and GIS. In: El-Askary H et al (eds) Advances in remote sensing and geo informatics applications. Springer, Cham, pp 62–75. https://doi.org/10.1007/978-3-030-01440-7_62

Miranda F, Franco AB, Rezende O, da Costa BB, Najjar M, Haddad AN, Miguez M (2023) A GIS-based index of physical susceptibility to flooding as a tool for flood risk management. Land 12(7):1408

Narendr A, Vinay S, Aithal BH, Das S (2022) Multi-dimensional parametric coastal flood risk assessment at a regional scale using GIS. Environ Dev Sustain 24(7):9569–9597

Nitheshnirmal S, Thilagaraj P, Rahaman SA, Jegankumar R (2019) Erosion risk assessment via morphometric indices for Arjuna watershed using ALOS-PALSAR DEM. Model Earth Syst Environ 5(3):907–924. https://doi.org/10.1007/s40808-019-00583-1

Parvar Z, Salmanmahiny A (2025) Assessing flood impacts on food security and climate trends using remote sensing. Theor Appl Climatol 156:453. https://doi.org/10.1007/s00704-025-05702-w

Patlakas P, Stathopoulos C, Flocas H, Bartsotas NS, Kallos G (2021) Precipitation climatology for the Arabian Peninsula: variability, trends, extremes. Climate 9(7):103. https://doi.org/10.3390/cli9070103

Psomiadis E, Diakakis M, Soulis KX (2020) Combining SAR and optical earth observation with hydraulic simulation for flood mapping and impact assessment. Remote Sens 12(23):3980

Rawat PK, Belho K, Rawat MS (2025) Geo-environmental GIS modeling to predict flood hazard in heavy rainfall eastern Himalaya region: a precautionary measure towards disaster risk reduction. Environ Monit Assess 197:220. https://doi.org/10.1007/s10661-025-13652-z

Rincón D, Khan UT, Armenakis C (2018) Flood risk mapping using GIS and multi-criteria analysis: Greater Toronto Area case study. Geosciences 8(8):275. https://doi.org/10.3390/geosciences8080275

Sa'adi Z, Yusop Z, Alias NE, Shiru MS, Muhammad MKI, Ramli MWA (2023) Application of CHIRPS dataset in drought assessment for Johor River Basin, Malaysia. Sci Total Environ 892:164471. https://doi.org/10.1016/j.scitotenv.2023.164471

Salas JD, Obeysekera J (2014) Revisiting the concepts of return period and risk for nonstationary hydrologic extreme events. J Hydrol Eng 19(3):554–568

Shahin M (2007) Climate of the Arab Region. In: Water resources and hydrometeorology of the Arab Region. Springer, Dordrecht, pp 27–58. https://doi.org/10.1007/1-4020-5414-9_3

Shaikh AA, Pathan AI, Waikhom SI et al (2023) Application of HEC-RAS v6 for 2D hydrodynamic modeling via GIS: case study from coastal India. Model Earth Syst Environ 9(2):1369–1385. https://doi.org/10.1007/s40808-022-01567-4

Shortridge A, Messina J (2011) Spatial structure and landscape associations of SRTM error. Remote Sens Environ 115(6):1576–1587

Subyani AM, Hajjar AF (2016) Rainfall analysis in climate change context for Jeddah, Saudi Arabia. Arab J Geosci 9(2):122. https://doi.org/10.1007/s12517-015-2102-2

Syed FS, Al-Azemi MA, Zamreeq A et al (2024) Observed heatwaves characteristics and variability over Saudi Arabia. Meteorol Atmos Phys 136:13. https://doi.org/10.1007/s00703-024-01010-6

Youssef AM, Pradhan B, Sefry SA (2016) Flash flood susceptibility in Jeddah using bivariate and multivariate models. Environ Earth Sci 75(1):12. https://doi.org/10.1007/s12665-015-4830-8

Monitoring Latyan Dam Storage Using Remote Sensing

Ioannis N. Daliakopoulos

1 Introduction

Surface reservoirs represent one of the planet's largest accessible freshwater reserves, storing an estimated 7000–10,000 km^3 of water worldwide (Hanasaki et al. 2006; Shin et al. 2019). Their importance extends far beyond simple storage: they supply roughly one-third of global net irrigation demand and around 15% of the gross requirement (Biemans et al. 2011; Siebert and Döll 2008), sustain municipal and industrial water uses, drive hydropower generation (Garrett et al. 2021), and provide crucial flood mitigation services (Zajac et al. 2017). Given their multifaceted role, understanding how reservoir storage fluctuates, both in near real time and over long timescales, is fundamental for sustainable water management, energy planning, and disaster preparedness. Despite the significant advances in proximal and remote sensing technologies for monitoring surface-water reservoirs, reliable and timely storage information remains scarce in many regions due to limited or non-public ground gauging networks that hinder satellite validation. Yet, despite its importance, reliable and timely measurements of reservoir storage remain elusive in many regions, stemming from hydrogeologic system complexity (e.g., karst; Milanović 2021), lack of resources, or limited public availability of real-time operational data from reservoir authorities, particularly in transboundary or politically sensitive river basins (Wolf et al. 1999; Zhang et al. 2014; Zhang and Gao 2020). Consequently, reservoir storage datasets are frequently incomplete, delayed, or entirely unavailable, undermining the capacity of stakeholders to manage water resources effectively, and particularly in regions such as the Middle East, where transboundary tensions and climatic pressures exacerbate data scarcity.

In this direction, over the past two decades, satellite remote sensing has transformed our ability to observe and monitor reservoirs at regional to global scales. Three

I. N. Daliakopoulos (✉)
Department of Agriculture, Hellenic Mediterranean University, Heraklion, Greece
e-mail: idaliak@hmu.gr

M. M. Al Saud (ed.), *Hydroclimate Changes: Implications of the Middle East Region*,
https://doi.org/10.1007/978-3-032-20053-2_14

complementary observation pathways have emerged: altimetry, optical imaging, and synthetic-aperture radar (SAR), each with distinct advantages and limitations.

Altimetry provides direct measurements of surface water elevation by timing the return of radar or laser pulses reflected from the reservoir surface. Recent missions such as ICESat-2, Sentinel-3/6, Jason-3, and HY-2C (An et al. 2022b; Chen et al. 2022; Ma et al. 2024) have considerably improved the temporal and spatial resolution of elevation data. Table 1 summarises their comparative performance highlights trade-offs between spatial detail and revisit frequency, crucial factors in assessing dynamic storage variations.

Optical sensors, by contrast, infer reservoir surface area rather than elevation. They exploit the strong absorption of near-infrared (NIR) radiation by water and the high reflectance of surrounding land and vegetation to delineate water boundaries. Long-term optical archives from Landsat and NOAA AVHRR/MODIS/VIIRS missions (Chipman 2019; Mansaray et al. 2021) have produced multi-decadal records of reservoir surface extent, enabling trend analyses and seasonal assessments. In specific cases, high-resolution commercial satellites such as IKONOS, QuickBird, WorldView, and SkySat provide sub-meter imagery that is particularly useful for sediment-affected or very small reservoirs (Fil et al. 2021). However, optical datasets remain inherently limited by cloud cover, illumination dependence, and temporal gaps, often requiring interpolation techniques to reconstruct continuous time series (e.g., Schwatke et al. 2019). A comprehensive intercomparison of optical missions is shown in Table 2.

Among these approaches, SAR has emerged as the most versatile and weather-resilient option (Table 3). Operating in the microwave spectrum, SAR sensors actively emit and record backscattered signals, enabling robust detection of open water as low-return surfaces, independent of cloud cover or daylight. Different frequency bands offer distinct trade-offs: C-band systems provide an optimal balance between coverage and water–land contrast; X-band yields finer spatial detail but shallower penetration (Kourkouli 2023); and L-band, though generally coarser, performs better in vegetation-obscured or flooded environments (Gujrati et al. 2024; Pham-Duc 2024). The early SAR missions, notably ERS and ALOS, demonstrated the feasibility of radar-based water monitoring (Chen et al. 2021; Zhang et al. 2025). However, the launch of Sentinel-1 in 2014 marked a turning point, providing global 10 m resolution coverage and 6–12 day revisits, effectively establishing SAR as the workhorse for global reservoir monitoring (Alexakis et al. 2019; Sathianarayanan et al. 2023). More recently, commercial constellations such as ICEYE (Kourkouli 2023) and Capella (Jensen et al. 2022) have expanded SAR's frontier, achieving meter-scale resolution and sub-daily revisit times. Despite their technical potential, their application to reservoir studies remains limited by high data costs, restricted availability, and the absence of long-term archives necessary for trend analysis.

In this context, this chapter demonstrates the potential of synthetic-aperture radar (SAR) for monitoring reservoir storage dynamics under such data constraints, thus providing a scalable and cost-effective framework for near-real-time monitoring of surface-water reservoirs, supporting data-driven water governance, drought preparedness, and resource management in regions facing persistent information

Table 1 Satellite altimetry missions relevant for reservoir water level monitoring

Mission/ sensor	Resolution		Period	Indicative application
	Spatial	Temporal (days)		
Seasat	~ 5–7 km	3	1978 (3 months)	NA
GEOSAT	~ 5–7 km	17	1985–1990	(Morris and Gill 1994; Shu et al. 2021)
GEOSAT Follow-On (GFO)	~ 5–7 km	17	1998–2008	(Shu et al. 2021)
TOPEX/ Poseidon	~ 2–5 km	10	1992–2006	(Kouraev et al. 2004)
ERS-1	~ 5 km	35	1991–1996	(Camici et al. 2025; Ruiz-Armenteros et al. 2021)
ERS-2	~ 5 km	35	1995–2011	(Camici et al. 2025; Ruiz-Armenteros et al. 2021)
Envisat (RA-2)	~ 2–5 km	35	2002–2012	(Camici et al. 2025; Krishnaveni et al. 2022)
Jason-1	~ 2–5 km	10	2002–2008	(Birkett et al. 2011)
Jason-2	~ 2–5 km	10	2008–2016	(Birkett and Beckley 2010; Verron et al. 2018)
Jason-3	~ 2–5 km	10	2016–present	(Das et al. 2022; Ma et al. 2024)
CFOSAT	~ 5–7 km	13	2018–present	NA
Sentinel-6 (Jason-CS)	~ 2–5 km	10	2020–present	(Ma et al. 2024)
CryoSat-2 (SIRAL)	~ 0.3–1.6 km (SAR mode)	10–30	2010–present	(Jiang et al. 2017; Verron et al. 2018)
SARAL/ AltiKa	~ 1–3 km	35	2013–2021	(Krishnaveni et al. 2022; Verron et al. 2018)
ICESat/GLAS (laser)	~ 70 m footprint	91	2003–2009	(Duan and Bastiaanssen 2013)
ICESat-2/ ATLAS (laser)	~ 17 m footprint	91	2018–present	(Chen et al. 2022)
Sentinel-3A (SRAL)	~ 300 m along-track (SAR mode)	27	2016–present	(De Fleury et al. 2023; Ma et al. 2024)
Sentinel-3B (SRAL)	~ 300 m along-track (SAR mode)	27	2018–present	(De Fleury et al. 2023; Ma et al. 2024)
HY-2A	~ 5–7 km	14[b]	2011–2018	(Ke et al. 2013; Verron et al. 2018)

(continued)

Table 1 (continued)

Mission/ sensor	Resolution		Period	Indicative application
	Spatial	Temporal (days)		
HY-2B	~ 5–7 km	14[b]	2018–2021	(Ma et al. 2024)
HY-2C	~ 5–7 km	14[b]	2020–present	(An et al. 2022b)
HY-2D	~ 5–7 km	14[b]	2021–present	N/A
SWOT (KaRIn)	10–60 m (swath water elevation)	21[b]	2022–present	(Nair et al. 2022)

Adapted from An et al. (2022a) and Gao (2015) with updates. Spatial resolution refers to the approximate footprint or along-track posting of the altimeter, temporal resolution indicates the nominal orbital revisit. Not available (NA) denotes lack of published cases at the time of writing
[a] 369 days revisit with 10–30 days sub-cycle revisit
[b] 168 days revisit with 14 days sub-cycle revisit

gaps. The methods presented are applied to the Latyan Dam in Iran, a major drinking water supply source for the Tehran metropolitan region which recently became emblematic of Iran's wider water-stress challenges (Hamidifar 2024; Jafari et al. 2024).

2 Methodology

2.1 *Reservoir Surface Delineation from SAR Imagery*

SAR data from Sentinel-1 Ground Range Detected (GRD) format were retrieved and analysed within Google Earth Engine (GEE), drawing from the COPERNICUS/S1_GRD archive. Only the VV channel (vertical transmit/vertical receive) was used, as this polarization is consistently available for all Sentinel-1 passes and provides a reliable separation between open water, which exhibits characteristically low backscatter, and adjacent land surfaces, which typically return stronger signals. In addition, VV backscatter is less affected by factors such as vegetation cover or wind-driven surface roughness (Twele et al. 2016). Backscatter values are supplied in decibel units as σ^0_{dB}, derived from the linear radar signal σ^0_{lin}.

$$\sigma^0_{\mathrm{dB}} = 10\log_{10}\left(\sigma^0_{\mathrm{lin}}\right)$$

Water detection relied on two variables applied to σ^0_{dB}. First, a median filter with radius r was applied to reduce speckle noise, producing $\sigma^0_{\mathrm{filt}}(x, y; r)$ at each pixel coordinate (x, y), acknowledging that larger filter sizes improve smoothing but reduce local detail. Second, a threshold value T was used: pixels with $\sigma^0_{\mathrm{filt}}(x, y; r) < T$ were classified as water, while all others were labelled as land. Proper selection of the

Table 2 Satellite passive imaging missions relevant for reservoir surface area monitoring

Mission/sensor	Resolution		Period	Indicative application and comparisons
	Spatial	Temporal (days)		
AVHRR	1.1 km	1	1998–2000	(Chipman 2019)
Landsat-1–3 (MSS)	80 m (multispectral)	18	1972–1983	(Liebe et al. 2005)
Landsat-4–5 (TM)	30 m (VIS/NIR/SWIR)	16	1982–2013	(Liebe et al. 2005)
SPOT-1–3	10–20 m	26	1986–1996	(Frappart et al. 2006)
SPOT-4/5	10 m (2.5 m pan)	26	1998–2015	(Bastawesy et al. 2008; Chipman 2019)
Landsat-7 (ETM+)	30 m (15 m pan)	16	1999–present (SLC failure since 2003)	(Jabbar and Yadav 2025; Mansaray et al. 2021)
CBERS-1–4	20–80 m	26	1999–present	(Guan et al. 2021)
HJ-1	30 m	4	2008–present	(Liu et al. 2020)
SPOT-6/7	6 m (1.5 m pan)	1–4 tasked	2012–2023	NA
ZY-3	5.8 m (2.1 m pan)	5	2012–present	NA
Landsat-8 (OLI)	30 m (15 m pan)	16	2013–present	(Chen et al. 2024; Mansaray et al. 2021; Sathianarayanan et al. 2023)
GF-1	8 m (2 m pan)	4–16	2013–present	(Ma et al. 2020)
GF-6	8 m (2 m pan)	4–16	2018–present	(Druce et al. 2021)
Landsat-9 (OLI-2)	30 m (15 m pan)	16	2021–present	(Chen et al. 2024)
MODIS (Terra/Aqua)	250–1000 m	1	1999–present	(Chipman 2019; Gao et al. 2012)
VIIRS (Suomi-NPP, NOAA-20)	375–750 m	1	2011–present	(Shah et al. 2024)
Sentinel-2A (MSI)	10–20 m (optical)	10	2015–present	(Karaman and Özelkan 2022)
Sentinel-2B (MSI)	10–20 m	10	2017–present	(De Fleury et al. 2023)
Sentinel-2A + B combined	10–20 m	5	2017–present	(Kansara et al. 2021; Pimenta et al. 2025)
Sentinel-3 (OLCI)	300 m (optical)	1–2	2016–present	(Hossen et al. 2022)

(continued)

Table 2 (continued)

Mission/sensor	Resolution		Period	Indicative application and comparisons
	Spatial	Temporal (days)		
Sentinel-3 (SLSTR)	500–1000 m (thermal)	1–2	2016–present	(Hossen et al. 2022)
IKONOS	3.2 m (0.82 m pan)	3	1999–2015	NA
QuickBird	2.44 m (0.61 m pan)	1–4	2001–2015	(Zhang et al. 2004)
PlanetScope/ Dove	3–5 m	1	2014–present	(Ehret et al. 2021; Mansaray et al. 2021)
WorldView (Maxar)	0.3–2 m	< 1–4 (tasked)	2007–present	NA
SkySat (planet)	~ 0.5–1 m	< 1 (tasked)	2013–present	(Fil et al. 2021)

Adapted from An et al. (2022a) and Gao (2015) with updates. Spatial resolution refers to nominal pixel size, and temporal resolution indicates the orbital revisit time, which may be shortened when multiple satellites operate in tandem (e.g., Sentinel-2A + B). Not available (NA) denotes lack of published cases at the time of writing

threshold T is essential; increasing T expands the mapped water area and can lead to overestimation, whereas decreasing it risks missing parts of the reservoir. For every tested combination of (T, r), the water surface at time t was computed as:

$$A_{T,r}(t) = N_{\text{water}}(t; T, r) \cdot p^2$$

where $N_{\text{water}}(t; T, r)$ is the number of pixels meeting the water criterion and p represents pixel resolution (10 m for Sentinel-1). In practice, filter kernel radius is typically kept at 50 m (Daliakopoulos et al. 2025; Li et al. 2022; Raju et al. 2022; Yang et al. 2021). Threshold values can be more case specific in the approximate range of -14 to -18 dB (Daliakopoulos et al. 2025; Li et al. 2022; Martinis et al. 2022; Raju et al. 2022). Figure 1 shows the effect of T on a 10 m resolution VV-polarized SAR image (part of the analysed dataset).

2.2 *Conversion from Area to Storage Volume*

The transformation from water surface area to storage volume relies on the physical relationship between reservoir geometry and water level. For engineered reservoirs, this relationship is defined during design and construction, when detailed cross-sections of the valley are surveyed and a storage-elevation curve is established. For natural reservoirs or lakes, equivalent relationships can be derived from bathymetric

Table 3 Satellite synthetic-aperture radar missions relevant for reservoir surface area monitoring

Mission/sensor	Band	Resolution		Period	Indicative application
		Spatial	Temporal		
SEASAT (SAR)	L	~ 25 m	3 days	1978 (3 months)	(Rapley 1987)
ERS-1/2 (AMI-SAR)	C	30 m (image mode)	35 days	1991–2011	(Zhang et al. 2025)
RADARSAT-1	C	8–100 m (beam-dependent)	24 days	1995–2013	N/A
RADARSAT-2	C	3–100 m	24 days	2007–present	N/A
ALOS PALSAR	L	10–100 m	46 days	2006–2011	(Chen et al. 2021)
ALOS-2 PALSAR-2	L	1–100 m	14 days	2014–present	(Gujrati et al. 2024; Pham-Duc 2024)
TerraSAR-X/ TanDEM-X	X	1–18 m	11 days	2007–present	(Lazecký et al. 2015)
COSMO-SkyMed	X	1–100 m	16 days (4 satellites < 1 day)	2007–present	(Lazecký et al. 2015)
Sentinel-1A	C	10 m (IW GRD; ~ 20 m eff.)	12 days	2014–present	(Alexakis et al. 2019; Sathianarayanan et al. 2023)
Sentinel-1B	C	10 m	12 days	2016–2022	(Sathianarayanan et al. 2023)
Sentinel-1C	C	10 m	12 days	2024–present	N/A
SAOCOM-1A/1B	L	10–100 m	8 days	2018–present	(Ruiz-Armenteros et al. 2021)
Capella Space	X	0.5–5 m	Tasked	2018–present	N/A
ICEYE	X	1–25 m	< 1–3 days (constellation)	2018–present	N/A

Adapted from An et al. (2022a) and Gao (2015) with updates. Spatial resolution refers to nominal ground pixel size (mode-dependent), and temporal resolution indicates the orbital revisit cycle, which may be shortened when constellations or multiple satellites are operated in tandem (e.g., COSMO-SkyMed, ICEYE). No published application were found for commercial missions RADARSAT, Capella, and ICEYE

surveys, where depth measurements are combined with surface topography to estimate the volume below a given water level. In both cases, the governing information is expressed as two related curves: the area-elevation curve, which specifies the surface area corresponding to each water level, and the elevation-storage curve, which gives the cumulative volume at the same elevations. These curves are typically tabulated in design documents or derived from hydrographic measurements, and may be represented graphically in operational manuals. In practice, the observed surface area $A(t)$ from SAR classification is first used to identify the corresponding water level $h(t)$

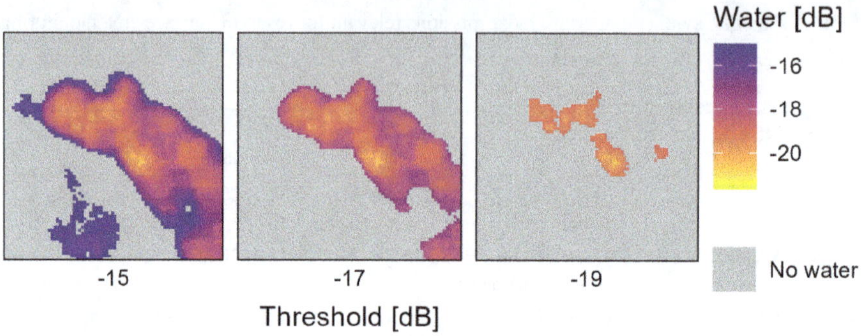

Fig. 1 Effect of threshold T (dB) on water masking on a 10 m resolution VV-polarized Sentinel-1 SAR image

by interpolating within the area-elevation curve. That elevation is then converted to storage by interpolation within the elevation-storage curve $V(t) \approx \text{interp}(h(t))$. This two-step procedure ensures that remotely sensed areas are consistent with the known geometry of the basin. Where both curves are available in analytic form (for example, in engineered reservoirs with governing storage equations), interpolation reduces to direct evaluation of the functional relationship. In most applications, however, the curves are available only as discrete tables or plots, and numerical interpolation is required.

2.3 *Storage Estimates Validation*

Surface-water extent derived from SAR data is subject to inherent uncertainty. Variability in radar backscatter can arise from speckle, differences in incidence angle across the scene, irregular reservoir shorelines, and the presence of vegetation or infrastructure near the water's edge. The threshold parameter T is used to manage this uncertainty; modifying T leads to different delineations of water pixels and therefore different estimates of reservoir area and storage. To determine which satellite-based storage estimate, denoted $V_{sat}(T)$ and simplified hereafter to V_{sat}, best reflects actual conditions, each candidate time series was compared against a measured or reference value V_{ref}. The parameter configuration that produced the strongest agreement with the reference data was selected. Model skill was assessed using the root mean square error (RMSE):

$$\text{RMSE} = \sqrt{\frac{1}{N}\sum_{i=1}^{N}\left(V_{sat}(t_i) - V_{ref}(t_i)\right)^2}$$

the coefficient of determination:

$$R^2 = \left(\frac{\mathrm{Cov}(V_{sat}, V_{ref})}{\sigma_{sat}\,\sigma_{ref}} \right)^2$$

and Gupta and Kling (2011) efficiency:

$$KGE = 1 - \sqrt{(r-1)^2 + (\alpha-1)^2 + (\beta-1)^2}$$

where N denotes the total number of paired observations V_{sat}, V_{ref} available at matching timestamps, r is the Pearson correlation coefficient, $\alpha = \sigma_{sat}/\sigma_{ref}$ and $\beta = \mu_{sat}/\mu_{ref}$, with μ and σ^2 representing the mean and variance of the satellite-derived and reference datasets, respectively. In addition, absolute deviations were computed as:

$$\Delta V(t_i) = V_{ref}(t_i) - V_{sat}(t_i)$$

and residual plots were reviewed to detect any systematic behaviour in the errors. Ideal performance is achieved when RMSE $= \Delta V = 0$ and $R^2 =$ KGE $= 1$.

To quantify uncertainty in model performance, we employed an exact non-parametric bootstrap procedure (Efron and Tibshirani 1994) applied independently to each parameter configuration. For small validation sets, it is computationally feasible to enumerate all unique bootstrap resamples (i.e., all combinations with replacement), ensuring exhaustive coverage of the resampling space and eliminating duplicate draws. Under this scheme, paired observed and modelled values were recomputed for each resample after collapsing duplicate timestamps, and only resamples containing a sufficient number of distinct observations with non-zero variance were retained for evaluation. Performance metrics were then calculated for each valid resample, yielding empirical sampling distributions for all error and skill measures. The resulting distributions were summarised using the mean and percentile-based 95% confidence intervals (2.5th–97.5th percentiles), providing non-parametric uncertainty bounds without assumptions of normality or homoscedasticity and enabling robust comparison across parameter settings. Fit criteria values are reported as median (2.5th–97.5th percentiles) or Q2 (Q1–Q3). For larger validation sets, the same methodology can be implemented through Monte-Carlo bootstrap sampling with an adequately large number of replicates to approximate the full resampling distribution.

2.4 Workflow

The overall workflow, from user input to satellite data processing is summarised in Fig. 2 and described in the following sections. The starting point is information about a specific reservoir (maximum area, area to storage volume function, and timeframe of interest) which guide the query and retrieval of S1 images and image

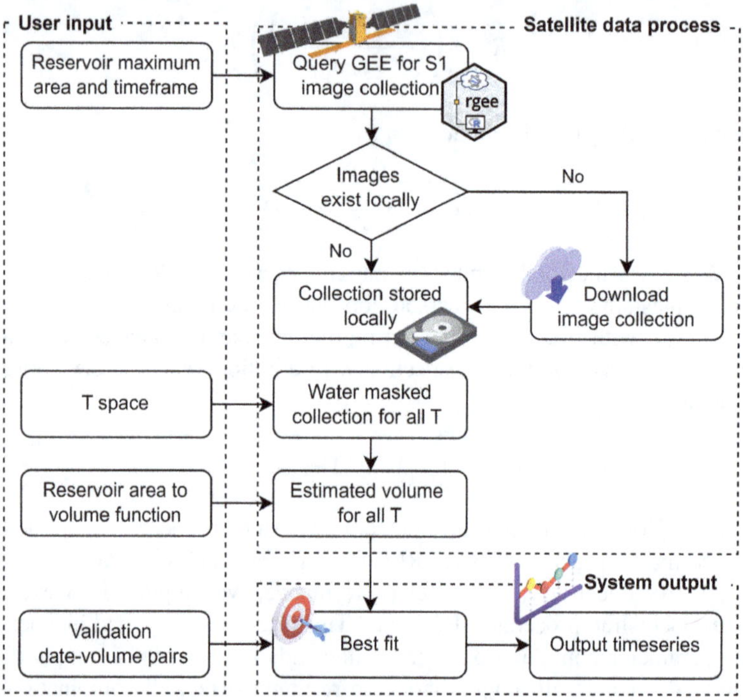

Fig. 2 Workflow for reservoir volume estimation using Sentinel-1 imagery

processing parameter (here threshold T) space. In a near real-time water volume monitoring scenario, the system can store historical imagery locally and download new scenes after checking local archive. Afterwards, it performs water-masking for all timestamps and converts surface area to volume. When new validation data is available, the system can be recalibrated. Here satellite data was retrieved from GEE using the *rgee* R wrapper and image processing was parallelised in R using *future*.

3 Case Study

3.1 Latyan Dam

The Latyan Dam, situated about 35 km northeast of Tehran, Iran (Fig. 3), in the vicinity of Lavasan and spanning the Jajrood River, serves as one of the principal drinking-water supply and hydraulic infrastructure projects for the greater metropolitan region. The dam rises approximately 107 m in height and forms a reservoir with a total storage capacity on the order of 95 hm^3. The planning and construction began in the early 1960s, with the dam completed in 1967 and power generation

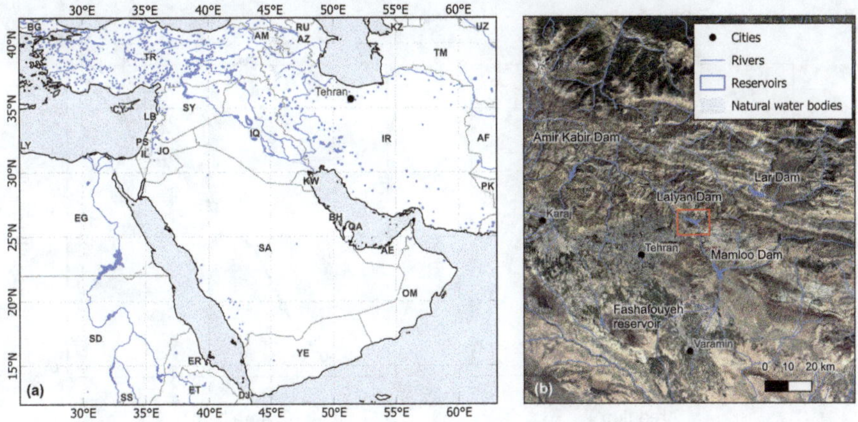

Fig. 3 Natural water bodies and reservoirs (Lehner et al. 2024) of the Middle East (**a**) and location of Latyan reservoir next in the suburbs of Tehran, Iran (**b**). *Satellite image source* Google Earth

commissioned in 1969 (with a later unit in 1987). As a multi-purpose facility, the project includes hydropower generation (45 MW installed capacity) alongside its role in urban water supply. Hydrologically, the reservoir is fed by flows from the Jajrood River basin and is integral to supplying surface water to Tehran's water system, while also contributing to regional flood mitigation and aquifer recharge through downstream infiltration. Since entering operation, the facility has become critical in securing water supply for Tehran's rapidly growing population.

Latyan Dam has been previously studied regarding sediment loading (Asiaei et al. 2021; Sadeghian et al. 2017), slope stability (Ahmadi and Eslami 2011), water quality (Asadollahfardi et al. 2012; Kakoei Dinaki et al. 2021), hydropower (Agheshlouei and Shourian 2021), and environmental flow (Naderi et al. 2020). More recently, however, it has become emblematic of Iran's wider water-stress challenges: current reservoir levels have dropped significantly amid prolonged drought and declining inflows. Monitoring of the dam and reservoir system continues to be important for both water-supply security and regional resilience in the face of climatic and infrastructure pressures.

3.2 *Satellite Data Acquisition*

The dataset used spans the period from 1 November 2020 to 31 October 2025, yielding a total of 445 Sentinel-1A SAR acquisitions over the study area. The collection includes 299 acquisitions in ascending orbit and 146 in descending orbit direction which determines the radar look angle and thus the incidence geometry with respect to the terrain and water body. For open water detection, these differences can introduce

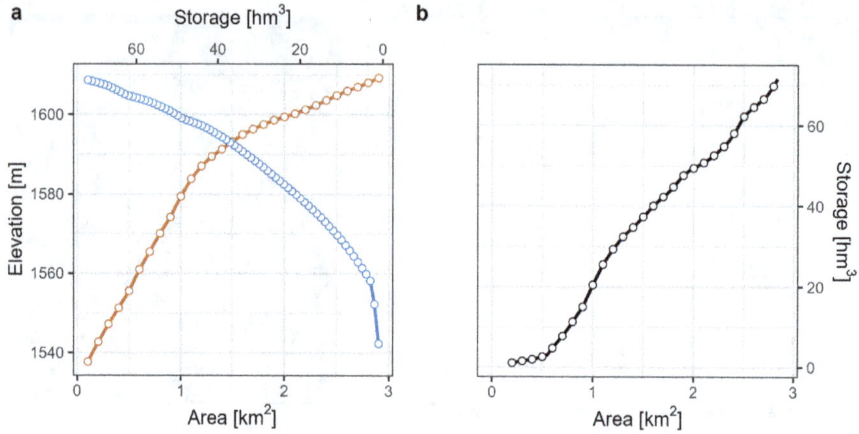

Fig. 4 **a** Design storage-area-elevation curve after Sadeghian et al. (2017) and **b** estimated area-storage curve for Latyan Dam

variations in backscatter, particularly along shorelines or in the presence of wind-roughened surfaces. While more sparce, here we use only the descending orbit that typically contains higher amount of backscatter (Rignot et al. 1994).

3.3 Dam Area and Storage

The transformation from water surface area to storage volume in Latyan Dam relies on the design storage-area-elevation relationship (Sadeghian et al. 2017). Figure 4 shows the original design curves; rearranging these relationships yields the direct area-storage curve (Fig. 4b), used to convert remotely sensed reservoir areas into storage estimates. In the absence of official data, date-storage validation pairs were recovered 7 validation points from internet sources (Table 4). The exact bootstrap therefore comprised all unique combinations-with-replacement of 7 samples, i.e. 1716 possible resamples. After applying the exclusion criteria (minimum three unique values and non-zero variance), 1583 valid resamples were retained and used to compute the distribution of performance metrics.

4 Results

The effect of varying detection threshold on model performance is illustrated in Fig. 5. Lower thresholds (below − 20 dB) generally resulted in poorer performance, with higher RMSE and lower efficiency scores. An intermediate parameter range produced the most favourable results, with a clear optimum around a threshold of T =

Table 4 Reported water storage volumes (hm^3) Latyan Dam across different dates from 2022 to 2025, as documented by various media sources

Date	Volume (hm^3)	Source
22-07-25	39	The Middle Eastern (2025)
08-03-25	9	Lulu (2025)
09-12-24	25	Energypress (2024)
22-07-24	65	The Middle Eastern (2025)
12-06-23	62	Financial Tribune (2023)
12-06-22	64	Financial Tribune (2023)
17-09-21	38	Iran International (2021)

− 18 dB. At this setting, RMSE was minimized, while R^2 and KGE were maximized, indicating a good balance between accuracy, correlation, and bias representation. More specifically T = − 18 dB scored the lowest KGE = 0.861 (0.640–0.946), also yielding high coefficient of determination $R^2 = 0.968$ (0.942–0.999) and low RMSE = 4.83 hm^3 (1.66–6.82) hm^3, indicating a strong overall fit (Table 5). The resulting V_{sat} time series is shown in Fig. 6 along with V_{ref}, Fig. 6 shows the Latyan Dam storage time series (hm^3) reconstructed from Sentinel-1 SAR between 2015 and 2025.

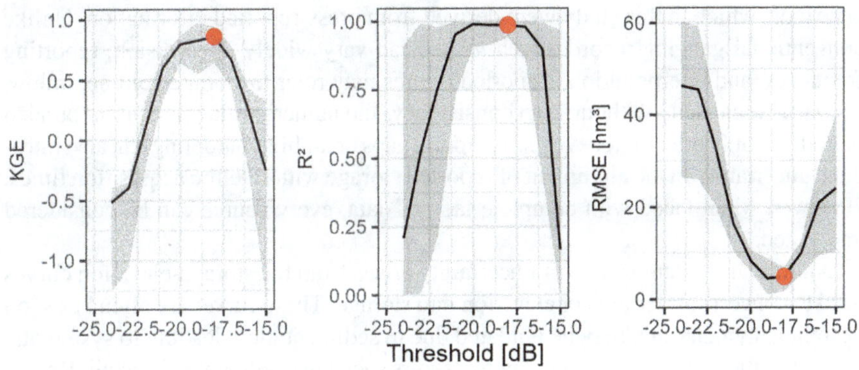

Fig. 5 Model performance as a function of threshold (dB). Panels show Kling-Gupta efficiency (KGE), coefficient of determination (R^2), and root mean square error (RMSE) (hm^3). The red dot indicates the parameter combination selected as optimal

Table 5 Media (Q2) and Q1–Q3 (25th–75th percentiles) for performance metrics (RMSE, R^2, KGE) for yielded for optimal threshold T = − 18 dB

Metric	Median (Q2)	Q1	Q3
RMSE (hm^3)	4.83	1.66	6.82
R^2	0.986	0.942	0.999
KGE	0.861	0.640	0.946

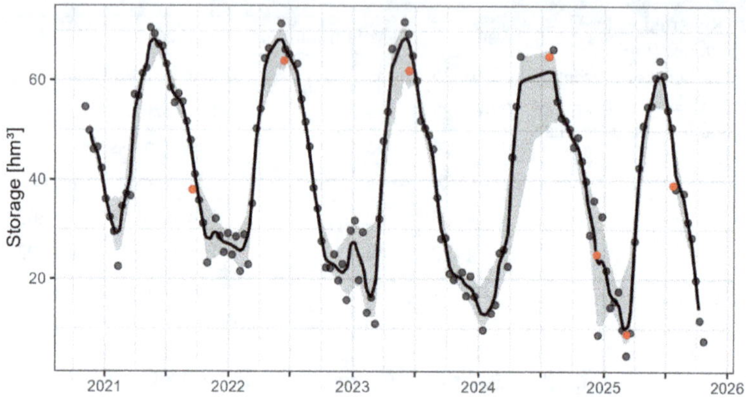

Fig. 6 Latyan Dam storage (hm^3) derived from Sentinel-1A SAR observations (black dots) and press-reported values (red dots) during 2020–2025. The black line represents the satellite-based smoothed storage trajectory

5 Limitations

The calibration of SAR imagery relies significantly on the validity of the reference data, which in this study were derived from press-reported information. Unlike conventional gauging records, such values can vary widely in precision, reporting frequency, and interpretation. Reported figures may refer to storage volume, inflow, or available capacity without strict consistency, and numeric values are often rounded or updated only after major events. Temporal gaps or ambiguous dating can also introduce uncertainty in the alignment of reported storage with satellite acquisition times. However, when faced with complete lack of data, every source can be considered and tested.

Another important source of uncertainty arises from the storage-elevation curves used to convert surface area or elevation into storage. These curves are often based on design bathymetry and may be outdated due to sedimentation, leading to systematic overestimation of capacity. In many reservoirs such curves do not exist at all. Recent approaches attempt to fill this gap either by deriving empirical hypsometric functions from satellite altimetry and surface water extents (e.g., Busker et al. 2019) or by generating synthetic storage–area–depth relationships through simplified geometric approximations at global scale (Yigzaw et al. 2018). Both approaches enable broader application of remote sensing for reservoir storage estimation, but they introduce additional uncertainties that must be considered when interpreting the results.

The comparison of RMSE, R^2, and KGE highlights the limitations of relying on a single performance metric. While the lowest RMSE is produced for T = – 19 dB, this statistic can primarily reflect overall fit and are strongly influenced by the magnitude of individual errors. By contrast, KGE, which balances correlation, variability, and bias, favoured $T = -18$ dB which also agrees better with the ranges found in literature (Li et al. 2022; Raju et al. 2022; Yang et al. 2021).

6 Conclusions

Besides limitations, the workflow developed here demonstrates that SAR-derived water extents can provide a robust and consistent basis for reconstructing multi-year storage variations in reservoirs where conventional measurements are sparse or inaccessible. Sentinel-1 imagery offers the spatial and temporal continuity required to monitor hydrological dynamics even under persistent cloud cover, while the processing framework ensures reproducible delineation of surface water across missions and acquisition geometries. The resulting storage series aligned closely with independent reference points, confirming that, when combined with existing design information, radar observations can serve as a dependable proxy for reservoir operation data. Such capability is particularly valuable in regions where storage information remains unavailable due to non-public gauging networks, institutional limitations, or the hydrogeologic complexity of karstic and semi-arid basins, conditions characteristic of many areas in the Middle East. By providing an independent, transparent, and reproducible means of assessing reservoir behaviour, SAR-based monitoring supports more informed water governance, enabling authorities and stakeholders to evaluate drought impacts, anticipate shortages, and plan adaptive management measures even where official data are lacking.

References

Agheshlouei A, Shourian M (2021) Probable maximum flood (PMF) estimation using a Bayesian model (case study: Latian Dam). In: cbip.org, pp 24–27

Ahmadi M, Eslami M (2011) A new approach to plane failure of rock slope stability based on water flow velocity in discontinuities for the Latian dam reservoir landslide. J Mount Sci 8(2):124–130. https://doi.org/10.1007/S11629-011-2088-5

Alexakis DD, Stavroulaki EG, Tsanis IK (2019) Using Sentinel-1A DInSAR interferometry and Landsat 8 data for monitoring water level changes in two lakes in Crete, Greece. Geocarto Int 34:703–721. https://doi.org/10.1080/10106049.2018.1434685

An C, Zhang F, Chan NW, Johnson VC, Shi J (2022a) A review on the research progress of lake water volume estimation methods. J Environ Manage 314:115057. https://doi.org/10.1016/J.JENVMAN.2022.115057

An Z, Chen P, Tang F, Yang X, Wang R, Wang Z (2022b) Evaluating the performance of seven ongoing satellite altimetry missions for measuring inland water levels of the great lakes. Sensors 22:9718. https://doi.org/10.3390/S22249718

Asadollahfardi G, Rahbar R, Fatemiaghda M (2012) Application of time series models to predict water quality of upstream and downstream of Latian dam in Iran. Univers J Environ Res Technol 26–35

Asiaei O, Karimaei Tabarestani M, Soltani A, Payan M (2021) Study on geotechnical properties of sediment deposited in Latian dam reservoir using field measurements. Iran J Soil Water Res 52:2457–2470. https://doi.org/10.22059/IJSWR.2021.326357.669010

Bastawesy MA, Khalaf FI, Arafat SM (2008) The use of remote sensing and GIS for the estimation of water loss from Tushka lakes, southwestern desert, Egypt. J Afr Earth Sci 52:73–80. https://doi.org/10.1016/J.JAFREARSCI.2008.03.006

Biemans H, Haddeland I, Kabat P, Ludwig F, Hutjes RWA, Heinke J, Von Bloh W, Gerten D (2011) Impact of reservoirs on river discharge and irrigation water supply during the 20th century. Water Resour Res 47:3509. https://doi.org/10.1029/2009WR008929

Birkett CM, Beckley B (2010) Investigating the performance of the Jason-2/OSTM radar altimeter over lakes and reservoirs. Mar Geodesy 33:204–238. https://doi.org/10.1080/01490419.2010.488983;WGROUP:STRING:PUBLICATION

Birkett C, Reynolds C, Beckley B, Doorn B (2011) From research to operations: the USDA global reservoir and lake monitor. Coast Altimetry 19–50. https://doi.org/10.1007/978-3-642-12796-0_2/FIGURES/7_2

Busker T, De Roo A, Gelati E, Schwatke C, Adamovic M, Bisselink B, Pekel JF, Cottam A (2019) A global lake and reservoir volume analysis using a surface water dataset and satellite altimetry. Hydrol Earth Syst Sci 23:669–690. https://doi.org/10.5194/HESS-23-669-2019

Camici S, Calmettes B, Daguzé JA, Piras F, Thibaut P, Amarnath G, Femenias P, Tarpanelli A (2025) Assessing the accuracy of satellite radar altimetry for inland water level monitoring. J Hydrol (Amst) 661:133556. https://doi.org/10.1016/J.JHYDROL.2025.133556

Chen X, Zheng Y, Peng J, Floris M (2021) Monitoring river water level using multiple bounces of bridges in SAR images. Adv Space Res 68:4016–4023. https://doi.org/10.1016/J.ASR.2021.08.006

Chen T, Song C, Luo S, Ke L, Liu K, Zhu J (2022) Monitoring global reservoirs using ICESat-2: assessment on spatial coverage and application potential. J Hydrol (Amst) 604:127257. https://doi.org/10.1016/J.JHYDROL.2021.127257

Chen J, Wang Y, Wang J, Zhang Y, Xu Y, Yang O, Zhang R, Wang J, Wang Z, Lu F, Hu Z (2024) The performance of Landsat-8 and Landsat-9 data for water body extraction based on various water indices: a comparative analysis. Remote Sens (Basel) 16:1984. https://doi.org/10.3390/RS16111984/S1

Chipman JW (2019) A multisensor approach to satellite monitoring of trends in lake area, water level, and volume. Remote Sens 11:158. https://doi.org/10.3390/RS11020158

Daliakopoulos I, Kadlec J, Skaloš J (2025) Reconstruction of reservoir water level and storage using Sentinel-1 C-SAR. https://doi.org/10.5194/EGUSPHERE-EGU25-19181

Das P, Hossain F, Khan S, Biswas NK, Lee H, Piman T, Meechaiya C, Ghimire U, Hosen K (2022) Reservoir assessment tool 2.0: stakeholder driven improvements to satellite remote sensing based reservoir monitoring. Environ Model Softw 157:105533. https://doi.org/10.1016/J.ENVSOFT.2022.105533

De Fleury M, Kergoat L, Grippa M (2023) Hydrological regime of Sahelian small waterbodies from combined Sentinel-2 MSI and Sentinel-3 synthetic aperture radar altimeter data. Hydrol Earth Syst Sci 27:2189–2204. https://doi.org/10.5194/HESS-27-2189-2023

Druce D, Tong X, Lei X, Guo T, Kittel CMM, Grogan K, Tottrup C (2021) An optical and SAR based fusion approach for mapping surface water dynamics over Mainland China. Remote Sens 13:1663. https://doi.org/10.3390/RS13091663

Duan Z, Bastiaanssen WGM (2013) Estimating water volume variations in lakes and reservoirs from four operational satellite altimetry databases and satellite imagery data. Remote Sens Environ 134:403–416. https://doi.org/10.1016/J.RSE.2013.03.010

Efron B, Tibshirani RJ (1994) An introduction to the bootstrap. https://doi.org/10.1201/9780429246593

Ehret T, Lajouanie S, Lefrançois V, de Franchis C (2021) Automatic monitoring of water level in small lakes using PlanetScope. In: International geoscience and remote sensing symposium (IGARSS), pp 3356–3359. https://doi.org/10.1109/IGARSS47720.2021.9554431

Energypress (2024) Tehran's dams' water reserves are only 20 percent. Iran Energy Press [WWW document]. Energypress, Iran. https://energypress.ir/en/?p=2670. Accessed 29 Oct 2025

Fil PP, Yurova AY, Dobrokhotov A, Kozlov D, Yurova PP, Dobrokhotov AY, Kozlov A, Cugueró-Escofet À, Puig V (2021) Estimation of infiltration volumes and rates in seasonally water-filled topographic depressions based on remote-sensing time series. Sensors 21:7403. https://doi.org/10.3390/S21217403

Financial Tribune (2023) Tehran dams 70% empty [WWW document]. Financial Tribune, Iran. https://financialtribune.com/articles/energy/118542/tehran-dams-70-empty. Accessed 29 Oct 2025

Frappart F, Do Minh K, L'Hermitte J, Cazenave A, Ramillien G, Le Toan T, Mognard-Campbell N (2006) Water volume change in the lower Mekong from satellite altimetry and imagery data. Geophys J Int 167:570–584. https://doi.org/10.1111/J.1365-246X.2006.03184.X/2/167-2-570-FIG013.JPEG

Gao H (2015) Satellite remote sensing of large lakes and reservoirs: from elevation and area to storage. Wiley Interdiscip Rev Water 2:147–157. https://doi.org/10.1002/WAT2.1065

Gao H, Birkett C, Lettenmaier DP (2012) Global monitoring of large reservoir storage from satellite remote sensing. Water Resour Res 48. https://doi.org/10.1029/2012WR012063;WEBSITE:WEBSITE:AGUPUBS;JOURNAL:JOURNAL:19447973;REQUESTEDJOURNAL:JOURNAL:19447973;WGROUP:STRING:PUBLICATION

Garrett K, McManamay RA, Wang J (2021) Global hydropower expansion without building new dams. Environ Res Lett 16:114029. https://doi.org/10.1088/1748-9326/AC2F18

Guan T, Xu Q, Chen X, Cai J (2021) A novel remote sensing method to determine reservoir characteristic curves using high-resolution data. Hydrol Res 52:1066–1082. https://doi.org/10.2166/NH.2021.035

Gujrati A, Pradhan R, Singh N, Jha VB, Gupta PK (2024) Adaptive water delineation algorithms for L- and C-band SAR imagery: a comparative analysis. Earth Sci Inform 17:5011–5025. https://doi.org/10.1007/S12145-024-01417-0/TABLES/5

Gupta HV, Kling H (2011) On typical range, sensitivity, and normalization of mean squared error and Nash-Sutcliffe efficiency type metrics. Water Resour Res 47:10601. https://doi.org/10.1029/2011WR010962;PAGE:STRING:ARTICLE/CHAPTER

Hamidifar H (2024) Water crisis in Iran: causes, consequences, and solutions. In: Water crises and sustainable management in the Global South, pp 85–109. https://doi.org/10.1007/978-981-97-4966-9_3

Hanasaki N, Kanae S, Oki T (2006) A reservoir operation scheme for global river routing models. J Hydrol (Amst) 327:22–41. https://doi.org/10.1016/J.JHYDROL.2005.11.011

Hossen H, Khairy M, Ghaly S, Scozzari A, Negm A, Elsahabi M (2022) Bathymetric and capacity relationships based on Sentinel-3 mission data for Aswan high dam lake, Egypt. Water 14:711. https://doi.org/10.3390/W14050711

Iran International (2021) Regional water facility manager: 60% of Latian dam reservoir is empty [WWW document]. Volant Media, UK. https://www.iranintl.com/en/20210917050961e. Accessed 29 Oct 2025

Jabbar Y, Yadav S (2025) Reservoir capacity estimation from Landsat 7 imagery using spatially parameterized bias correction model. Hydrol Sci J. https://doi.org/10.1080/02626667.2025.2531995;WGROUP:STRING:PUBLICATION

Jafari M, Andik B, Mianabadi H, Ghorbani A, Nasri Fakhredavood S, Choobchian S, Mirhashemi Dehkordi SS (2024) Water conflict analysis in a changing context: a qualitative systematic review of trends, patterns, and approaches in Iran. World Water Policy 10:1292–1326. https://doi.org/10.1002/WWP2.12212;JOURNAL:JOURNAL:23806540;WGROUP:STRING:PUBLICATION

Jensen K, De S, Hughes L, Yalla G (2022) Flood monitoring with X-band and C-band SAR: a case study of the 2021 British Columbia floods. In: International geoscience and remote sensing symposium (IGARSS), July 2022, pp 5535–5538. https://doi.org/10.1109/IGARSS46834.2022.9884604

Jiang L, Nielsen K, Andersen OB, Bauer-Gottwein P (2017) Monitoring recent lake level variations on the Tibetan Plateau using CryoSat-2 SARIn mode data. J Hydrol (Amst) 544:109–124. https://doi.org/10.1016/J.JHYDROL.2016.11.024

Kakoei Dinaki F, Cheraghi M, Lorestani B, SobhanArdakani S, Chamani A (2021) Investigating the potential of native plants in phytoremediation of heavy metals and nutrients in the Latian and Lar dam reservoirs. Iran-Water Resour Res 17:130–141

Kansara P, Li W, El-Askary H, Lakshmi V, Piechota T, Struppa D, Sayed MA (2021) An assessment of the filling process of the grand Ethiopian renaissance dam and its impact on the downstream countries. Remote Sens 13:711. https://doi.org/10.3390/RS13040711

Karaman M, Özelkan E (2022) Comparative assessment of remote sensing-based water dynamic in a dam lake using a combination of Sentinel-2 data and digital elevation model. Environ Monit Assess 194:1–20. https://doi.org/10.1007/S10661-021-09703-W/FIGURES/14

Ke X, Heguang L, Jingshan J (2013) Active and passive microwave remote sensing technology of the HY-2A ocean satellite mission. Strateg Study Chin Acad Eng 15:4–11

Kouraev AV, Zakharova EA, Samain O, Mognard NM, Cazenave A (2004) Ob' river discharge from TOPEX/Poseidon satellite altimetry (1992–2002). Remote Sens Environ 93:238–245. https://doi.org/10.1016/J.RSE.2004.07.007

Kourkouli P (2023) Natural disaster monitoring using ICEYE SAR data. In: Geoinformatics for geosciences: advanced geospatial analysis using RS, GIS and soft computing, pp 163–170. https://doi.org/10.1016/B978-0-323-98983-1.00010-7

Krishnaveni AS, Jayaram C, Chowdary VM, Jha CS, Jayaram C (2022) Reservoir monitoring using satellite altimetry: a case study over Mayurakshi reservoir, pp 295–310. https://doi.org/10.1007/978-3-030-90479-1_18

Lazecký M, Perissin D, Zhiying W, Ling L, Yuxiao Q (2015) Observing dam's movements with spaceborne SAR interferometry. In: Engineering geology for society and territory. volume 5: urban geology, sustainable planning and landscape exploitation, pp 131–136. https://doi.org/10.1007/978-3-319-09048-1_26/FIGURES/7

Lehner B, Beames P, Mulligan M, Zarfl C, De Felice L, van Soesbergen A, Thieme M, Garcia de Leaniz C, Anand M, Belletti B, Brauman KA, Januchowski-Hartley SR, Lyon K, Mandle L, Mazany-Wright N, Messager ML, Pavelsky T, Pekel J-F, Wang J, Wen Q, Wishart M, Xing T, Yang X, Higgins J (2024) The Global Dam Watch database of river barrier and reservoir information for large-scale applications. Sci Data 11:1069. https://doi.org/10.1038/s41597-024-03752-9

Li M, Zhang T, Tu Y, Ren Z, Xu B (2022) Monitoring post-flood recovery of croplands using the integrated Sentinel-1/2 imagery in the Yangtze-Huai River basin. Remote Sens (Basel) 14:690. https://doi.org/10.3390/RS14030690/S1

Liebe J, van de Giesen N, Andreini M (2005) Estimation of small reservoir storage capacities in a semi-arid environment: a case study in the Upper East Region of Ghana. Phys Chem Earth Parts a/b/c 30:448–454. https://doi.org/10.1016/J.PCE.2005.06.011

Liu H, Zheng L, Jiang L, Liao M (2020) Forty-year water body changes in Poyang Lake and the ecological impacts based on Landsat and HJ-1 A/B observations. J Hydrol (Amst) 589:125161. https://doi.org/10.1016/J.JHYDROL.2020.125161

Lulu (2025) [@upuouo] Tehran is facing the worst water crisis in over a century [WWW document]. X (formerly Twitter). https://x.com/upuouo/status/1898146013342491050. Accessed 29 Oct 2025

Ma X, Wang L, Chen C, Du J, Sun S (2020) Simulation of the dynamic water storage and its gravitational effect in the head region of Three Gorges Reservoir using imageries of Gaofen-1. Remote Sens 12:3353. https://doi.org/10.3390/RS12203353

Ma A, Chao N, Qin P, Hwang C, Zhu C, Chen G, Wang Z, Wang S (2024) Performance assessment of Sentinel-3/6 altimeter data for marine gravity recovery. Front Mar Sci 11:1440845. https://doi.org/10.3389/FMARS.2024.1440845/BIBTEX

Mansaray AS, Dzialowski AR, Martin ME, Wagner KL, Gholizadeh H, Stoodley SH (2021) Comparing PlanetScope to Landsat-8 and Sentinel-2 for sensing water quality in reservoirs in agricultural watersheds. Remote Sens 13:1847. https://doi.org/10.3390/RS13091847

Martinis S, Groth S, Wieland M, Knopp L, Rättich M (2022) Towards a global seasonal and permanent reference water product from Sentinel-1/2 data for improved flood mapping. Remote Sens Environ 278:113077. https://doi.org/10.1016/J.RSE.2022.113077

Milanović P (2021) Dams and reservoirs in karst? Keep away or accept the challenges. Hydrogeol J 29:89–100. https://doi.org/10.1007/S10040-020-02273-0/FIGURES/7

Morris CS, Gill SK (1994) Variation of Great Lakes water levels derived from Geosat altimetry. Water Resour Res 30:1009–1017. https://doi.org/10.1029/94WR00064

Naderi M, Alioghli S, Jahandideh O, Rajabizadeh Y, Salarijazi M (2020) Determination of optimal and desirable environmental flow release from Latian dam reservoir with consideration of ecohydraulic, hydrological and hydromorphological characteristics to protect the habitat of the Jajrood River. Iran J Irrig Drain 14:1277–1300

Nair AS, Verma K, Karmakar S, Ghosh S, Indu J (2022) Exploring the potential of SWOT mission for reservoir monitoring in Mahanadi basin. Adv Space Res 69:1481–1493. https://doi.org/10.1016/J.ASR.2021.11.019

Pham-Duc B (2024) Comparison of synthetic aperture radar Sentinel-1 and ALOS-2 observations for lake monitoring. Vietnam J Earth Sci 46:322–338. https://doi.org/10.15625/2615-9783/20639

Pimenta J, Fernandes JN, Azevedo A (2025) Remote sensing tool for reservoir volume estimation. Remote Sens 17:619. https://doi.org/10.3390/RS17040619

Raju RT, Thampi SG, Sathish Kumar D (2022) Flood mapping using Sentinel-1 SAR data. Lect Notes Civ Eng 234:577–590. https://doi.org/10.1007/978-981-19-0304-5_42/FIGURES/6

Rapley CG (1987) Exploratory study of inland water and land altimetry using SEASAT data. ESA CR (P)

Rignot E, Way JB, McDonald K, Viereck L, Williams C, Adams P, Payne C, Wood W, Shi J (1994) Monitoring of environmental conditions in Taiga forests using ERS-1 SAR. Remote Sens Environ 49:145–154. https://doi.org/10.1016/0034-4257(94)90051-5

Ruiz-Armenteros AM, Delgado JM, Bakon M, Sousa JJ, Lamas-Fernández F, Marchamalo-Sacristán M, Sánchez-Ballesteros V, Papco J, González-Rodrigo B, Lazecky M, Perissin D (2021) ReMoDams: monitoring dams from space using satellite radar interferometry. In: International geoscience and remote sensing symposium (IGARSS), July 2021, pp 5331–5334. https://doi.org/10.1109/IGARSS47720.2021.9553375

Sadeghian MS, Daneshkhah M, Heydari M (2017) A comparison between the amount of reservoir sediment load prediction during design period and the operational period. Int J Ground Sediment Water 6:277–292

Sathianarayanan M, Saraswat A, Athick ASMA, Lin HM (2023) Intercomparison between Sentinel-1, Sentinel-2, and Landsat-8 on reservoir water level estimation. Sustain Water Resour Manag 9:1–12. https://doi.org/10.1007/S40899-023-00974-4/FIGURES/8

Schwatke C, Scherer D, Dettmering D (2019) Automated extraction of consistent time-variable water surfaces of lakes and reservoirs based on Landsat and Sentinel-2. Remote Sens 11:1010. https://doi.org/10.3390/RS11091010

Shah D, Zhang S, Sarkar S, Davidson C, Zhang R, Zhao M, Devadiga S, Noojipady P, Román MO, Gao H (2024) Transitioning from MODIS to VIIRS global water reservoir product. Sci Data 11:1–17. https://doi.org/10.1038/S41597-024-03028-2

Shin S, Pokhrel Y, Miguez-Macho G (2019) High-resolution modeling of reservoir release and storage dynamics at the continental scale. Water Resour Res 55:787–810. https://doi.org/10.1029/2018WR023025

Shu S, Liu H, Beck RA, Frappart F, Korhonen J, Lan M, Xu M, Yang B, Huang Y (2021) Evaluation of historic and operational satellite radar altimetry missions for constructing consistent long-term lake water level records. Hydrol Earth Syst Sci 25:1643–1670. https://doi.org/10.5194/HESS-25-1643-2021

Siebert S, Döll P (2008) The global crop water model (GCWM): documentation and first results for irrigated crops. Research report

The Middle Eastern (2025) [@TMiddleEastern] What is the current state of Tehran's water reservoirs? [WWW document]. X (formerly Twitter). https://x.com/TMiddleEastern/status/1947622823343596017. Accessed 29 Oct 2025

Twele A, Cao W, Plank S, Martinis S (2016) Sentinel-1-based flood mapping: a fully automated processing chain. Int J Remote Sens 37:2990–3004. https://doi.org/10.1080/01431161.2016.1192304

Verron J, Bonnefond P, Aouf L, Birol F, Bhowmick SA, Calmant S, Conchy T, Crétaux JF, Dibarboure G, Dubey AK, Faugère Y, Guerreiro K, Gupta PK, Hamon M, Jebri F, Kumar R, Morrow R, Pascual A, Pujol MI, Rémy E, Rémy F, Smith WHF, Tournadre J, Vergara O (2018) The benefits of the Ka-band as evidenced from the SARAL/AltiKa altimetric mission: scientific applications. Remote Sens 10:163. https://doi.org/10.3390/RS10020163

Wolf AT, Natharius JA, Danielson JJ, Ward BS, Fender JK (1999) International river basins of the world. Int J Water Resour Dev 15:387–427. https://doi.org/10.1080/07900629948682

Yang Z, Wei J, Deng J, Gao Y, Zhao S, He Z (2021) Mapping outburst floods using a collaborative learning method based on temporally dense optical and SAR data: a case study with the Baige landslide dam on the Jinsha River, Tibet. Remote Sens 13:2205. https://doi.org/10.3390/RS13112205

Yigzaw W, Li HY, Demissie Y, Hejazi MI, Leung LR, Voisin N, Payn R (2018) A new global storage-area-depth data set for modeling reservoirs in land surface and earth system models. Water Resour Res 54:10372–10386. https://doi.org/10.1029/2017WR022040;PAGEGROUP:STRING:PUBLICATION

Zajac Z, Revilla-Romero B, Salamon P, Burek P, Hirpa F, Beck H (2017) The impact of lake and reservoir parameterization on global streamflow simulation. J Hydrol (Amst) 548:552–568. https://doi.org/10.1016/J.JHYDROL.2017.03.022

Zhang S, Gao H (2020) Using the digital elevation model (DEM) to improve the spatial coverage of the MODIS based reservoir monitoring network in South Asia. Remote Sens 12:745. https://doi.org/10.3390/RS12050745

Zhang J, Xu K, Watanabe M, Yang Y, Chen X (2004) Estimation of river discharge from non-trapezoidal open channel using QuickBird-2 satellite imagery. Hydrol Sci J 49:247–260. https://doi.org/10.1623/HYSJ.49.2.247.34831;WGROUP:STRING:PUBLICATION

Zhang S, Gao H, Naz BS (2014) Monitoring reservoir storage in South Asia from multisatellite remote sensing. Water Resour Res 50:8927–8943. https://doi.org/10.1002/2014WR015829

Zhang Z, Ma P, Wang X, Hou J, Zhang Q, Guo Y, Xu Z, Wang Y, Abdulhamid K (2025) Surface elevation dynamics of Lake Karakul from 1991 to 2020 inversed by ICESat, CryoSat-2 and ERS-1/2. Remote Sens 17:2816. https://doi.org/10.3390/RS17162816

Index

M. M. Al Saud (ed.), *Hydroclimate Changes: Implications of the Middle East Region*, https://doi.org/10.1007/978-3-032-20053-2

E

www.ingramcontent.com/pod-product-compliance
Lightning Source LLC
Chambersburg PA
CBHW062102180526
45290CB00003B/26
9783032200525